SCALABLE GREEN CHEMISTRY

edited by

Stefan G. Koenig

SCALABLE GREEN CHEMISTRY

Case Studies from the Pharmaceutical Industry

PAN STANFORD PUBLISHING

Published by

Pan Stanford Publishing Pte. Ltd.
Penthouse Level, Suntec Tower 3
8 Temasek Boulevard
Singapore 038988

Email: editorial@panstanford.com
Web: www.panstanford.com

British Library Cataloguing-in-Publication Data
A catalogue record for this book is available from the British Library.

Scalable Green Chemistry: Case Studies from the Pharmaceutical Industry

ISBN 978-981-4316-49-1 (Hardcover)
ISBN 978-981-4364-81-2 (eBook)

Printed in the USA

For Lena

Contents

Preface

This work aims to engage students and practitioners of chemistry and to inspire them to take part in the green chemistry movement by providing real-world examples, demonstrated on large, even commercial, scales. This is important on two levels: (1) that the technologies or projects presented actually work reproducibly and serve as a practical validation of the field, and (2) that the pharmaceutical companies, academic institutions, and funding agencies behind this work support these efforts, as there would be little point in doing so if green chemistry were regarded as a waste of time and money. Making chemical processes more environmentally friendly is a winning proposition for all stakeholders.

The book is divided roughly into three sections. Chapters 1–4 introduce green chemistry and feature topics that relate broadly to the pharmaceutical industry. Chapters 5–9 describe focused studies of specific pharmaceutical projects, from investigational compounds to marketed products. The final chapters (10–13) review the latest technologies and provide a perspective on the future of green chemistry. In total, an incredible overview of the field is provided, with case studies illustrating a myriad of improvements. I thank the chapter authors for contributing to this valuable endeavor, creating a work that will inform both beginners and practicing chemists alike to adopt the principles of green chemistry.

Chapter 1 initiates the presentation with valuable background information and key definitions. The author is an advocate whose work with the American Chemical Society Green Chemistry Institute's Pharmaceutical Roundtable demonstrates that established pharma and affiliated companies are committed to conducting their business in a more sustainable manner. Chapter 2 provides excellent historical examples to showcase the impact on important and established therapeutic products with improved manufacturing routes with reduced environmental burden. Biocatalysis is introduced in Chapter 3, demonstrating how enzymatic transformations can impact synthetic chemical processes. A multitude of commercialized medicines are discussed to describe the efficiency with which certain raw materials, intermediates and active substances can be targeted with biocatalytic transformations. Chapter 4 provides an in-depth discussion of how green chemistry demonstrates its value by the application of metrics. The focus is on a series of synthetic strategies to the important anti-influenza treatment, oseltamivir phosphate (Tamiflu®). Different chemical syntheses are evaluated to compare the efficiency of each route.

In Chapter 5, extensive process development of an antidepressant candidate is described. The chemical route was thoroughly investigated to produce a streamlined synthesis and ready access to the active substance. The approach to green chemistry in the generics industry is discussed in Chapter 6. Though this segment of the pharmaceutical industry is less visible in contributions to sustainable manufacturing, the case is made that efficiency gains make established products more lucrative, illustrated by the development of a less wasteful route to a widely-prescribed diabetes medication. Chapter 7 describes the replacement of the original route to an important antibiotic with a more sustainable chemical process. This second generation provides better access to a treatment for drug-resistant bacterial infections with significant environmental benefits for a medicine produced on metric ton scale. In Chapter 8, the synthesis of a key, water-soluble intermediate in a manufacturing process is presented. Practical considerations are dutifully investigated to carefully select solvents and avoid the use of water to enable isolation of this highly polar synthetic intermediate. Chapter 9 depicts process development work to a recently approved therapeutic for rheumatoid arthritis. A robust commercial route was designed, guided by green chemistry considerations.

Chapter 10 presents continuous processing as a revolutionary manufacturing technique, particularly for hazardous reactions that would not be feasible in the traditional batch mode. It is described by field experts, with an overview of the benefits and illustration with a case study. Chapter 11 introduces real-time analytical technology—

as applied to spectroscopy, calorimetry or crystallization—to show how real-time data improves decision-making and worker safety, while also providing energy and material efficiency gains. Chapter 12 discusses the merits of microwave technology and the application of this tool to chemical transformations, which are now becoming possible at kilo-scale. Finally, Chapter 13 reflects on international green chemistry efforts and offers recommendations to meet global challenges as they relate to the impact of pharmaceutical products and their manufacture.

Green chemistry is an accepted philosophy in placing a chemical undertaking on a more sustainable path. The pharmaceutical industry has adopted the 12 Principles to achieve its goal of developing important medicines to benefit patients. The principles lead to safer, more efficient and sustainable chemical routes to important therapies. Only by respecting and preserving the environment will the industry's mission of improving health worldwide truly become reality. Chemists should practice green chemistry because it offers a safer work environment and a healthier community for all.

Stefan G. Koenig
Summer 2013

Acknowledgments

As with any project of this scope, I would first like to thank my family and friends for their support in enabling me to see this book through to completion. Life presents so many challenges, and meeting them is only possible in a mutually supportive environment. Several people contributed their perspectives and valuable discussions during the project. I would like to specifically thank Yvette Wild Koenig, Linda Koenig, Edgar Wild, and Ruben Savizky for their exceptional efforts in assisting me to create a cohesive theme.

I would like to express my gratitude to the authors for their chapters. It is often very difficult to craft a compelling story about a project that involves so many different aspects. I believe that the chapters in this book convey important information in a readable format and am thankful for the hard work the authors put in. In addition, I would like to thank many members of the American Chemical Society Green Chemistry Institute—past and present—for their support and contributions to this project. Lastly, I would like to thank all chemists that have made a commitment to incorporating green chemistry into their practice and instruction. Only through you will the transformation to a more sustainable enterprise take place.

Chapter 1

Introduction to Green Pharmaceutical Science: Fact, Fiction, and Future

Julie B. Manley
Guiding Green, LLC, 457 E. Mier Road, Sanford, MI 48657, USA
juliemanley@guidinggreen.com

Modern chemistry can be traced back to the eighteenth century, when Antoine Lavoisier, who is now recognized as the father of modern chemistry, disproved phlogiston theory by establishing that oxygen was the component of air that combines with substances as they burn.[1] Two centuries later, the first Nobel Prize in Chemistry, among other fields, was awarded in 1901. Over the next 100 years, the field of chemistry advanced to increase the understanding of chemical processes and their contribution to modern technology. In 2005, the Nobel committee recognized metathesis, which enables the development of synthesis reactions that are more efficient, simpler to use, and environmentally friendlier, as a significant achievement for the development of pharmaceuticals and lauded it as a "great step forward for green chemistry."[2]

Green chemistry, defined as the design of chemical processes and products to minimize the use and/or generation of hazardous materials,[3] has its foundation in the 1990s. The Pollution Prevention

Scalable Green Chemistry: Case Studies from the Pharmaceutical Industry
Edited by Stefan G. Koenig

ISBN 978-981-4316-49-1 (Hardcover), 978-981-4364-81-2 (eBook)
www.panstanford.com

Act of 1990 in the United States created a precedent with the waste management hierarchy, recognizing that it was better to prevent waste than to treat it. The act itself recognized the economic value, in addition to the environmental, health, and safety benefits inherent to reducing waste at the source.[4] In the mid-1990s, Paul T. Anastas, then chief of the Industrial Chemistry Branch of the US Environmental Protection Agency, and John C. Warner, then in exploratory and media research at Kodak Corporation, established the formal definition of green chemistry and enumerated its 12 principles (Table 1.1), the framework to design chemicals and processes to be inherently less hazardous. Green chemistry is the integration of source and hazard reduction into the core of chemistry.

Green chemistry is inherent in the mission of the pharmaceutical industry: to develop medicines to help patients live longer, healthier, and more productive lives.[5] It would be paradoxical to develop pharmaceuticals in a manner that threatens the health and safety of those that the industry aims to protect. It is more than just corporate responsibility: 88% of respondents to a pharma-manufacturing survey indicated their commitment to green is primarily motivated by cost savings.[6]

As of 2010, the pharmaceutical industry has received the Presidential Green Chemistry Challenge Award—the United States' highest environmental award designed to recognize innovative chemical technologies that prevent pollution—a total of nine times since the award's inception in 1996 (see Table 1.2). Winning technologies across all industries are responsible for reducing the use or generation of more than 198 million pounds of hazardous chemicals (enough to fill a freight train nearly 11 miles long), saving 21 billion gallons of water (enough to meet the annual needs of over 2,100 people), and eliminating 57 million pounds of carbon dioxide releases to the air (equal to taking 6,000 automobiles off the road).[7] Seven different pharmaceutical companies have been honored with the award, which has recognized processes from research and development (R&D) through commercialization, demonstrating that green chemistry is relevant to the broader industry and the breadth of the development phases.

Table 1.1 The 12 principles of green chemistry[4]

1.	**Prevention** It is better to prevent waste than to treat or clean up waste after it has been created.
2.	**Atom Economy** Synthetic methods should be designed to maximize the incorporation of all materials used in the process into the final product.
3.	**Less Hazardous Chemical Syntheses** Wherever practicable, synthetic methods should be designed to use and generate substances that possess little or no toxicity to human health and the environment.
4.	**Designing Safer Chemicals** Chemical products should be designed to effect their desired function, while minimizing their toxicity.
5.	**Safer Solvents and Auxiliaries** The use of auxiliary substances (e.g., solvents, separation agents) should be made unnecessary wherever possible and innocuous when used.
6.	**Design for Energy Efficiency** Energy requirements of chemical processes should be recognized for their environmental and economic impacts and should be minimized. If possible, synthetic methods should be conducted at ambient temperature and pressure.
7.	**Use of Renewable Feedstocks** A raw material or feedstock should be renewable rather than depleting, whenever technically and economically practicable.
8.	**Reduce Derivatives** Unnecessary derivatization (use of blocking groups, protection/deprotection, temporary modification of physical/chemical processes) should be minimized or avoided, if possible, because such steps require additional reagents and can generate waste.
9.	**Catalysis** Catalytic reagents (as selective as possible) are superior to stoichiometric reagents.
10.	**Design for Degradation** Chemical products should be designed so that at the end of their function, they break down into innocuous degradation products and do not persist in the environment.
11.	**Real-Time Analysis for Pollution Prevention** Analytical methodologies need to be further developed to allow for real-time, in-process monitoring and control prior to the formation of hazardous substances.
12.	**Inherently Safer Chemistry for Accident Prevention** Substances and the form of a substance used in a chemical process should be chosen to minimize the potential for chemical accidents, including releases, explosions, and fires.

Table 1.2 Pharmaceutical industry recipients of the Presidential Green Chemistry Challenge Award (source: www.epa.gov)

Pharmaceuticals: 9 technologies

Year	Winner	Description of the winning technology in relation to the topic area
2010	Merck & Co., Inc.; Codexis, Inc.	Sitagliptin, the active ingredient in Januvia™, a treatment for type II diabetes, manufactured using an evolved, highly stereoselective transaminase
2006	Codexis, Inc.	The key chiral building block for atorvastatin calcium (the active ingredient in Lipitor® used to lower cholesterol), synthesized by three biocatalysts greatly improved by directed evolution
2006	Merck & Co., Inc.	Sitagliptin, the active ingredient in Januvia™, used to treat type II diabetes, made by a novel green synthesis for ß-amino acids
2005	Merck & Co., Inc.	Aprepitant, the active ingredient in Emend®, used to treat chemotherapy-induced nausea and vomiting, made by a convergent, highly atom-economical safer synthesis that also saves water
2004	Bristol-Myers Squibb Company	Paclitaxel, the active ingredient in Taxol®, used to treat ovarian and breast cancer, synthesized by plant cell fermentation
2002	Pfizer, Inc.	Sertraline, the active ingredient in Zoloft®, used to treat depression, synthesized by a process that eliminates waste, reduces solvents, and doubles overall product yield
2000	Roche Colorado Corporation	Ganciclovir, the active ingredient in Cytovene®, a potent antiviral agent, synthesized by the guanine triester process, eliminating 2 hazardous solid waste streams and 11 chemicals
1999	Lilly Research Laboratories	A drug candidate for the treatment of epilepsy, synthesized by a process including a yeast-mediated asymmetric reaction that eliminates chromium waste and large volumes of solvent
1997	BHC Company (now BASF Corporation)	Ibuprofen, the active ingredient in Advil™, Motrin™, and other over-the-counter pain relievers, synthesized in three catalytic steps with virtually no wasted atoms

Yet even with the demonstrated success of green chemistry in the pharmaceutical industry to date, future opportunities far outweigh the implementation achieved thus far. The field of green chemistry is still considered in its infancy, and it takes time for full integration, application, and even the most fundamental first step, comprehension, to be absorbed. Understanding green chemistry, how it impacts the business of pharmaceutical development, and how to implement its principles are questions being addressed throughout the industry. The remainder of this chapter will be devoted to addressing the most important facts, dispelling common myths, and exploring future opportunities for green chemistry in the pharmaceutical industry.

Fact 1: There is no universally accepted definition of *green*.

Greenness is not an absolute; it is a continuous endeavor. Science continues to evolve. What is environmentally responsible today may not be considered *green* enough tomorrow. Similarly, an efficient, low-hazard batch process may be replaced in the future with an even more efficient, safer, and environmentally preferable continuous process. The lack of a finite endpoint for *greenness* should not limit the ability or interest in striving for it. Instead, incremental improvements should invigorate the desire to innovate toward more economical, efficient, and less hazardous syntheses.

To illustrate this continuous improvement, one single product has won the award on two occasions. In 2005, Merck & Co., Inc., received the Presidential Green Chemistry Challenge Award for its catalytic synthesis for sitagliptin, the active ingredient in Januvia™, a treatment for type II diabetes. As compared to the original process, the new process eliminated 220 pounds of waste for each pound of active pharmaceutical ingredient (API) manufactured and increased the yield by nearly 50%.[8] Although the process was recognized for its successful demonstration of the principles of green chemistry, there remained opportunities for improvement, namely, an inadequate reaction stereoselectivity necessitating a crystallization step and high-pressure reaction conditions requiring expensive, specialized manufacturing equipment and an expensive rhodium catalyst. Collaboration between Merck and Codexis, Inc., led to an improved, greener route for the manufacture of sitagliptin. The new enzymatic process eliminated the high-pressure hydrogenation, all metal reagents from the previous award-winning technology, and the chiral purification step, resulting in a 56% improvement in productivity with existing equipment, a 10–13% overall increase in yield, and a 19% reduction in overall waste generation.[7] Merck and Codexis were acknowledged for this achievement in 2010 with the Presidential Green Chemistry Challenge Award, proving that what is green today could become even greener tomorrow.

With regard to individual compounds, even the apparently safest chemical can be toxic in large doses. Illustrations of Paracelsus' sentiment of "the dose makes the poison" would include a substance as ubiquitous as water. It is similarly difficult to definitively state that chemical X or process Y is *green*. The 12 principles and related metrics can be used to make relative determinations; although it is often difficult to develop a process that adheres to all 12 principles. For example, an efficient catalytic process controlled by process analytical technology could increase product yield, reduce the number of process steps, minimize the cycle time, utilize less solvent and energy, and reduce waste generation (principles 1, 2, 3, 5, 6, 8, 9, and 12). Most would consider this a very *green* process. However, notably absent from the list of benefits is that renewable feedstocks were not used in the development of this pharmaceutical ingredient, which is inherently toxic in order to be efficacious, and it does not degrade innocuously in the environment (principles 4, 7, and 10). The limited availability of renewable materials is a challenge in the ability to meet principle 4. However, this is beginning to change with more materials becoming commercially available and viable. Examples include solvents such as 2-methyltetrahydrofuran and cyclopentyl methyl ether, as well as nonpyrophoric reagents such as alkali silica gels for reductions or DABAL-Me for methylations. It is more challenging to design a pharmaceutical product that is efficacious and, upon excretion, degrades benignly in the environment, but it is not impossible.

Most importantly, it should become an accepted fact that there is no universally accepted definition of *green* and there is no perfect application of the 12 principles. To recognize this not as a hindrance, but as an opportunity to innovate, will allow *greener* to become attainable. Don't let perfect get in the way of possible.

Fact 2: The pharmaceutical industry generates a disproportionate amount of waste per kilogram of API produced.

With the many synthetic advances over the years and the realization that waste has long-term impacts, it is no longer acceptable to

Table 1.3 E-factor for sectors of the chemical industry

Industry sector	Product tonnage	E-factor
Oil refining	10^6–10^8	ca 0.1
Bulk chemicals	10^4–10^6	<15
Fine chemicals	10^2–10^4	5–50
Pharmaceuticals	10^1–10^3	25–100+

generate tremendous quantities of waste to produce high-value products, even life-saving drugs. The amount of waste generated per kilogram of product produced, a metric commonly known as the E-factor, is orders of magnitude higher in the pharmaceutical industry than in other segments of the chemical enterprise (Table 1.3).[9]

$$\text{E-factor} = \frac{\text{By-products (kg)}}{\text{Product (kg)}}$$

A benchmarking study performed by the American Chemical Society's (ACS) Green Chemistry Institute® (GCI) Pharmaceutical Roundtable suggests that these E-factors are reasonably consistent with current industry data. The roundtable is a partnership between the ACS GCI and member pharmaceutical companies dedicated to the integration of green chemistry and green engineering in the global pharmaceutical industry. The roundtable chose to benchmark against the process mass intensity (PMI), a measure of the material used per kilogram of API produced.

$$\text{PMI} = \frac{\text{Total mass in a process or process step (kg)}}{\text{Mass of product (kg)}}$$

As compared to the E-factor, the PMI is focused on the efficient use of materials in the process, rather than a calculation of the waste generation. Although the roundtable chose to benchmark against the PMI, the conclusions were consistent with the earlier E-factor reports. The median PMI across the seven companies participating in the study was 120 kg material used/kg API.[10] The study also recognized that the four highest PMI values were from three different companies, indicating that the values are not a reflection of one company's contribution. Although complexity was not specifically evaluated in the study, it was also noted that the process with the highest PMI was a complex process with greater than eight steps.

The study broke down the materials used into categories: solvent, reactant, process water, and others, concluding that ~80% of the PMI was related to solvent and water use. Utilizing that baseline, calculations based on annual sales with some assumptions about average dose and daily selling price suggest that as much as 3 million kilograms of solvent waste could be co-produced in the manufacture of the API.[11] Even making the most conservative assumption that all of the waste could be incinerated at a low cost, there is a tremendous potential for significant financial savings if the industry were able to minimize the amount and hazard of the solvents used.

Process complexity (number of transformations) and strict quality requirements have been a logical part of the industry's defense against these inefficient processes. Assuming these constraints are a fact of doing business in the pharmaceutical industry, green chemistry can still substantially improve the E-factor or the PMI. The median PMI of 120 kg/kg has been demonstrated to be reduced by 80% or more with the application of the green chemistry principles. Pfizer improved the synthesis of pregabalin, the active ingredient in Lyrica, an anticonvulsant drug used to treat chronic pain, reducing its E-factor from 86 to 9. In 2007, Pfizer switched from a classical chemical enantiomeric resolution to an enzyme-based resolution engineered as a continuous process. The improvements minimized solvent use by an estimated 185,000 tons of solvent (a 90% reduction) and 15,000 tons of starting material (a 50% reduction) based on production through 2020.[12]

Bristol–Myers Squibb received the Presidential Green Chemistry Challenge Award in 2004 for their plant cell fermentation (PCF) technology process for paclitaxel,[7] the active ingredient in the anticancer drug Taxol®. PCF replaced the conventional process that extracted a paclitaxel building block from the leaves and twigs of the European yew tree. The conventional process required 11 chemical transformations, 7 isolations, 13 solvents, and 13 organic reagents and other materials. Compared to the conventional process, the PCF process had no chemical transformations and eliminated six intermediates. During its first five years, the PCF process was estimated to eliminate ~71,000 pounds (32 metric tons) of hazardous chemicals and other materials. In addition, the PCF process eliminated 10 solvents and 6 drying steps, saving a considerable amount of energy.

The pharmaceutical industry does generate a disproportionate amount of waste per kilogram of product produced, but it is also true that green chemistry is the mechanism to turn this statement into a falsehood. The pharmaceutical industry can become competitive with bulk and fine chemicals through the use of green chemistry. The important takeaway is that green chemistry has demonstrated the potential to radically change the face of the pharmaceutical industry and will continue to do so.

Fiction 1: We are already doing green chemistry; we just don't call it that.

Process chemistry and green chemistry have common goals; the distinction between the two is the intention. Both aim to develop processes that are robust and efficient. Characteristics of a high-quality, cost-efficient, scalable process are consistent with the principles of green chemistry. A reduction in the number of transformations generally translates to a more energy-efficient process. Improved process quality is generally consistent with waste reduction with less off-specification material, etc. At times these environmental improvements are the aim of the process improvement to minimize regulatory liability or improve operational flexibility, but by and large they are generally more a by-product of good process chemistry. Green chemistry, in contrast, is by definition the intentional design of a process or product to be less hazardous. The intentional application of the 12 principles yields the highest efficiency potential that exists for a given process.[13]

Green chemistry operates at the intersection of a crossroads of decision points on cost, product performance, and environmental responsibility (see Fig. 1.1). To be considered green chemistry, the process or product designed according to the 12 principles must not only be environmentally favorable but also must perform equally or better than its current alternative and must be economically viable.

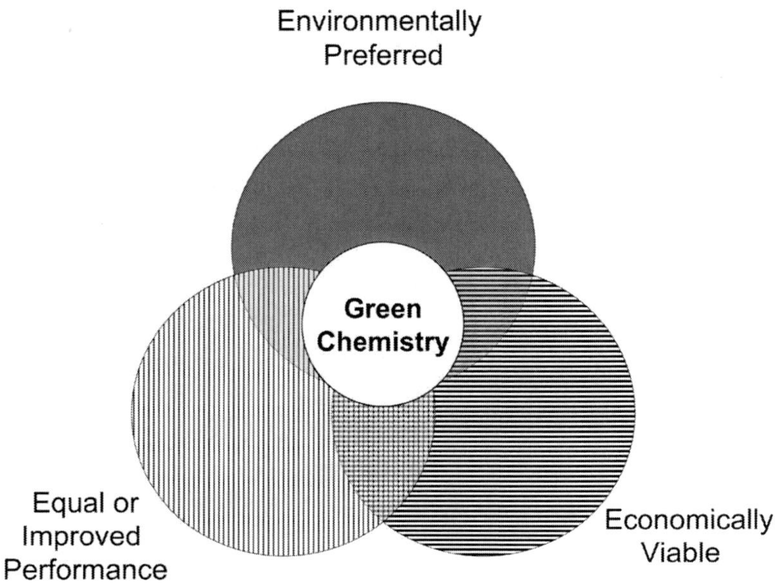

Figure 1.1 Green chemistry at the intersection.

This is best explained by an example outside the pharmaceutical industry: I purchase a cost-competitive, environmentally preferred cleaning product for everyday use in my home. Upon using it, I notice that it does not clean as well as the previous product, and I need to use a larger amount of the product to clean the intended area. In this case, the product is obviously not as effective as the current alternative, which translates into the product not really being as environmentally preferred, because it takes more product (and all of its associated production material and energy impact) to meet the intended need. Similarly, if the most effective, environmentally preferred cleaner is too expensive to purchase, then there is no market and the environmental impacts of the product go unrealized. Within the pharmaceutical industry, companies have yet to utilize environmental claims as a market differentiator. However, the same rationale applies. The US Food and Drug Administration (FDA) will not approve a product that has a lower quality profile or is less efficacious than its alternative even though it may be environmentally preferred.

Louis Pasteur once said, "Chance favors the prepared mind."[14] Empowering and enabling scientists to make informed green chemistry decisions will help improve the likelihood of developing the highest-efficiency process. Many pharmaceutical companies, including AstraZeneca, GlaxoSmithKline, Eli Lilly, Merck, Pfizer, and others, have developed training programs for their scientists to share the fundamentals of green chemistry and to provide them with the tools available to implement the 12 principles. Solvent selection guides, reagent selection guides, metrics, research lectures from academics making breakthrough innovations, etc., all help to enable chemists to make informed decisions and to integrate the 12 principles into their daily responsibilities. Ideally, training programs for current scientists would evolve to be less introductory and more focused on company-specific resources. To facilitate this evolution, green chemistry, toxicology, and other relevant topics should be introduced earlier during undergraduate and graduate degree programs. Similarly, the industry should recognize the importance of this knowledge when hiring new employees and request this green chemistry skill set as a preferred job qualification. Ultimately, the most sought-after, highly qualified candidates will be prepared to implement green chemistry as an integrated component of good process chemistry.

Perhaps in the future, this fictional statement could become a truth.

Fiction 2: Green chemistry costs money.

The pharmaceutical industry has faced increasing pressure to develop quality medicines faster and for less money, while facing ever greater generic competition. In 2000, generic drugs accounted for 49% of prescriptions filled as compared to a resounding 74% in 2010. Recent research also shows the time until new medicines face competition within their class has decreased from 10.2 in the 1970s to 2.5 years in 2000–2003. With the cost to develop a new drug up to $1.3 billion and a development timeline of 10–15 years,[15] it stands to reason that one would be hesitant to implement green chemistry if it had the potential to challenge these numbers even further.

So, the question really is, how can green chemistry be implemented in a way that not only preserves the current budget but provides added value back to the bottom line? If existing resources, namely, personnel, instrumentation, and equipment, can be used to implement green chemistry, it can be demonstrated that the application of green chemistry saves significant money, and a substantial amount at that. Reports of annual savings of $1–10 million per year are arguably an average. When additional capital has been required, examples have routinely demonstrated a favorable return on investment.

Boehringer Ingelheim had a process for a macrocyclic hepatitis C protease inhibitor that was too difficult and costly to run at a large scale because of the energy used and waste produced. In addition, they estimated it would require a multimillion dollar capital investment to run the process at scale. By redesigning the process to include a modified ring-closing metathesis (RCM), the process could be run in existing equipment, negating the need for further capital expenditures. Additional benefits included a 20-fold higher concentration, 50 times less catalyst, and $1/50^{th}$ the reaction time, with an E-factor of 28.[16]

Merck reported that a new process for imipenem, a component of the broad-spectrum antibiotic drug PRIMAXIN, eliminated the use of methylene chloride, increased product yield, and improved product quality, while adding manufacturing flexibility. The cost of the process change was $34 million, yet with $14 million per year in cost savings generated by the change, the investment was deemed justifiable.[17]

It is fairly uncommon for the industry to explicitly state the financial impact of a process change; it tends to favor a description of the qualitative and quantitative changes to yield, cycle time, waste generation, and raw material use. However, even with the generalities, one can calculate the estimated savings on average of $5–10 million per year, using conservative assumptions. Abbott Laboratories reportedly saved $12–13 million per year on changes to its clarithrymycin process.[18] Bristol–Myers Squibb reported saving $7.5 million per year in association with their green chemistry improvements on irbesartan.[19] A GlaxoSmithKline process for manufacturing a diabetes drug improved yield by 37%, used 81% less solvent, 30% less water, and less than half the energy, resulting in a raw material and waste disposal estimated cost savings of $175 million per year at production scale.[16]

Certainly, the financial profile for green chemistry in existing equipment is better than one requiring capital expenditure. However, even those examples requiring additional outlays have been successful in demonstrating a favorable return on investment. The cost benefits are even greater when the 12 principles are integrated at the earliest stages of development, allowing the beneficial results to be realized throughout the product life cycle.

Fiction 3: Green chemistry cannot be implemented in discovery/medicinal chemistry.

Recent reports show that only 2 in 10 medicines recoup the investment made by the companies in developing them.[15] It is fully appreciated that there could be a hesitance to integrate anything into discovery that has the potential to distract from the purpose of identifying promising novel compounds in large numbers as quickly as possible. The 12 principles of green chemistry should be implemented in R&D as early as possible to better align their design intent with company goals in order to fully capitalize on the innumerable benefits of higher efficiency and reduced impact throughout all phases of development. Ultimately, it becomes more a question of how to implement green chemistry rather than of its overall merit.

It could similarly be argued that one should not implement green chemistry during late-stage development or manufacturing for fear of jeopardizing the quality profile of the drug and its regulatory filing. Like the example of the Boehringer Ingelheim process for the macrocyclic hepatitis C protease inhibitor described earlier, it could take significant time, and possibly money, to achieve a scalable process. Yet designing a process with the 12 principles in consideration from the outset would greatly minimize downstream hurdles and improve the cost profile and time to market.

Medicinal chemists need to have resources available to them to enable the implementation of the 12 principles in a manner that does not hinder the ability to meet their primary job functions. These resources may appear to be different than those used in process chemistry organizations, but they are designed with the same information; the display just changes. For example, a solvent selection guide may provide detail for a process chemist to understand the environmental impacts in air, water, and waste, to enable them to make informed decisions with regard to available equipment, manufacturing location, and regulatory limitations. The medicinal chemist may use a simplified guide that summarizes relative solvent *greenness* to enable faster decisions.

Pfizer developed a solvent selection guide based on the assessment of worker safety, process safety, and environmental and regulatory considerations and then translated the outcome into a simple guide for the medicinal chemists. Results showed a 50% reduction in chlorinated solvent use across research divisions over a three-year period.[20] Even sites with increases in the number of chemists during that period were able to report a 50% reduction in chlorinated solvent use.

The ACS GCI Pharmaceutical Roundtable developed a reagent selection guide based on the design of one developed by Pfizer. For each reaction in the guide, there is a visual representation in the form of a Venn diagram of the various reagents and how they compare on three measures: scalability, wide utility, and greenness. The ideal reagent would fall at the intersection of all three. This visual guide has received favorable feedback from the members of the roundtable and has been utilized by scientists within these respective corporations.

"Restraint rather than constraint" was a theme first shared by Pfizer and echoed throughout the industry when discussing how to integrate green chemistry in medicinal chemistry organizations. There should be a focus on awareness and providing options to chemists to make informed decisions quickly, without interfering in the ability to meet performance objectives. A chromatography example[21] demonstrates this point well. Chromatography generates large amounts of silica waste that is not easily recycled. To minimize this waste, silica cartridges may be reused multiple times following a polar eluent flush to "clean" the column and minimize the risk of cross contamination. Educating chemists to make them aware of the potential to recycle the cartridges up to five times, depending on the manufacturer's recommendations, would not only minimize waste but also avoid the double economic penalty of disposing of waste and purchasing new materials.

Integration of green chemistry into medicinal chemistry organizations is still in its infancy. Yet companies have recently begun moving past concerns about whether it is appropriate to pursue this integration and on to the question of how to do so in an appropriate manner. In many cases it is the medicinal chemists themselves that are interested in implementing the 12 principles into the discovery paradigm. Capitalizing on this interest within the organization should enable initiating an informed, deliberate program based on restraint rather than constraint.

Future 1: The application of green chemistry in the design and manufacture of biopharmaceuticals is of increasing importance.

There is a misconception that biopharmaceuticals are inherently *green* because of the biological nature of the drugs and the heavy reliance on water in their manufacture. It is true that the manufacturing process typically requires fewer organic solvents and less hazardous materials. However, water use, energy consumption, and waste generation are areas that also are in need of significant improvement.

A study by Ho *et al.*[22] analyzed waste streams from the manufacture of therapeutic proteins by fermentation and concluded that these processes required approximately 10 to 100 times more water per kilogram of product compared to small-molecule drugs. The manufacturing processes utilized large amounts of water and more water still for the supportive operations, such as water for injection, equipment cleaning and sterilization, waste processing, etc. This translated into significant energy consumption as well and common compounds such as NaCl and other acids and bases ending up as salts in the aqueous waste. One-time-use consumables are quite common and generate substantial waste. Minimizing downstream processing steps, improving yield, and recycling water and other processing materials would improve the *greenness* of therapeutic proteins.

The expansion of the industry into biologics, predicted to grow at double-digit rates, twice as fast as traditional pharmaceuticals,[23] emphasizes the opportunity to direct attention to the integration of the 12 principles into this growing area. Improvements will not just make a positive impact on the process environmental and cost profile. Water minimization is of increasing importance as water scarcity is a global concern that could affect future operations and corporate citizenship.

According to the World Health Organization, one in three people around the world has insufficient clean water to meet daily needs.[24] Managing freshwater to ensure all global needs are met, including health, cleanliness, food production, industrial use, and energy, is essential, and countries have the right to regulate accordingly.[25] Beyond corporate citizenship, business impacts could range from potential business interruption resulting from a limitation and potential delay in the supply chain, government regulations, and others. Investors[26] are reportedly beginning to assess reliance on water resources and the vulnerability of their portfolios to the problems of water availability and pollution.

There is not a more direct example of the life cycle impacts of green chemistry than that presented by water scarcity. The pharmaceutical industry develops medicines to meet the needs of those suffering from diseases, including those caused by unhealthy and unhygienic conditions related to the problem of water scarcity. Green chemistry is inherent in the mission of the pharmaceutical industry, which is to develop medicines to help patients live longer, healthier, and more productive lives. Developing them in a manner that threatens the health and safety of those patients is not an effective means to meet this goal. With the increasing importance of biopharmaceuticals, understanding and minimizing water usage, waste generation, and energy use will have a positive effect not only on the process itself but also on more global challenges.

Future 2: Enhancement of the green toolbox will enable the pharmaceutical industry and the broader chemical enterprise to integrate the 12 principles across the product life cycle.

As innovative as the pharmaceutical industry is known to be, it is somewhat restricted in its ability to *be green* by the limited tools available to it. It is not the direct responsibility of the industry to develop *greener* alternatives to commonly used reactions, to develop renewable solvents and other raw materials, or to educate future generations of scientists, etc. These components comprise the "green toolbox," which is largely empty at this point. Academia, the broader chemical industry, the government, and others have a role in enlarging the toolbox to enable the pharmaceutical and other industries to more effectively integrate green chemistry into their operations.

Pharmaceutical companies have taken leadership in this area through their work in the ACS GCI Pharmaceutical Roundtable where they collaboratively address noncompetitive challenges with respect to integrating green chemistry into the industry. The strategic priorities are to inform and influence the research agenda, develop tools to help scientists make informed decisions, educate current and future scientists on the 12 principles and their application, and do all of this globally, recognizing the industry is not limited in geography.[27] The priorities developed in 2005 are still relevant today, not because of a lack of achievement, but simply because there is so much more to do. To meet the needs of pharmaceutical research, development, and manufacturing, the toolbox would have several categories, or drawers, to fill. Some examples are summarized next:

- Chemistry
  Nobel Prize–winning metathesis is one chemistry advance that has been added to the toolbox for the pharmaceutical industry and others. Impactful chemistries do not have to be prize-winning technologies to make a demonstrated change for the industry.

One of the first accomplishments of the roundtable was to identify and publish a list of 12 key green chemistry research areas.[28] These are 10 reactions and 2 more general areas that are commonly used in the industry and in need of greener alternatives. By narrowing down the scope of work to a dozen areas, there may be a greater chance for impactful change in the short term. The roundtable has also awarded over $1 million to academic institutions worldwide to accelerate research in these areas. As research continues to evolve, the list of reactions will need to be reevaluated.

- Engineering

  Advancements in engineering have the potential to revolutionize how the industry operates and change its environmental profile. For example, the industry primarily operates in batch mode compared to the alternative of continuous processing, which is just being introduced. Predictions are that microreactors will be in widespread use in the pharmaceutical sector by 2017.[29] For the appropriate synthesis, continuous processing has the potential to improve efficiency and energy consumption, while minimizing the use of solvents and other raw materials as well as waste generation.

  Further application of process analytical technology (PAT) will reap environmental benefits as well. The goal of PAT is to understand and control the process, which is fundamental to green chemistry. PAT represents an essential tool to allow for real-time, in-process monitoring and control prior to the formation of hazardous substances (green chemistry principle 11).

  As new technologies are developed to improve quality and efficiency, the industry will benefit from engineers and chemists working collaboratively to integrate these new endeavors into the business.

- Biocatalysis

  As demonstrated earlier with the collaboration between Merck and Codexis on the award-winning enzymatic process for sitagliptin, biocatalysis is an important tool in the green toolbox. Enzymes are highly efficient and selective and allow reactions to be run in an aqueous environment under ambient conditions, minimizing energy use and organic solvents.[30] Biocatalysis can help achieve many of the 12 principles and significantly minimize process metrics such as E-factor or the PMI.

- Renewable materials

  The application of green chemistry principle 7, use of renewable materials, is extremely limited at this time by the availability of the materials themselves. Development and subsequent marketing of materials made from renewable resources are an essential first step, which is outside the scope of the pharmaceutical industry. Renewable materials are somewhat like biologics in that there is an assumption of greenness. However, to truly understand the relative greenness of the respective material, a life cycle analysis should be completed to compare the two materials side by side. Not all cases will show a favorable outcome to the renewably sourced option.

- Less hazardous materials

  Solvents comprise a majority of the PMI of an API. Engineering, PAT, catalysts, etc., all have the potential to minimize the volume of solvent used. However, even with these technologies, solvents will continue to be a large contributor to the environmental footprint of the industry. The availability of less hazardous alternatives is an essential resource for chemists.

  The ACS GCI Pharmaceutical Roundtable has placed an emphasis on minimizing the use and inherent hazard of solvents in recent years. Collaborations such as the roundtable have the opportunity to raise awareness of this need and encourage chemical manufacturers to engage in this area. A group of corporations speaking on behalf of their industry has more impact and is more demonstrative of the market potential than individual companies trying to achieve the same objective independently.

- Guides
  As previously mentioned, tools such as solvent selection and reagent guides enable chemists and engineers to make more informed decisions. Guides can be company specific, industry wide, or even broader. They should be readily accessible, be understandable, and allow the user to make a quick and informed decision in the best interest of the process. The types of guides are limited only in our ability to design them. Perhaps in the future there will be broad chemistry and/or engineering guides to help determine which reaction, equipment, or technology may be most efficient and *greener*.

Green chemistry, with its 12 principles spanning the life cycle of a chemical or process from cradle to grave, is so broad that one industry cannot accomplish it all successfully without collaboration among industry sectors, academia, the government, and other nongovernmental groups. Further enhancement of the green toolbox will enable the pharmaceutical industry and the broader chemical enterprise to integrate the 12 principles across the product life cycle.

Future 3: Improved collaboration through the supply chain is essential for complete integration of green chemistry.

The development of a single API is not restricted to one plant, one company, or one geographic region any more. Whether the focus is on R&D or manufacturing, some major corporations are planning to outsource as much as 40% of their API production needs.[31] Others hire third parties to assemble many of the chemistry components for drug discovery and development. Biology outsourcing is a newer but growing trend in the drug industry.[32] With the continuing trend of outsourcing in the pharmaceutical industry, it is becoming necessary to work through the supply chain to effectively integrate green chemistry.

It is routine practice for quality and environmental, health, and safety organizations to audit third parties to assure compliance with company policy and applicable regulations. Less common is the sharing of environmental metrics, such as the PMI, although companies have started to request this information from their suppliers. To get an accurate calculation of the PMI, it is necessary to have data on the manufacture of all of the intermediates, many of which are now done outside company walls. The PMI benchmarking study by the roundtable presented in 2007 recognized the difficulty in obtaining a complete set of data from commonly available starting materials through to the final, dried form of the API. The roundtable study repeated two years later had a noticeable improvement in the number of processes with complete data. To improve the ability of pharmaceutical companies in obtaining data from their third-party suppliers, the roundtable began a project to standardize the PMI request. The outcome of the project should facilitate the communication between the supplier and the pharmaceutical company, resulting in an increase in nonproprietary information shared between entities. One could envision a future where the PMI, or another measure of *greenness*, could be used as a distinguishing factor between suppliers.

Conclusion

The purpose of this chapter was to provide an introduction to green chemistry in the pharmaceutical industry and to concisely address important facts and myths while also highlighting opportunities for the future. In summary:

- There is no universally accepted definition of *green*, and there is no perfect application of the 12 principles. To recognize this is not a hindrance but an opportunity to innovate.
- The pharmaceutical industry generates a disproportionate amount of waste per kilogram of product produced, and green chemistry is the mechanism to change that.
- Process chemistry and green chemistry have common goals; the distinction between the two is the intention.

- Green chemistry may cost money, depending on the application, but the return on investment is favorable in justifying the investment.
- Green chemistry can and should be implemented in discovery/medicinal chemistry organizations, with an emphasis on restraint, not constraint.
- The application of green chemistry in the design and manufacture of biopharmaceuticals is of increasing importance.
- Enhancement of the *green* toolbox will enable the pharmaceutical industry and the broader chemical enterprise to integrate the 12 principles across the product life cycle.
- Improved collaboration through the supply chain is essential for the complete integration of green chemistry.

Green chemistry as a field is still young, having only been defined in the late 1990s. Yet it has already been explicitly recognized in a Nobel Prize–winning discovery, and with that award a technology has been introduced that greatly benefits the pharmaceutical industry, among others. The strong link between green chemistry and the industry is a demonstration of the great potential to impact the triple bottom line (economic, environmental, and social success) of a corporation dedicated to developing products to help patients live longer, healthier lives. There has been much success to date, as demonstrated in this chapter as well as in those following. The excitement lies in the opportunities yet to be realized and the innovations yet to be dreamed.

References

1. www.acs.org
2. Nobelprize.org. (14 November 2010). *Press Release: The Nobel Prize in Chemistry 2005.* http://nobelprize.org/nobel_prizes/chemistry/laureates/2005/press.html
3. Anastas, P. T.; Warner, J. C. (1998). *Green Chemistry: Theory and Practice.* New York, NY: Oxford University Press.
4. 42 U. S. C. 13101 Pollution Prevention Act of 1990.
5. www.phrma.org

6. Thomas, P. (15 October 2009). *Pharma's green evolution*. http://www.pharmamanufacturing.com/articles/2009/149.html
7. The Presidential Green Chemistry Challenge Awards Recipients 1996-2010, US Environmental Protection Agency, Office of Pollution Prevention and Toxics, 744K10003, June 2010.
8. The Presidential Green Chemistry Challenge Awards Program: Summary of 2006 Entries and Recipients, US Environmental Protection Agency, Office of Pollution Prevention and Toxics, EPA 744-R-06-003, June 2006.
9. Sheldon R. A. (1994). *Chem. Tech.*, **24**, 38–47.
10. Henderson, R. K.; Kindervater, J.; Manley, J. B. (June 2007). *Lessons Learned through Measuring Green Chemistry Performance—the Pharmaceutical Experience*, paper presented at the 11th Annual Green Chemistry and Engineering Conference, Washington, DC. http://portal.acs.org/portal/fileFetch/C/CTP_005585/pdf/CTP_005585.pdf
11. Cue, B. W.; Berridge, J.; Manley, J. B. (2009). "PAT & Green Chemistry: The intersection of Benign by Design and Quality by Design," *Pharm. Eng.*, **29**(2), 8–20.
12. Ritter, S. K. (2010). *Chem. Eng. News*, **88**(43), 45–47.
13. Tucker, J. L. (2006). *Org. Proc. Res. Dev.*, **10**, 315–319.
14. http://www.lhup.edu/~dsimanek/sciquote.htm
15. Pharmaceutical Research and Manufacturers of America (March 2010). *Pharmaceutical Industry Profile 2010*. Washington, DC: PhRMA.
16. Thayer, A. M. (2010). *Chem. Eng. News*, **87**(29), 19–22 (and references therein).
17. www.p2pays.org/ref/34/33215.pdf, Pennsylvania DER, 1994.
18. CHI Process R&D Summit, Philadelphia 2005.
19. PR Chem 2006, Rio Grande, Puerto Rico 2006.
20. Alfonsi, K.; Colberg, J.; Dunn, P. J.; Fevig, T.; Jennings, S.; Johnson, T. A.; Kleine, H. P.; Knight, C.; Nagy, M. A.; Perry, D. A.; Stefaniak, M. (2008). *Green Chem.*, **10**, 31–36.
21. Richter, D. T. (June 2010). *Green Chromatography: Small is Big!*, paper presented at the 14th Annual Green Chemistry & Engineering Conference, Washington, DC.
22. (a) Ho, S. V. (30 September–1 Oct 2010). *Pfizer, Environmental Considerations in Biologics Manufacturing*, paper presented at the 3rd International Symposium on Green Processing in the Pharmaceutical and Fine Chemical Industries, University of Massachusetts, Boston, MA;

(b) Ho, S. V.; McLaughlin, J. M.; Cue, B. W.; Dunn, P. J. (2010). *Green Chem.*, **12**, 755–766.

23. Mullin, R. (2010). *Chem. Eng. News*, **88**(43), 26–28.
24. http://www.who.int/features/factfiles/water/en/
25. 1 Oct 2010 UN News Service. http://www.un.org/apps/news/story.asp?NewsID=36308&Cr=water&Cr1=#
26. *The Bubble Is Close to Bursting: A Forecast of the Main Economic and Geopolitical Water Issues Likely to Arise in the World during the Next Two Decades*, draft for discussion at the World Economic Forum Annual Meeting, 2009.
27. ACS GCI Pharmaceutical Roundtable Business Plan, June 2009.
28. Constable, D. J. C.; Dunn, P. J.; Hayler, J. D.; Humphrey, G. R.; Leazer, J. L., Jr., Linderman, R. J.; Lorenz, K.; Manley, J.; Pearlman, B. A.; Wells, A.; Zaks, A.; Zhang, T. Y. (2007). *Green Chem.*, **9**, 411–420.
29. Bush, N.; Trybus, S. (1 October 2010). *Systematic Quantification of Micro Reactor Potential in Pharma and Specialty Chemicals*, paper presented at the 3rd International Symposium on Green Processing in the Pharmaceutical & Fine Chemical Industries, Boston, MA.
30. Ran, N.; Zhao, L.; Chen, Z.; Tao, J. (2008). *Green Chem.*, **10**, 361–372.
31. Thayer, A. M. (2010). *Chem. Eng. News*, **88**(6), 21–26.
32. McCoy, M. (2010). *Chem. Eng. News*, **88**(40), 24–25.

Chapter 2

Green Chemistry in Drug Development

Apurba Bhattacharya,[a] Siddhartha Akasapu,[a] and Rakeshwar Bandichhor[b]

[a] *Department of Chemistry, Texas A&M Kingsville, Kingsville, Texas 78363, USA*
[b] *Center of Excellence, Research & Development, Integrated Product Development, Dr. Reddy's Laboratories Ltd., Survey Nos. 42, 45, 46, & 54 Bachupally, Qutubullapur, Ranga Reddy Dist 500072, Andhra Pradesh, India*
kfab002@tamu.edu and rakeshwarb@drreddys.com

2.1 Introduction

The steady growth of the pharmaceutical industry in the past half century, coupled with the more recent and rapid growth in the generic business, has been of great benefit to human health. Today there are 4,500 drugs in development, including experimental ones, which comprise about 70% of the total. Nearly 44% of Americans are taking at least one prescription drug.[1,2] The annual value of US sales was estimated at $330 billion for 2010, showing an annual growth rate over many decades of nearly 8.5%. From 1950 to 2010, the total resident population of the U.S. increased from 151 million to 305 million people, representing an average annual growth rate of 1.2%.[3–5] This expansion in population and the accompanying

Scalable Green Chemistry: Case Studies from the Pharmaceutical Industry
Edited by Stefan G. Koenig

ISBN 978-981-4316-49-1 (Hardcover), 978-981-4364-81-2 (eBook)
www.panstanford.com

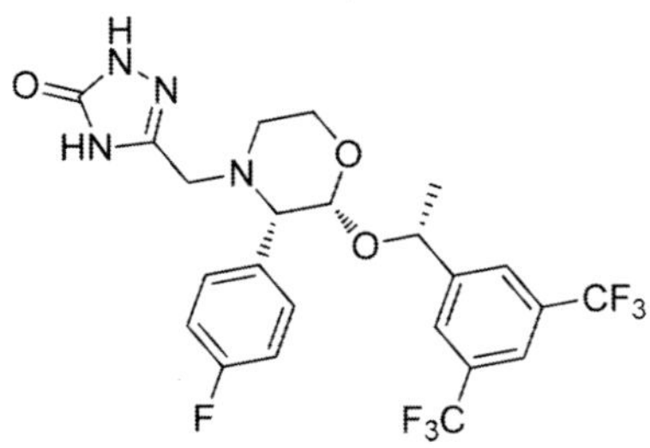

1 (Aprepitant)

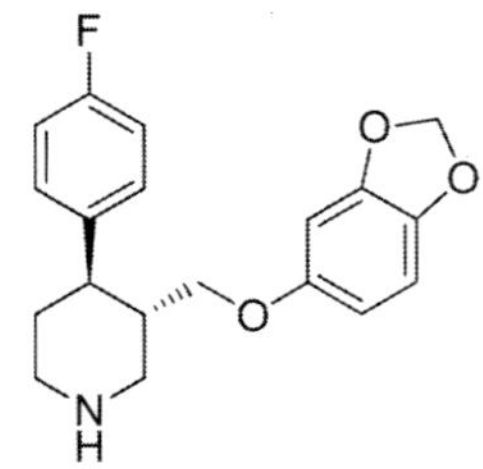

2 (Paroxetine)

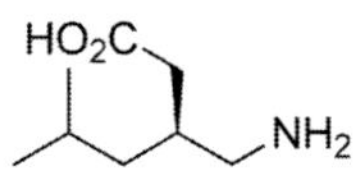

3 (Pregabalin)

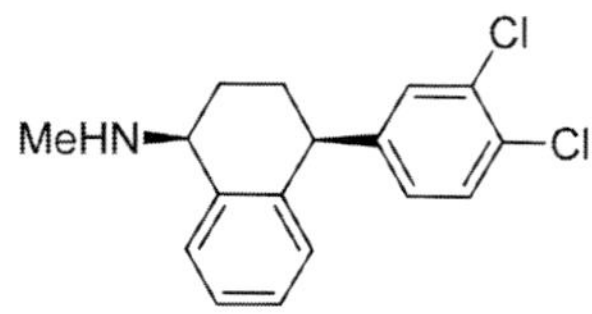

4 (Sertraline)

Figure 2.1 Structure of APIs.

growth of pharmaceutical production to meet customers' needs came with a significant environmental cost. Unfortunately several industries are relying on outdated methods to produce active pharmaceutical ingredients (APIs), making it the most wasteful chemical producer as measured by the E-factor, a relation of waste produced versus desired material obtained.[6–9] Yet times are changing, as the examples in this chapter demonstrate. In their drive to harness waste, many pharmaceutical producers have adopted the 12 green chemistry principles and redesigned their chemical processes to important medicines. Eyeballing innovative ways to reformulate their current manufacturing process has led them to trim down their impact on the environment. The present chapter confers about the evolution in production from initial strategy to greener approaches for several significant pharmaceutical APIs (Fig. 2.1).

2.2 Aprepitant

Aprepitant **1**, marketed as Emend®, belongs to a class of neurokinin-1 (NK-1) receptor antagonists developed by Merck.[10]

2.2.1 *Medicinal Chemistry Route*

Synthesis of aprepitant **1** involved reduction of morpholinone **5** with *L*-selectride, followed by quenching with 3,5-bis(trifluoromethyl)benzoyl chloride to afford **8** in 79% yield. This intermediate was then subsequently converted to aprepitant **1** using a series of reactions (Scheme 2.1).[11] The route was environmentally unfriendly due to stoichiometric titanium metal-mediated olefination and a protecting group strategy.

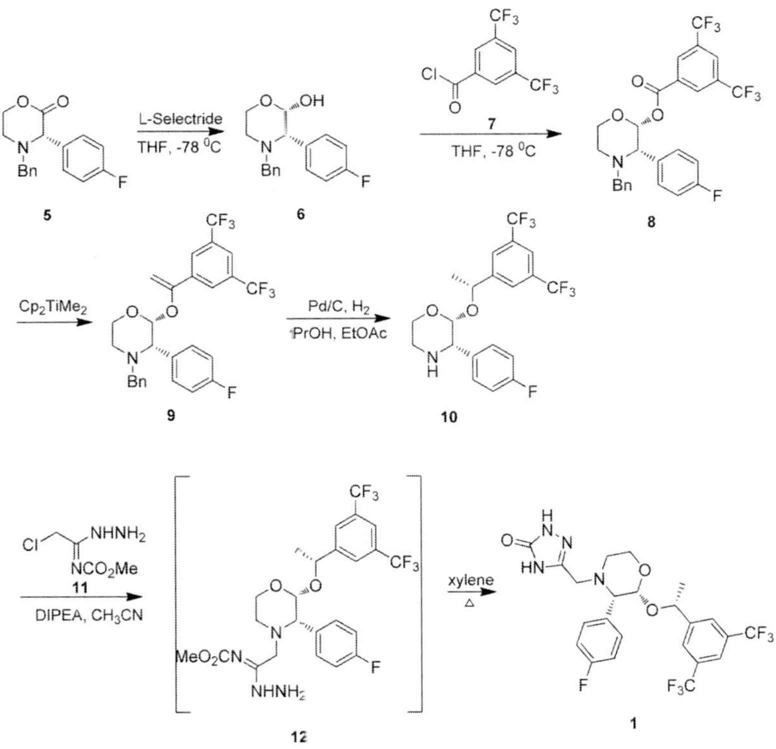

Scheme 2.1 Medicinal chemistry route for aprepitant **1**.

2.2.2 *Early Development Route*

To embark on a new route, *p*-fluorobenzaldehyde **13** (Scheme 2.2) was converted to nitrile derivative **14**. After several subsequent steps, **14** led to intermediate *trans*-**6**, which was found to be suitable as a key material for **1** with respect to cost, convenience, and scalability.

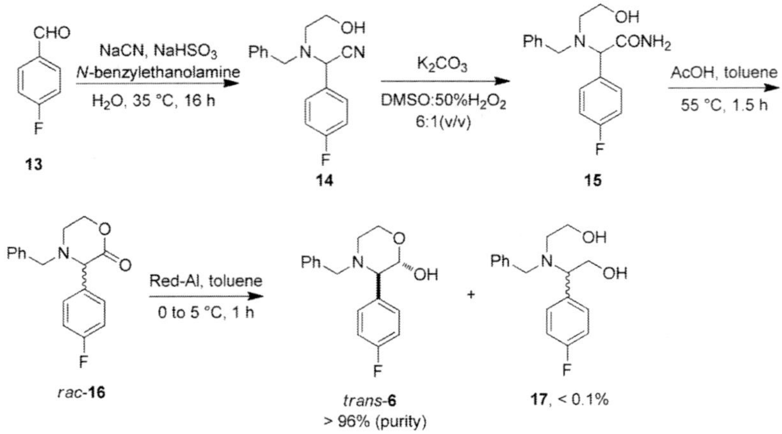

Scheme 2.2 Early development route for morpholinol derivative *trans*-**6**.

Reduction of acetophenone derivative **18** gave the corresponding hydroxy compound **19** as a crystalline solid. Subsequent treatment with HBr/H_2SO_4 gave the desired bromo compound **20** (Scheme 2.3).

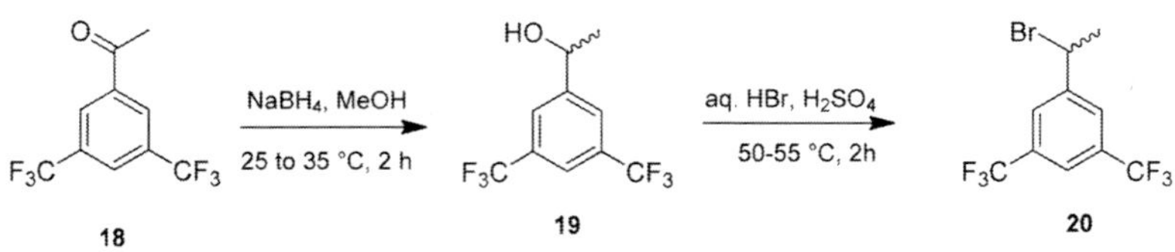

Scheme 2.3 Process chemistry route for a bromo derivative.

Treatment of racemic *trans*-**6** with bromide **20** furnished acetal derivative **21** (obtained as a 2:1 mixture of racemic diastereomers; Scheme 2.4). Hydrogenolysis of the hydrochloride salt of **21** in the presence of Pd-C/H_2 produced amine **10a**, with the same diastereomeric ratio. The major pair of amine enantiomers **10b** was isolated by crystallization from isopropyl alcohol, with subsequent

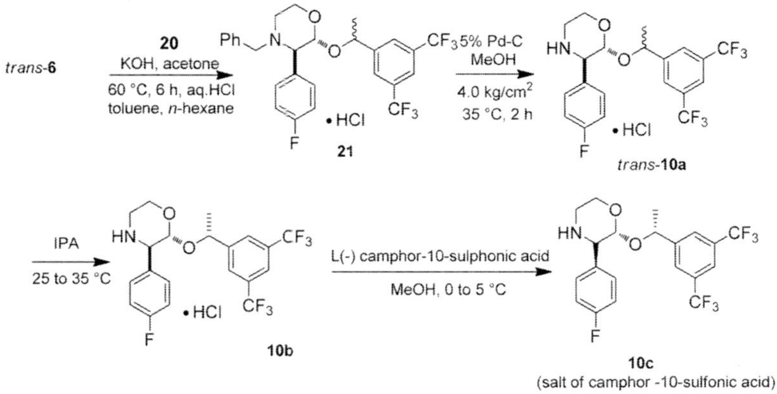

Scheme 2.4 Early development route for camphor-10-sulfonic acid salt of advanced intermediate **10c**.

resolution being accomplished using *L*-(-)-camphorsulfonic acid to afford the lone stereoisomer **10c**.

Conversion of **10c** into **10** was carried out in the manner according to Zhao *et al.*[12] (Scheme 2.5). Imine **22** was reduced with *cis*-stereoselectivity to afford **10**. Subsequent alkylation with chloroacetamidrazone, following a known process,[13] and cyclization of the resulting intermediate **12** afforded **1**.[14] The route

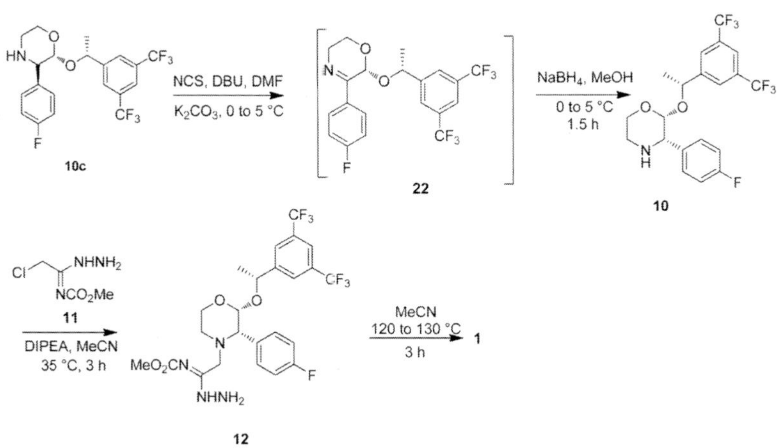

Scheme 2.5 Endgame process strategy for **1**. *Abbreviations*: DBU, 1,8-diazabicyclo[5.4.0]undec-7-ene; DMF, dimethylformamide.

avoided the olefination strategy but still employed a wasteful *N*-chlorosuccinimide (NCS)-mediated transformation, as well as resolution and protection-deprotection steps.

2.2.3 *A Greener Approach*

Embarking on a new approach, Merck researchers condensed **23** with **24** at ambient temperature to afford **25** in high yield.[15,16] The resultant **25** rearranged to **27** in a mixture of water and tetrahydrofuran (THF), accompanied by decarboxylated by-product **26**. Nevertheless, heating **23** with aqueous glyoxylic acid in THF regenerated the rearranged starting material **25** at the expense of evolving formaldehyde. To compensate for the loss of substrate **24** due to thermal decomposition, excess glyoxylic acid (2.3 eq.) was employed to obtain a better yield of **27** (76%), as shown in Scheme 2.6.[10]

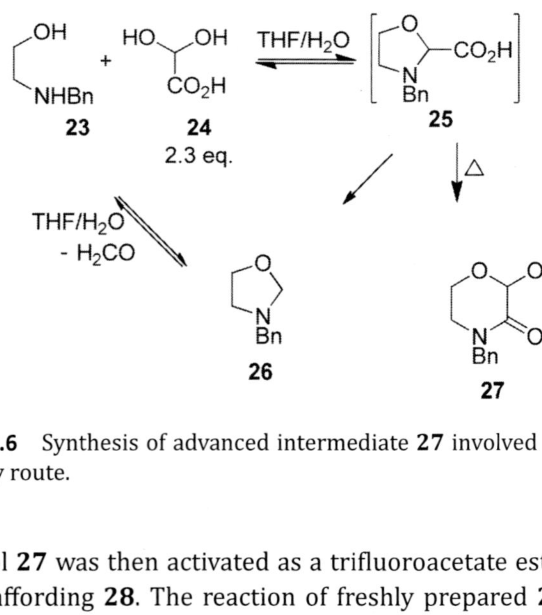

Scheme 2.6 Synthesis of advanced intermediate **27** involved in the green chemistry route.

Lactol **27** was then activated as a trifluoroacetate ester without a base, affording **28**. The reaction of freshly prepared **28** with **29** catalyzed by $BF_3 \cdot Et_2O$ subsequently yielded a 55:45 mixture of **30a** and **30b** (Scheme 2.7). This process was efficient and scalable, while the dimeric impurity **31** did not present much concern.[10]

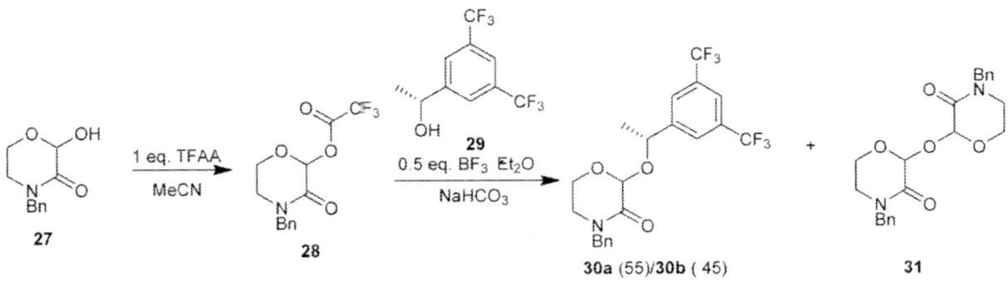

Scheme 2.7 Synthesis of advanced intermediate **31** involved in the green chemistry route.

A robust one-pot process for intermediate **10** was developed to align with the earlier work (Scheme 2.8). After the completion of the Grignard reaction with **30a**, the mixture was quenched and subsequently hydrogenated using Pd/C catalyst. The hydrogenation selectivity in favor of the desired *cis* isomer **10** increased greatly when the quenched Grignard reaction mixture was acidified with *p*-toluenesulfonic acid (TSA) before hydrogenation. The nature and amount of acid were perilously important in achieving the best reaction performance,[10] though morpholine derivative **10** was isolated from the reaction mixture as its hydrochloride salt.

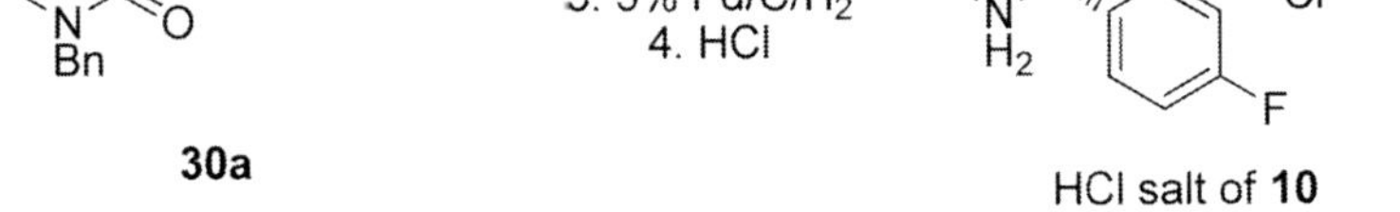

Scheme 2.8 Synthesis of advanced intermediate **10** involved in the green chemistry route.

Conversion of **10** · HCl to aprepitant **1** was achieved either by the reaction with chloroamidrazone **11,** as shown in Schemes 2.1 and 2.5 or with chloromethyltriazolinone **32** (Scheme 2.9).[17]

The Merck route involved a convergent approach to achieve the synthesis in three steps. Chiral alcohol **29** was used to set

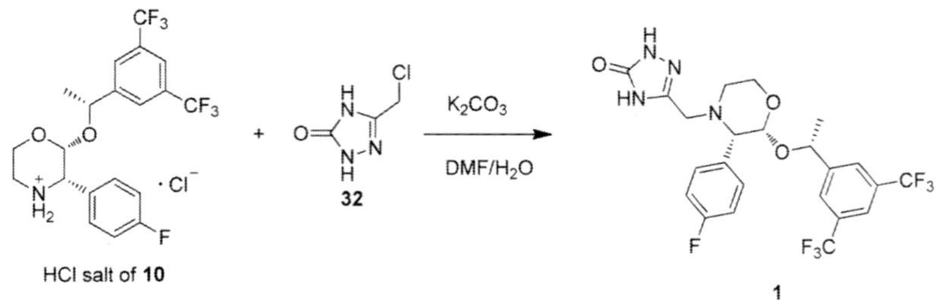

Scheme 2.9 Completion of synthesis of **1**.

the stereochemistry at the remaining pro-chiral centers in two subsequent steps. This process was significantly better from an environmental standpoint and avoided the operational hazards inherent with certain reagents, for example, sodium cyanide, dimethyl titanocene, and gaseous ammonia. By practicing this atom-economical green approach, the energy requirements were lowered significantly and approximately 41,000 kg of waste per 1,000 kg of aprepitant **1** were eliminated.

2.3 Paroxetine

Paroxetine **2**, developed by GlaxoSmithKline under the trade names Seroxat and Paxil®, is found to be effective as a first-line therapy for generalized anxiety disorder.[18]

2.3.1 *Medicinal Chemistry Route*

The reaction of methyl ester **33** and 4-fluorophenylmagnesium bromide **34** afforded a mixture of 1-methyl-4-(4′-fluorophenyl)-piperidine-3-carboxylic acid methyl ester **35a/35b**. The base-hydrolysis/equilibration of **35a/35b** gave the corresponding *trans* carboxylic acid derivative **36**, which after conversion to acid chloride, esterification, resolution, reduction, chlorination, and alkylation with 3,4-methylenedioxyphenoxide (sesamol) afforded **40**. Finally, demethylation via treatment with chlorourethane and decarbamoylation afforded **2** (Scheme 2.10).[19] The route was not eco-friendly for several reasons. It was lengthy, involving protection-deprotection as well as resolution steps, and further, the recycling of undesired diastereomer was not possible.

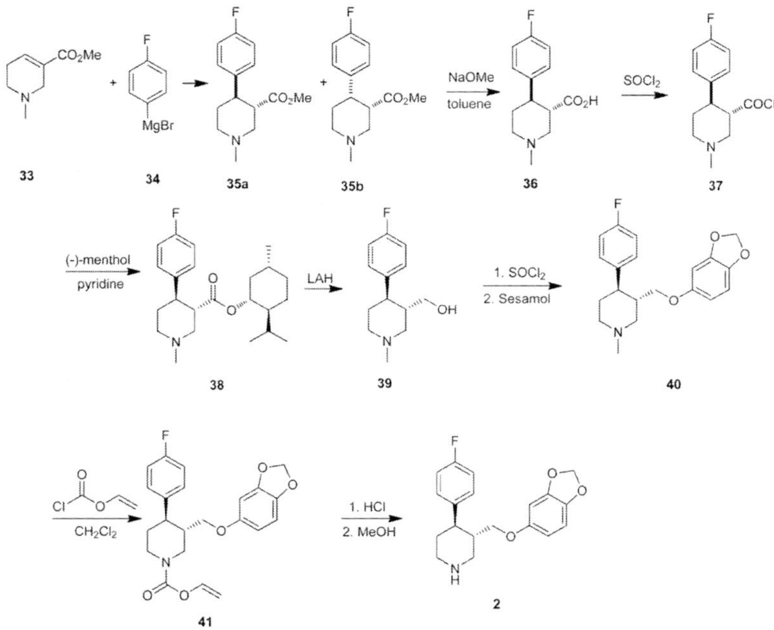

Scheme 2.10 Medicinal chemistry route for **2**.

2.3.2 *Early Development Route*

Bis-ester **43** was prepared by the reaction of *p*-fluorobenzaldehyde **42** with ethyl acetoacetate, followed by esterification (Scheme 2.11). The hydrolytic resolution of **43** employing pig liver esterase afforded acid ester **44**. To invert the stereochemistry at the C-3 center and achieve reduction of the ester functionality of **44** simultaneously, the reactions were strategically performed, that is, first deprotonation and second reduction to afford alcohol **45**. Intermediate **45** was mesylated and treated with benzylamine to provide lactam **46**. Homologation to **47** followed by reduction gave amidoalcohol **48**. Etherication with sesamol afforded the penultimate precursor **49**, which was subjected to hydrogenolysis to yield paroxetine **2**.[20]

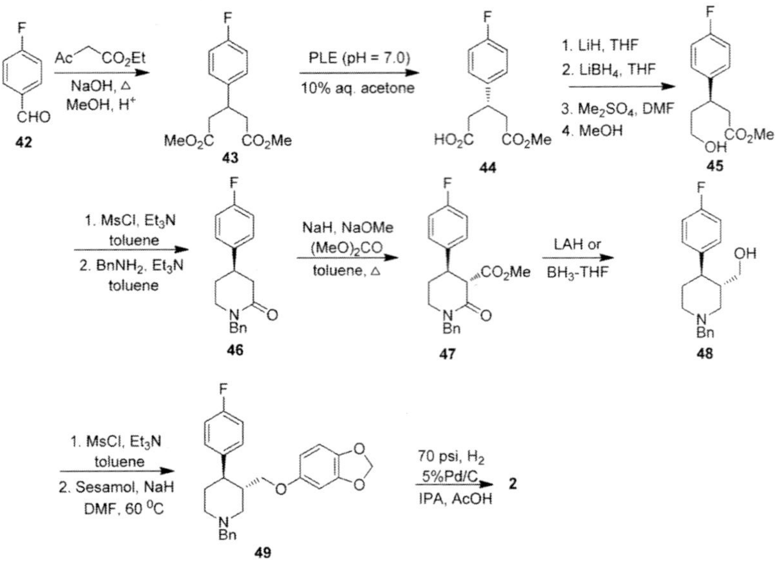

Scheme 2.11 Early development route for **2**.

The route benefited from a biocatalytic desymmetrization to avoid chemical resolution but did create significant amounts of aqueous waste during workup.

2.3.3 *A Greener Approach*

A more efficient, enzymatic desymmetrization strategy was recently reported, as shown in Scheme 2.12.[21] The key step involved a protease-catalyzed desymmetrization of meso diester **50** utilizing subtilisin Carlsberg, a member of a family of serine endopeptidases isolated from *Bacillus subtilis*. Meso diester **50** was synthesized from **42** via a series of reactions, namely, Knoevenagel condensation, Michael addition, and intramolecular cyclization, to afford (3*R*, 4*S*)-ester **52** as a single enantiomer after decarboxylation of intermediate **51**. Demethylation and global reduction followed by etherification with sesamol afforded **2**. This route was also based on the enzymatic desymmetrization concept; however, the yield of the overall transformation was almost double that of the process

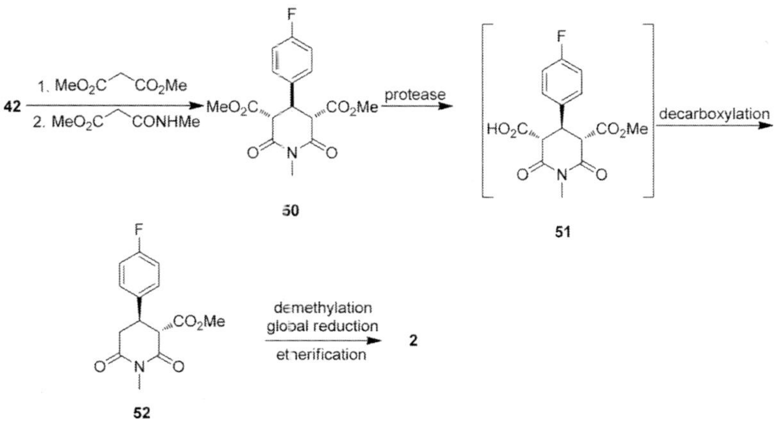

Scheme 2.12 Greener route for **2**.

chemistry route, resulting in a greener, shorter, and more cost-efficient synthesis for **2**.

2.4 Pregabalin

Pregabalin **3**, sold as Lyrica© by Pfizer, is prescribed for the management of neuropathic pain and epilepsy and it is one of the bestselling candidates.[22]

2.4.1 *Medicinal Chemistry Route*

4-Methylpentanoic acid **53** was converted to the corresponding acid chloride with thionyl chloride and further reacted with Evans's chiral auxiliary to afford intermediate **54**. Thereafter, a cascade of reactions afforded **3**, as shown in Scheme 2.13.[22] The route involved the use of a chiral auxiliary and protection-deprotection steps and was notably waste producing.

2.4.2 *Early Development Route*

Condensation of diethyl malonate with isovaleraldehyde **58** gave **59**, followed by Michael addition of potassium cyanide to afford cyanodiester **60**. Conversion of **60** to **61** followed by resolution

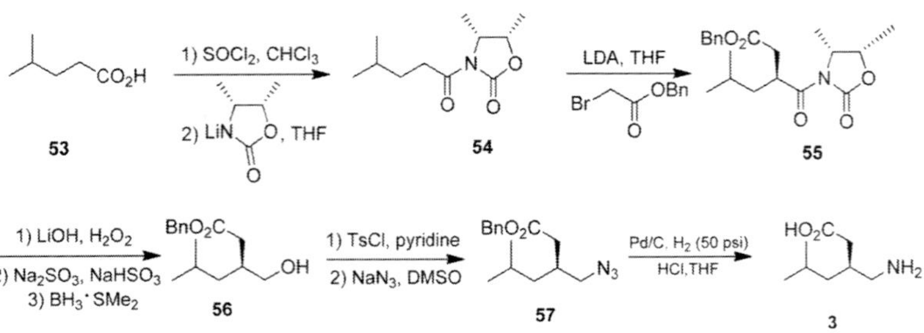

Scheme 2.13 Medicinal chemistry route for **3**.

produced pregabalin **3** (Scheme 2.14).[23] The route is operationally simple, high yielding, and cost effective. However, it is not green since the undesired enantiomer (*ent*-**3**) could not be recycled.

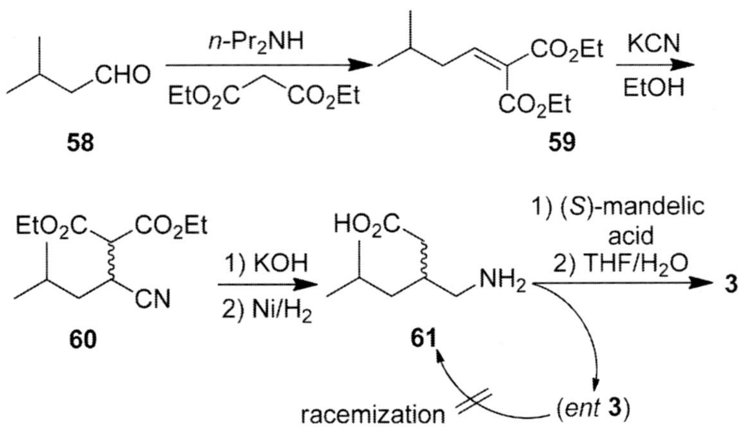

Scheme 2.14 Early development route for **3**.

2.4.3 *A Greener Approach*

Enzyme-catalyzed resolution of cyano diester **60** with a special lipase, sold as Lipolase™, afforded the desired (*S*)-mono acid enantiomer (in solution, i.e., not isolated), which was subjected to decarboxylation, hydrolysis, and hydrogenation to afford **3** (Scheme 2.15).[24] Recycling of the undesired (*R*)-cyano diester **62** rendered the process high yielding and very eco-friendly, with a reduction

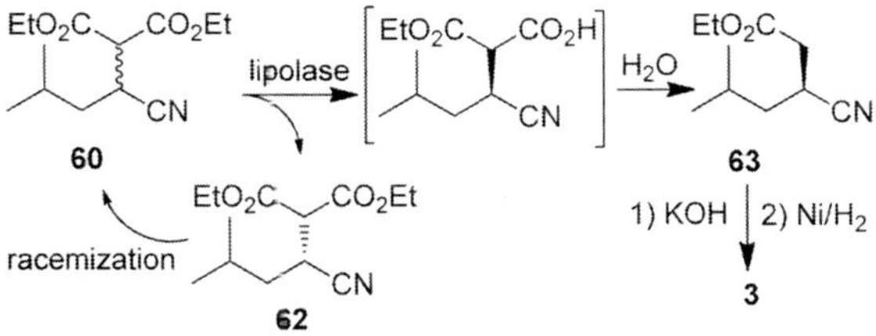

Scheme 2.15 Green chemistry route for **3**.

in the E-factor from 86 to 17. Almost 38 million liters of alcoholic organic solvents and nearly 2,000 metric tons of raw materials (mandelic acid, cyano diester **63**, and Ni reagent) were eliminated on an annual basis.

2.5 Sertraline

Sertraline hydrochloride 4, better known in the U.S. as Zoloft®, is a selective inhibitor of serotonin reuptake and is used for the management of depression.[25–27]

2.5.1 *Medicinal Chemistry Route*

3,4-Dichlorobenzophenone **64** reacted with diethyl succinate **65** to afford intermediate **66**, which was subjected to HBr-glacial acetic acid to obtain intermediate **67**. Hydrogenation of **67** afforded intermediate **68**. Subsequent acid chloride formation followed by acylation afforded intermediate **69**. Dehydration then led to Schiff base **70**, followed by further steps, including a classical resolution, to afford **4** (Scheme 2.16).[26] The route generated a significant amount of titanium dioxide waste and involved a wasteful borohydride-mediated imine reduction to obtain **71**.

2.5.2 *Early Development Route*

Condensation of intermediate **69** with methylamine followed by reduction of the intermediate **70** with Pd/C under an H_2 atmosphere afforded a *syn*- and *anti*-diastereomeric mixture (6:1) of the amine

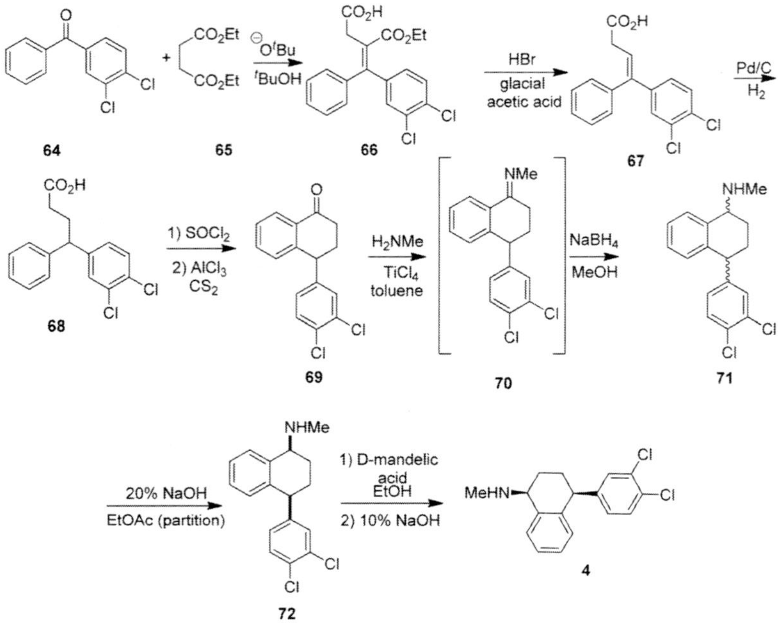

Scheme 2.16 Medicinal chemistry route for **4**.

71, isolated as the hydrochloride salt. The desired *syn* amine **72** was then selectively crystallized and resolved to produce the **4** · *D*-(-)-mandelic acid salt. This route, though efficient, employed environmentally unfriendly and waste-producing reagents (TiO_2 and $MeNH_2{\cdot}HCl$) (Scheme 2.17).[28] The route utilized waste-producing and hazardous titanium tetrachloride. Substituting the sodium borohydride reduction with Pd/C-catalyzed hydrogenation of imine **70** rendered the process marginally better from an environmental perspective.

2.5.3 *A Greener Approach*

Intermediate **69** was converted to **70** by using methylamine in a suitable pressure-rated vessel, avoiding the use of any titanium reagents In the reduction step to obtain **71**, Pd/$CaCO_3$ was found to be the catalyst of choice, offering an improved *cis*/*trans* ratio (20:1). Without further purification, intermediate **71** was treated with

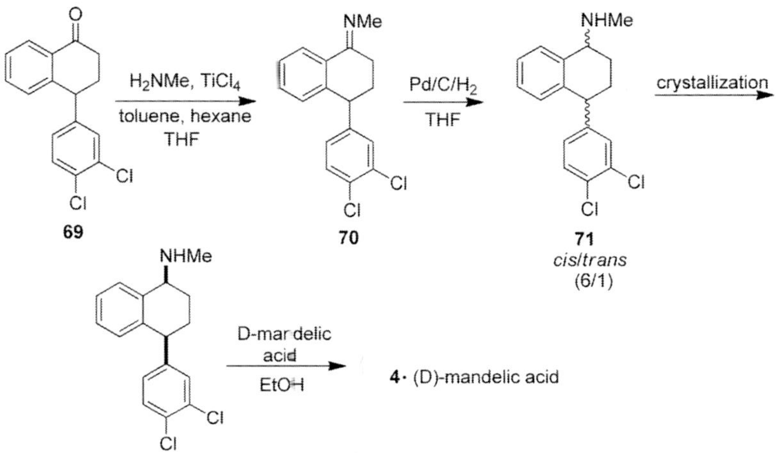

Scheme 2.17 Early development route for **4**.

D-(-)mandelic acid to obtain **4**·*D*-(-)-mandelic acid salt (Scheme 2.18).[28]

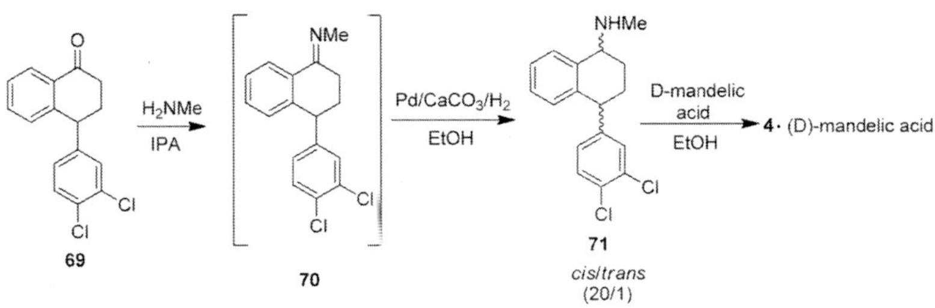

Scheme 2.18 Green chemistry route for **4**.

Intermediates **70** and **71** were carried forward without isolation. The number of solvents was reduced from five to two and total volumes reduced to 24% of the original process. This process became more robust, operationally simple and cost effective. With more than 100 metric tons of sertraline hydrochloride manufactured every year, 440 tons/year of TiO_2-$MeNH_2$·HCl wet-cake waste and over 40 tons of the unwanted *trans* isomer waste were eliminated. Additionally, worker safety was improved due to the absence of over 140 metric tons of $TiCl_4$ per year.[28]

2.6 Conclusions

Consideration of environmental issues is best introduced during the early stages of development to achieve maximum sustainability in processes and products. Accordingly, incorporation of green chemistry principles into synthetic route design has evolved from intuitive efforts by select individuals toward an institutionalized practice among major pharmaceutical and biotechnology companies. This chapter presented several brief examples of successful green chemistry in drug manufacturing. These processes evolved from initial discovery routes through nascent development efforts to deliver solid green chemistry syntheses in the production of pharmaceutical APIs.

The development of drug molecules by employing green technologies is the foremost scientific pursuit gaining momentum within the drug industry, all with an eye toward reducing the environmental burden. Developing processes that avoid functional group protection/deprotection, limit the usage of solvents, allow reactions to progress at ambient temperature and facilitate catalytic C–C or C-heteroatom bond formation has proven to be extremely challenging. Thus, integration of green chemistry, green engineering and biocatalysis ought to be an essential consideration at the earliest stages of development.

This review chapter focused on evolving trends in the development of important drug molecules. The featured compounds were initially produced with lower process efficiency and higher manufacturing costs but are now being manufactured with minimized waste at lower cost after remarkable development efforts. These efforts were made to improve the lifetime impact of these medicines and improve conditions in the world. It is important to realize that we have borrowed this planet from our children and owe it to them to leave it a better place than we found it. Making medicines without creating unnecessary waste helps improve everyone's well-being.

Acknowledgments

We thank Dr. Reddy's Laboratories for supporting this work.

References

1. Wishart, D. S.; Knox, C.; Guo, A. C.; Cheng, D.; Shrivastava, S.; Tzur, D.; Gautam, B.; Hassanali, M. (2008). *Nucl. Acids Res.*, **36**, D901–D906.
2. US Department of Health and Human Resources. (2004). *Health*. Hyattsville: MD National Center for Health Statistics.
3. (2005). *IMS Health Reports*. Norwalk, CT: IMS Health.
4. Class, S. (2004). *Chem. Eng. News*, **82**(49), 18–29.
5. US Department of Health and Human Resources. (2006). *Health*. Hyattsville: MD National Center for Health Statistics.
6. Sheldon, R. A. (1994). *CHEMTECH*, **24**(3), 38–47.
7. Sheldon, R. A. (2000). *C R Acad. Sci. Ser. IIc. Chim.*, **3**, 541–551.
8. Sheldon, R. A. (2005). *Green Chem.*, **7**, 267–278 (and references therein).
9. Anastas, P. T.; Warner, J. C. (1998). *Green Chemistry Theory and Practice*. New York: Oxford University Press.
10. Brands, K. M. J.; Payack, J. F.; Rosen, J. D.; Nelson, T. D.; Candelario, A.; Huffman, M. A.; Zhao, M. M.; Li, J.; Craig, B.; Song, Z. J.; Tschaen, D. M.; Hansen, K.; Devine, P. N.; Pye, P. J.; Rossen, K.; Dormer, P. G.; Reamer, R. A.; Welch, C. J.; Mathre, D. J.; Tsou, N. N.; McNamara, J. M.; Reider, P. J. (2003). *J. Am. Chem. Soc.*, **125**, 2129–2135.
11. Hale, J. J.; Mills, S. G.; MacCoss, M.; Finke, P. E.; Cascieri, M. A.; Sadowski, S.; Ber, E.; Chicchi, G. G.; Kurtz, M.; Metzger, J.; Eiremann, G.; Tsou, N. N.; Tattersall, D.; Rupniak, M. J.; Williams, A. R.; Rycroft, W.; Hargreaves, R.; MacIntyre, D. E. (1998). *J. Med. Chem.*, **41**, 4607–4614.
12. Zhao, M. M.; McNamara, J. M.; Ho, G. J.; Emerson, K. M.; Song, Z. G. J.; Tschaen, D. M.; Brands, K. M. J.; Dolling, U. H.; Grabowski, E. J. J.; Reider, P. J.; Cottrell, I. F.; Ashwood, M. S.; Bishop, B. C. (2002). *J. Org. Chem.*, **67**, 6743–6747.
13. Dorn, C. P.; Finke, P. E.; Hale, J. J.; MacCoss, M. M.; Sander, G. S.; Shrenik, K. C.; Mark, S. H.; Timothy, L.; Tamara, W.; Brian, J. (1998). US Patent 5719147.
14. Elati, C. R.; Kolla, N.; Gangula, S.; Naredla, A.; Vankawala, P. J.; Avinigiri, M. L.; Chalamala, S.; Sundaram, V.; Mathad, V. T.; Bhattacharya, A.; Bandichhor, R. (2007). *Tetrahedron Lett.*, **48**, 8001–8004.
15. Woodward, R. B.; Gosteli, J.; Ernest, I.; Friary, R. J.; Nestler, G.; Raman. H.; Sitrin, R.; Suter, C.; Whitesell, J. K. (1973). *J. Am. Chem. Soc.*, **95**, 6853–6855.
16. Harwood, L. M.; Vines, K. J.; Drew, M. G. B. (1996). *Synlett*, 1051–1053.

17. Cowden, C. J.; Wilson, R. D.; Bishop, B. C.; Cottrell, I. F.; Davies, A. J.; Dolling, U. H. (2000). *Tetrahedron Lett.*, **41**, 8661–8664.
18. Johnson, T. A.; Curtis, M. D.; Beak, P. (2001). *J. Am. Chem. Soc.*, **123**, 1004–1005.
19. Crowe, D.; Jones, D. A.; Ward, N. (2001). WO 2001017966.
20. Yu, M. S.; Lantos, I.; Peng, Z. Q.; Yu, J.; Cacchio, T. (2000). *Tetrahedron Lett.*, **41**, 5647–5651.
21. Tao, J.; Xu, J. H. (2009). *Curr. Opin. Chem. Bio.*, **13**, 43–50.
22. Yuen, P. W.; Kanter, G. D.; Taylor, C. P.; Vartanian, M. G. (1994). *Bioorg. Med. Chem. Lett.*, **4**, 823–826.
23. Hoekstra, M. S.; Sobieray, D. M.; Schwindt, M. A.; Mulhern, T. A.; Grote, T. M.; Huckabee, B. K.; Hendrickson, V. S.; Franklin, L. C.; Granger, E. J.; Karrick, G. L. (1997). *Org. Process Res. Dev.*, **1**, 26–38.
24. Martinez, C. A.; Hu, S.; Dumond, Y.; Tao, J.; Kelleher, P.; Tully, L. (2008). *Org. Process Res. Dev.*, **12**, 392–398.
25. MacQueen, G.; Born, L.; Steiner, M. (2001). *CNS Drug Rev.*, **7**, 1–24.
26. Welch, W. M.; Kraska, A. R.; Sarges, R.; Koe, B. K. (1984). *J. Med. Chem.*, **27**, 1508–1515.
27. McRae, A. L., Brady, K. T. (2001). *Expert Opin. Pharmacother.*, **2**, 883–892.
28. Taber, G. P.; Pfisterer, D. M.; Colberg, J. C. (2004). *Org. Process Res. Dev.*, **8**, 385–388.

Chapter 3

Development of Green-by-Design, Practical Biocatalytic Processes

Gjalt W. Huisman[a] and Steven J. Collier[b]

[a]*Codexis Inc., 200 Penobscot Drive, Redwood City, CA 94063, USA*
[b]*Codexis Laboratories Singapore Pte Ltd., 61 Science Park Road, Singapore 117525, Singapore*
gjalt.huisman@codexis.com and steve.collier@jmusa.com

3.1 Introduction

There is a growing need for more sustainable approaches to the manufacture of fine chemicals for pharmaceutical and agrochemical agents.[1] Of particular note is the challenge associated with the manufacture of chiral chemicals in an environmentally sound way. Traditional approaches to the asymmetric synthesis of chiral compounds often involve the use of resolving agents, stochiometric chiral reagents (e.g., chiral hydrides), auxiliaries, or catalysts dependent upon chiral ligands and heavy metals such as rhodium and ruthenium. The use of such reagents to induce asymmetry in a molecule is not truly sustainable, and such processes often suffer from less-than-perfect enantioselectivities, giving products that require yield-eroding upgrading via recrystallization, or demand

Scalable Green Chemistry: Case Studies from the Pharmaceutical Industry
Edited by Stefan G. Koenig

ISBN 978-981-4316-49-1 (Hardcover), 978-981-4364-81-2 (eBook)
www.panstanford.com

reaction conditions that require specialized plant equipment for operation, for example, high pressures and/or temperatures. In contrast, biocatalysts offer an attractive combination of exquisite selectivity and true sustainability, being biodegradable proteins produced from renewable resources. Directed evolution technologies developed in the early 1990s enabled the optimization of enzymes to function optimally in the desired process and provided a boost to the field.[2] This chapter aims to first introduce biocatalysis and directed evolution technologies and then give specific case studies on the development of efficient, sustainable, green-by-design biocatalytic processes.

3.2 Introduction on Biocatalysis and Directed Evolution

Biocatalysts have been used in synthetic chemistry for decades, and the technology is typically associated with hydrolytic enzymes—for example, for the production of (-)-lactam for abacavir,[3] or whole-cell fermentation processes such as remote hydroxylation, for example, for pravastatin.[4] Although it is generally accepted that enzymes are highly selective catalysts, they are also often considered to be active only in a tight process operating range in terms of temperature and co-solvents and require relatively high catalyst loadings and low substrate concentrations. The reason for such perceived[5,6] shortcomings is that enzymes have evolved in nature over millions of years to support *the fitness of the host*. They have developed as part of complex metabolic systems to provide certain levels of a product but to shut down before the levels of the product become toxic to the host organism. As a result, the activity of enzymes is often regulated via substrate and/or product inhibition. Enzymes that are capable of producing products at concentrations toxic to the host will negatively impact the fitness of the host, resulting in extinction.[7]

In contrast, for commercial applications, the volumetric productivity of a process is critical and industrial enzymes need to be capable of operating at substrate and product concentrations vastly higher than they would encounter in a natural environment. Product titers, a measure of the concentration of a compound in solution, of at least 10% w/w are typically required for commercially attractive chemical processes. In addition, enzymes should optimally be stable to aggressive conditions, tolerating organic co-solvents to solubilize hydrophobic substrates and products, and be able to withstand high temperatures needed to accelerate reaction rates and improve solubilities. To develop enzymes that can not only survive but be commercially viable under such challenging conditions, advanced enzyme-engineering technologies are leveraged.[7] Directed evolution utilizes state-of-the-art enzyme-engineering techniques to develop diverse but targeted, high-quality libraries of variant enzymes and employs high-throughput screening (HTS) technologies to sift through the diversity generated to identify mutations that are beneficial under the chosen screening conditions (e.g., high temperature, high concentrations of the substrate and product as well as co-solvent).[8,9] A schematic representation of directed evolution is shown in Fig. 3.1.

A key consideration in developing industrially viable enzymes through directed evolution is the need to screen under conditions

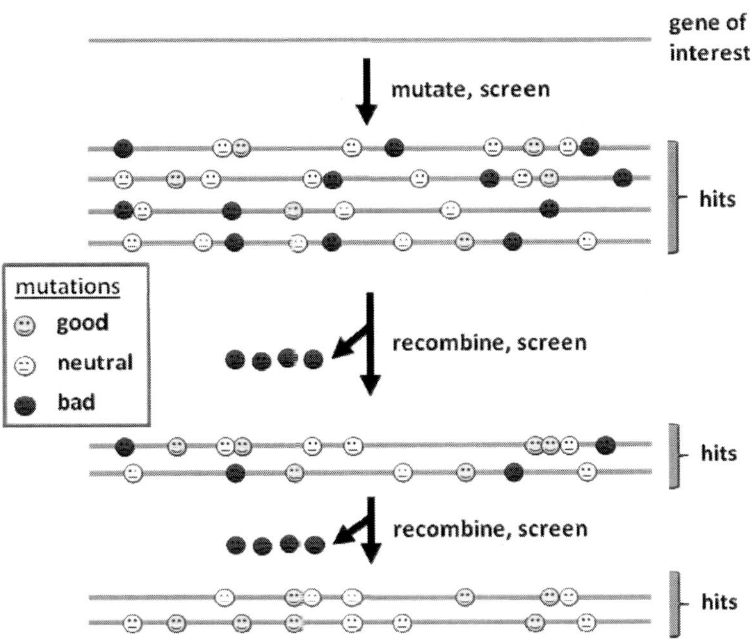

Figure 3.1 Directed evolution schematic.

that are representative of the desired commercial process.[10] Because the variants in these libraries are subjected to evolutionary pressure, and selected by their ability to survive under such stress, if the screening conditions are not truly representative, variants may be selected that are improved for function under the screening conditions and not the commercial-scale process conditions. Ultimately this boils down to what is sometimes called the first law of directed evolution: "You get what you screen for." Thus, it is essential to evolve the enzyme in the right direction, that is, under process-representative conditions. Following this approach, iterative rounds of directed evolution can provide truly process-viable biocatalysts by repeatedly adding beneficial mutations to a parent enzyme.

However, there is a need to develop biocatalysts more rapidly, and this has driven further developments in enzyme evolution technologies. To this end, collections of "preevolved" enzymes have been generated (Codex™ panels) on the basis of a recombinant backbone that is robust under process-type conditions.[11] The manufacture of such biocatalysts can be readily scaled up in large-scale fermentation of the recombinant *Escherichia coli* (*E. coli*) strain. These collections are phenotypically diverse and, as a population, contain variants with different combinations of mutations known to impart a range of substrate specificities and deliver different stereoisomers of a product. The amino acid sequence of each individual variant is known, and the data obtained from screening the collection can be processed using a protein sequence–activity relationship (ProSAR) algorithm to deconvolute and identify beneficial and deleterious mutations and thus guide new combinations of this diversity that can be expected to give further improvements in enzyme performance.[12] Examples of such collections include the ketoreductase (KRED), transaminase (TA), ene-reductase, nitrilase, acylase, monoamine oxidase, and halohydrin dehalogenase Codex™ panels.[13]

3.3 Biocatalysis and Green Chemistry

Before discussing the application of enzyme-engineering techniques to the development of scalable manufacturing processes of important pharmaceutical intermediates and active pharmaceutical ingredients (APIs), it is worth emphasizing that biocatalysts have very strong green credentials as compared to more traditional chemical approaches. If one considers the 12 principles of green chemistry, as formulated by Paul Anastas and John Warner,[14] it is easy to see how biocatalysis can meet many of these criteria. Biocatalysis and the 12 principles of green chemistry (Anastas and Warner):[14]

1. *Prevention: It is better to prevent waste than to treat or clean up waste after it has been created:* Biocatalysts are usually highly selective for a given transformation and as such do not produce many unwanted by-products. In addition, enzymes can be engineered to operate at high substrate loads under process-relevant concentrations to minimize solvent use.
2. *Atom economy: Synthetic methods should be designed to maximize the incorporation of all materials used in the process into the final product:* Biocatalysts, particularly the widely used ketoreductases and emerging transaminases, are more atom economical than many of the competing stoichiometric reducing or aminating agents.
3. *Less hazardous chemical syntheses: Wherever practical, synthetic methods should be designed to use and generate substances that possess little or no toxicity to human health and the environment:* Many traditional chemical transformations, including reductions, reductive aminations and oxidations, use hazardous stoichiometric redox agents (e.g., alkylboranes, metal borohydrides, pressurized hydrogen gas, or potent inorganic oxidants). In contrast, the stoichiometric agents used by biocatalysts are easily handled and considerably less hazardous (e.g., glucose, isopropanol, isopropylamine, or air/oxygen).
4. *Designing of safer chemicals: Chemical products should be designed to effect their desired function, while minimizing their toxicity:* Given that biocatalysts are typically used to manufacture known molecules, this principle is not as applicable.

5. *Safer solvents and auxiliaries: The use of auxiliary substances (e.g., solvents, separation agents) should be made unnecessary wherever possible and innocuous when used:* Biocatalytic reactions are highly selective, giving high-purity products with low levels of by-products, thus circumventing the need for auxiliary separation agents. In addition, enzymes operate effectively in aqueous environments (and water is an ideal green solvent) and can be engineered to perform in the presence of safe organic co-solvents, such as isopropanol.
6. *Design for energy efficiency: Energy requirements of chemical processes should be recognized for their environmental and economic impacts and should be minimized. If possible, synthetic methods should be conducted at ambient temperature and pressure:* Because biocatalysts operate efficiently under ambient or slightly elevated temperatures, reaction processes can avoid the requirements for extreme (high or low) temperature and/or high-pressure reactions.
7. *Use of renewable feedstocks: A raw material or feedstock should be renewable rather than depleting, whenever technically and economically practical:* Enzymes are themselves renewable materials as they are formed via overexpression of the corresponding gene in a host organism that consumes simple renewable feedstocks and synthesizes the enzymes from metabolic intermediates. In contrast, many traditional metallic catalysts are derived from precious metals, which are mined from finite resources. In addition, the reagents and by-products from biocatalytic reactions, such as KREDs or TAs, are renewable and readily recycled.
8. *Reduction of derivatives: Unnecessary derivatization (use of blocking groups, protection/deprotection, temporary modification of physical/chemical processes) should be minimized or avoided, if possible, because such steps require additional reagents and can generate waste:* Enzymes are highly selective catalysts and can react with pinpoint accuracy at a specific reactive center in a molecule, bearing multiple reactive positions. This superb regioselectivity can obviate the need for protecting groups.

9. *Catalysis: Catalytic reagents (as selective as possible) are superior to stoichiometric reagents:* Enzymes are, by definition, highly selective catalytic systems. In addition, through appropriate enzyme and process design, the requirement for stoichiometric co-factors can be avoided and these, too, can be operated catalytically.
10. *Design for degradation: Chemical products should be designed so that at the end of their function, they break down into innocuous degradation products and do not persist in the environment:* Although the products are designed for therapeutic use and may not be designed to degrade innocuously, enzymes themselves are proteins, made from common amino acids, and are readily degraded under standard waste treatment protocols.
11. *Real-time analysis for pollution prevention: Analytical methodologies need to be further developed to allow for real-time, in-process monitoring and control prior to the formation of hazardous substances:* Some biocatalytic reactions can be monitored by real-time analysis. For instance, ketone reductions that use a glucose/glucose dehydrogenase co-factor-recycling system need to be run in pH-stat mode to maintain a neutral reaction pH. This is achieved by feeding of an appropriate base, the addition of which is monitored continuously.
12. *Inherently safer chemistry for accident prevention: Substances and the form of a substance used in a chemical process should be chosen to minimize the potential for chemical accidents, including releases, explosions, and fires:* Enzymes catalyze reactions under benign conditions, avoiding high temperatures and/or pressures, and use low hazard donor molecules. For example, KREDs use isopropanol (or other alcohols) as stoichiometric reducing agents, generating only acetone (or another oxidized by-product). This is in contrast to the hazard associated with high-pressure hydrogenations or pyrophoric hydride reducing agents. Similarly, TAs utilize isopropylamine as an amine donor to avoid the use of high-pressure reductive aminations or the need to reduce imines with hydrides.

3.4 Ketoreductase Processes for Chiral Alcohols

Chiral secondary alcohols are common intermediates for pharmaceutical APIs. Many commercial approaches to these intermediates utilize stoichiometric and hazardous borane-based reagents, which may be variously irritants, corrosive, flammable, unstable, and/or water reactive. A second approach involves asymmetric hydrogenations, which are dependent upon heavy metal catalysts and chiral ligands. Where these approaches fail, the common fallback is a classical resolution procedure and a potential loss of >50% in yield as the undesired stereoisomer is removed (although racemization and reuse of the undesired isomer may improve process economics). Biocatalytic reductions enabled by KRED enzymes offer a greener, more sustainable alternative. KREDs (also known as carbonyl reductases and alcohol dehydrogenases) are ubiquitous in nature, and the sequences of more and more natural KRED genes are being reported. KRED-mediated reductions require the use of a co-factor (nicotinamide adenine dinucleotide [NADH] or reduced nicotinamide adenine dinucleotide phosphate [NADPH]) to provide the reducing equivalents needed, and enzymes are also available for the efficient regeneration of these expensive co-factors. Most commonly, co-factor regeneration is achieved through coupling with glucose dehydrogenase (GDH), which converts glucose to gluconic acid (essentially an irreversible reaction), or through using a KRED that also accepts isopropyl alcohol (IPA) as a substrate, giving acetone as the oxidized by-product (Fig. 3.2). In the latter case, the overall reaction is a coupled equilibrium (ketone + IPA gives chiral alcohol + acetone), but in the few cases where high conversion is not naturally achieved, the equilibrium can be driven toward the product by removal of the acetone by-product via distillation. Overall, both coupled systems use renewable materials as stoichiometric reducing agents, and the catalysts themselves are also renewable, readily obtainable by fermentation of *E. coli* strains overexpressing the gene of interest. There are a number of published reviews focusing on KREDs, which include discussions of the mechanism and substrate range of these enzymes,[15] applications in the fine chemical and pharmaceutical industries, and as part of

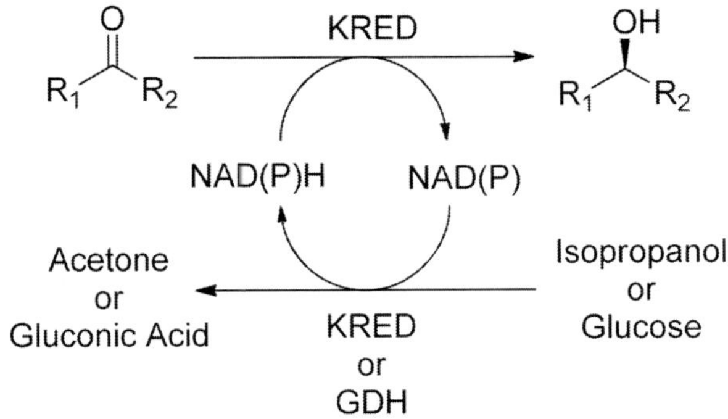

Figure 3.2 KRED-catalyzed ketone reduction coupled to co-factor regeneration.

more wide-ranging coverage of biocatalysts for pharmaceutically interesting targets.[16]

The use of KRED-catalyzed reduction is now an established strategy to achieve catalytic ketone reduction to provide a single enantiomer or diastereomer of product in very high purity.[17,18] However, with some substrates natural enzymes are not sufficiently active or not capable of delivering the product in high enough chiral purity. In such cases, the product thus obtained requires upgrading, resulting in concomitant loss of yield. The use of directed evolution technologies allows for fully tunable systems, in which the enzyme is optimized to provide high activity and outstanding selectivity for products that previously were produced with poor selectivity or were even inaccessible with natural enzymes. Simultaneously, the catalysts can be engineered to withstand the rigors of a commercial manufacturing environment, allowing them to withstand conditions intolerable for many natural KREDs. The following section details a number of examples of KRED-enabled processes developed by Codexis.

3.4.1 *Montelukast: Hydroxyester Intermediate*

Montelukast (developed and marketed as Singulair® by Merck) is an orally active leukotriene receptor antagonist used to control and relieve the symptoms of asthma and seasonal allergies. The molecule has a complex architecture and is formed from the pivotal alcohol

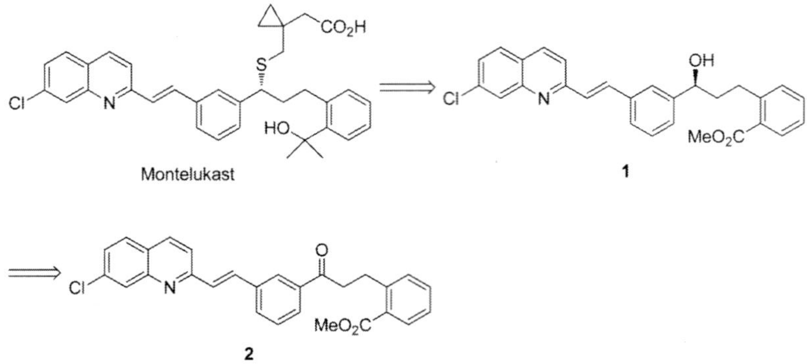

Figure 3.3 Retrosynthesis of montelukast.

ester intermediate **1** (Fig. 3.3). The original commercial route to the chiral alcohol involved reduction of the corresponding ketone **2** using a significant excess of diisopinocampheylborane (DIP-Cl) in tetrahydrofuran (THF) at low temperature. Postreaction, the crude product required upgrading via crystallization to increase the chiral purity from 97.5% e.e. to 99.5% e.e., resulting in isolated yields of 65–87%. Clearly, there was scope for improvement in the approach to this chiral alcohol, and it represented an attractive target for biocatalysis. As noted by Merck scientists,[19] a screen of our natural enzyme collection failed to provide an enzyme capable of producing the desired *S*-product. However, screening our collection of preevolved variants (based on *Lactobacillus kefir*) did identify an enzyme that could deliver the required product with perfect enantiopurity. The activity of this variant was far from commercial utility, so process development and enzyme optimization via directed evolution were initiated.

The process was developed to be a slurry-to-slurry reaction and operated at 45°C. Using a co-solvent mixture of isopropanol (acting both as co-solvent and as co-factor regenerant), toluene and buffered water provided a substrate solubility of ~0.01 g/L. As a result, the biocatalyst was evolved for increased robustness to organic co-solvent (>60% organics), increased thermostability (up to 50°C), and increased tolerance to the acetone co-product, a competitive inhibitor. The final enzyme variant was improved over 2,000-fold compared to the initial enzyme variant that exhibited

Table 3.1 Summary of the biocatalytic process for manufacture of the chiral alcohol for montelukast

Reaction conditions	100 g/L substrate load; 3.3% catalyst load; 1:5:3 PhMe:IPA:water; 45°C
Chiral purity of product (crude)	>99.5% e.e.
Isolated yield	>90%
Comments	Isolation via direct filtration

detectable activity. The fully developed process had high volumetric productivity, and a high-purity product was obtained simply via direct filtration of the reaction mixture. The enantiopurity of the product was extremely high as the undesired enantiomer was not detected in the product. Table 3.1 summarizes the process currently used on a manufacturing scale to supply the intermediate for generic montelukast manufacture.[20]

3.4.2 *Duloxetine: Aminoalcohol Intermediate*

Duloxetine (developed by Lilly and marketed as Cymbalta®) is a dual serotonin and norepinephrine reuptake inhibitor and is used in the treatment of major depressive disorder. The chiral motif in the final API is an (*S*)-thienylpropanolamine moiety and duloxetine can be prepared from either the corresponding monomethyl aminoalcohol, MMAA (**3a**, R = H), or dimethyl aminoalcohol (DMAA) (**3b**, R = Me) and fluoronaphthalene, with a late-stage *N*-demethylation being required for the latter strategy (Fig. 3.4). The route to the chiral aminoalcohols traditionally involved resolution of racemic alcohols, which is inherently wasteful.

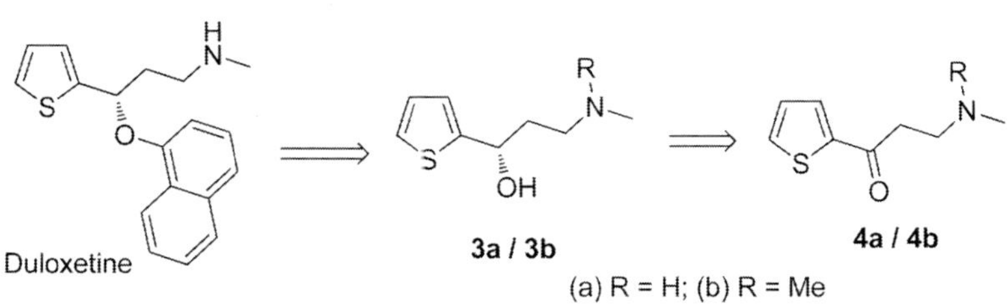

Figure 3.4 Retrosynthetic analysis of duloxetine.

A direct, asymmetric route to each aminoalcohol would offer distinct advantages, and we screened our enzyme collection for variants capable of achieving this transformation using ketone **4** (also either mono- or dimethylated). None of the natural enzymes in our collection had the desired activity, but a preengineered enzyme from a previous evolution program did exhibit the requisite activity and enantioselectivity.

Manufacture of both monomethyl aminoalcohol (MMAA) and DMAA was targeted through enzyme evolution, using IPA as a stoichiometric reductant. Interestingly, the evolutionary path of the enzymes for the respective alcohols diverged due to the differing stabilities of the two substrates at different pH. Dimethylaminoketone **4b** (R = Me) was found to be unstable under neutral conditions, undergoing elimination of dimethylamine, and as such, it was required that the process be run at pH $\geq$9 with a large excess of IPA co-solvent to minimize decomposition. In contrast, the monomethyl aminoketone **4a** (R = H) was unstable under basic and neutral conditions and the process required that an acidic buffer (pH 6) be used, and the enzyme was evolved to operate efficiently under these conditions. Thermostability was also required as it was necessary to distill off the acetone co-product to drive the reactions to completion. In both cases, subjecting the KRED libraries to the required process conditions in parallel evolution programs allowed identification of relevant diversity to increase the activity and stability of the enzymes. Recombination of this respective diversity ultimately delivered different families of enzymes capable of operating under commercially viable conditions to produce either MMAA[21] or DMAA.[22] These biocatalytic processes delivered the products with very high enantiopurity and allowed the process to DMAA to go into commercial operation to supply generic duloxetine manufacturers. The performance of the two processes is tabulated later in Table 3.2.

3.4.3 *Chiral Alcohol for Ezetimibe*

Ezetimibe is a cholesterol-lowering drug developed by Schering-Plough (now Merck) and marketed as a monotherapy under the trade name Zetia® or in combination with simvastatin as Vytorin®. The commercial route to ezetimibe proceeds via the chiral benzylic

Table 3.2 Summary of biocatalytic processes for manufacture of the chiral alcohols for duloxetine

	MMAA	DMAA
Conditions	150 g/L substrate load; 1.5% catalyst; 90% IPA/water; 25°C, ~50 mm Hg	150 g/L substrate load; 1% catalyst; 50% IPA/water; 40°C
Chiral purity (crude)	>99.5% e.e.	>99.5% e.e.
Isolated yield	>75%	87%
Comments	Process run at pH 5 Facile extractive workup	Process run at pH 9 Facile extractive workup

alcohol **5**, which is typically prepared via asymmetric reduction of the corresponding ketone **6** using stoichiometric borane (as its THF complex) in the presence of a chiral boron catalyst, (*R*)-methyl-CBS (Fig. 3.5). Both the reagent and the catalyst are hazardous materials, and postreaction large quantities of borate waste are produced. Furthermore, the product obtained from the reduction has less-than-perfect diastereopurity and requires upgrading.[23]

Codexis investigated the potential for replacing the boron chemistry with a KRED-mediated reduction and screened the Codex™ panel plates to this effect. Although the wild-type parent of the panel enzymes had no detectable activity, active variants were found on the panel itself. The sequence activity data from the panel screen was used to design a library of new variants, incorporating diversity predicted to improve the performance of the enzyme. This resulted in a new variant with improved activity and stereoselectivity, which was then used as a backbone for further evolution.[24] The process was designed to operate as a biphasic system in a mixture of toluene and buffered water, using a glucose/GDH-based co-factor-recycling system. Evolution delivered

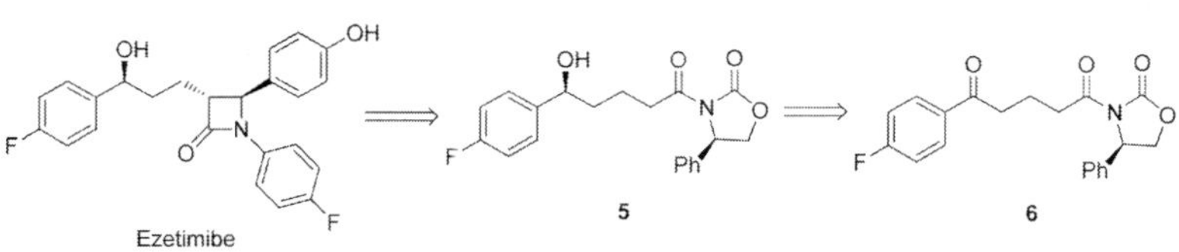

Figure 3.5 Retrosynthesis of ezetimibe.

Table 3.3 Summary of the biocatalytic process for manufacture of the chiral alcohol for ezetimibe

Process conditions	100 g/L substrate load; 2% KRED, 0.8% GDH; 20% toluene/water; glucose/GDH recycling; 30°C, pH 7.0
Chiral purity (crude)	>99.9% d.e.
Reaction yield	>95%
Comments	Facile extractive workup suitable for telescoping

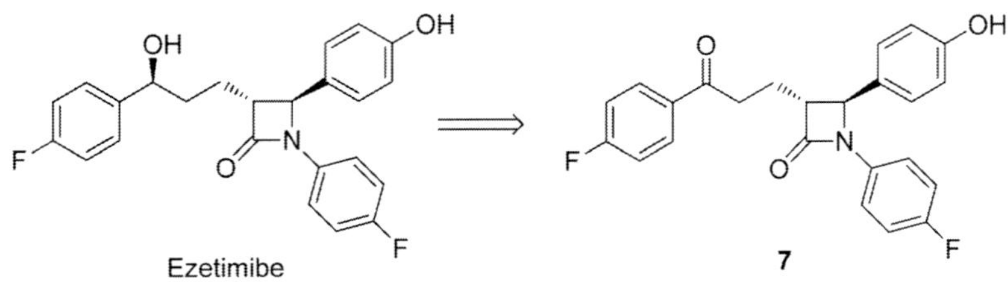

Figure 3.6 Late-stage ketone reduction for synthesis of ezetimibe.

a 12,000-fold improvement required to meet commercial targets. Performance of the biocatalytic process is tabulated next (Table 3.3).

Codexis subsequently developed a second approach to ezetimibe, targeting the API itself using a late-stage reduction of the pro-ezetimibe ketone **7** (Fig. 3.6). Typically, this intermediate was made using asymmetric hydrogenation or with chiral borane reagents, often of a benzyl-protected precursor, of which both approaches were suboptimal. The starting point for this program was one of the variants developed for the earlier ezetimibe approach. In this case, the process was developed using IPA co-factor recycling (rather than glucose/GDH), and the enzyme was evolved to tolerate high IPA concentrations and elevated temperatures to allow full solubilization of the substrate and to drive the reaction to high conversion. The final process characteristics are summarized in Table 3.4.

3.4.4 *Carbapenems: AOSA Precursor*

Acetoxysilyloxyazetidinone (AOSA), or more formally (3*R*, 4*R*)-4-acetoxy-3-[(*R*)-(*tert*-butyldimethylsilyloxyethyl]-2-azetidinone) **8**,

Table 3.4 Summary of the biocatalytic process for manufacture of an advanced chiral alcohol for ezetimibe

Conditions	100 g/L substrate load; 2% catalyst load; 65% IPA/water; 30°C
Chiral purity (crude)	>99.9% d.e.
Isolated yield	90% (unoptimized)
Comments	Isolated by direct filtration

is a key intermediate in the synthesis of many synthetic beta-lactam antibiotics, including meropenem **9** (Fig. 3.6). One of the most successful approaches to this intermediate was developed by Takasago and proceeds via syn-hydroxyester **10**, which was manufactured using an innovative reductive dynamic kinetic resolution of ketoester **11** (Fig. 3.7).[25] The reduction required very high pressure (100 atm H_2) and a Ru-(*R*)-BINAP catalyst in superheated (50°C) dichloromethane. Despite these demanding conditions, the reaction gave a product of imperfect stereopurity, and a 96:4 syn:anti ratio of diastereoisomers was obtained requiring further upgrading. Another approach to the intermediate developed by Kaneka involved the use of chlorosulfonyl isocyanate, a highly toxic, corrosive, and violently water-sensitive reagent.[26]

Codexis investigated the possibility of performing the diastereoselective reduction under milder reaction conditions to give a product of higher chiral purity in a KRED-based process using IPA for cofactor recycling. Screening of the Codex™ panels identified a wide range of variants that were active on the substrate, and selection of the appropriate enzyme allowed predominant formation of any one of the four possible diastereoisomers. The best variant identified from the panel gave a 90:10 ratio of diastereoisomers and was completely selective for the 3*R*-alcohol but less specific for the 2*S* position. This variant required further evolution to improve its diastereoselectivity and a ~250-fold activity improvement to meet the required process targets.

The data from the panel screen was analyzed using ProSAR to identify a series of important mutations, which were used to prepare a new combinatorial library of enzyme variants. One of these variants provided the desired product in analytically perfect diastereoselectivity, and after incorporation of a further three

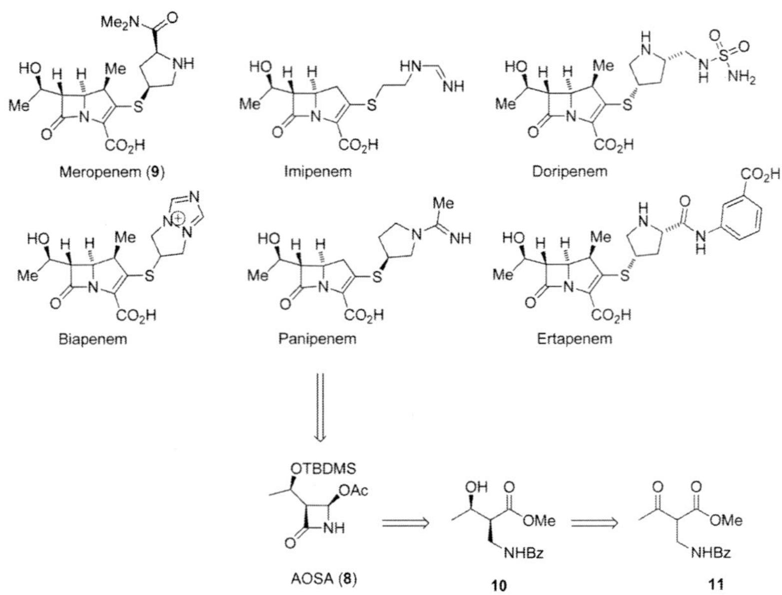

Figure 3.7 Retrosynthetic analysis of penem antibiotics.

mutations (targeting stability, co-factor binding, and expression), the resulting enzyme exceeded all of the desired process targets.[27] Notably, the biocatalyst was developed using only mutations that were previously identified in our evolution projects and the enzyme optimization was completed in only three months. The progress of the evolution program and final performance statistics are given in Table 3.5.

Table 3.5 Optimization of a KRED for commercial-scale manufacture of AOSA for penem antibiotics

Parameter	Target	Wild type	Panel hit	Shuffled hit	Final catalyst
[Substrate] g/L	>200	20	20	250	250
[Enzyme] g/L	<1	5	5	2	1
[NADP] g/L	≤0.1	0.1	0.1	0.02	0.01
Time (h)	≥24	24	24	24	24
% Conversion	>90	<5	17	96	100
Stereopurity	>99	80	90	>99	>99.9

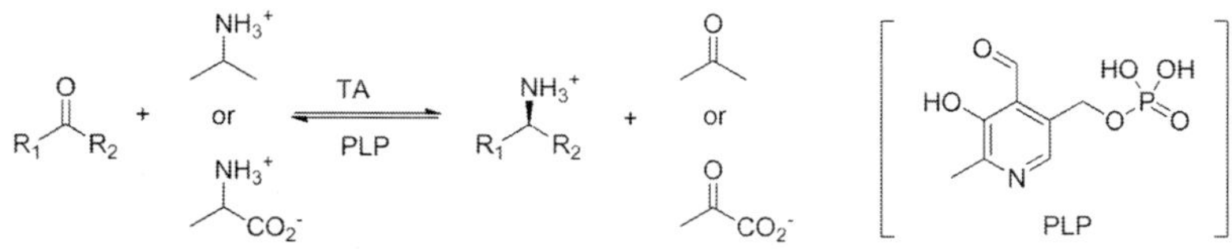

Figure 3.8 Schematic of a transaminase reaction using isopropylamine or alanine as an amine donor.

3.5 Transaminase Processes to Chiral Primary Amines

TAs convert ketones into chiral amines using an amine donor and pyridoxal phosphate (PLP) as a co-factor.[28] Often-used amine donors include alanine and isopropylamine, which give pyruvate and acetone as co-products, respectively (Fig. 3.8). The TA reaction is a powerful synthetic tool as it allows the formation of high-value chiral amine intermediates under mild conditions and typically provides products in exquisite enantiopurity.

The thermodynamics of TA reactions are such that the equilibrium typically needs to be pulled to the desired product and it is often necessary to remove the product or the co-product from the reaction to allow high conversions. As such, the preferred and most simple approach to achieving high conversion involves use of isopropylamine as the donor and removal of the acetone co-product via distillation.[29] The next section discusses the application of these enzymes to a large-scale API-manufacturing program.

3.5.1 *Sitagliptin*

Sitagliptin (marketed by Merck as Januvia® and as Janumet® in co-formulation with metformin) is an important new drug for the treatment of type II diabetes, a disease that affects over 240 million individuals worldwide. Demand for this drug is expected to increase significantly in the coming years, and a sustainable and efficient manufacturing route is highly desirable. Sitagliptin contains a chiral aminoamide unit, and the traditional approach to this intermediate involved an asymmetric hydrogenation of the corresponding enamide **12** using a rhodium Josiphos catalyst, as shown in Fig. 3.9. Although the hydrogenation was a successful strategy for

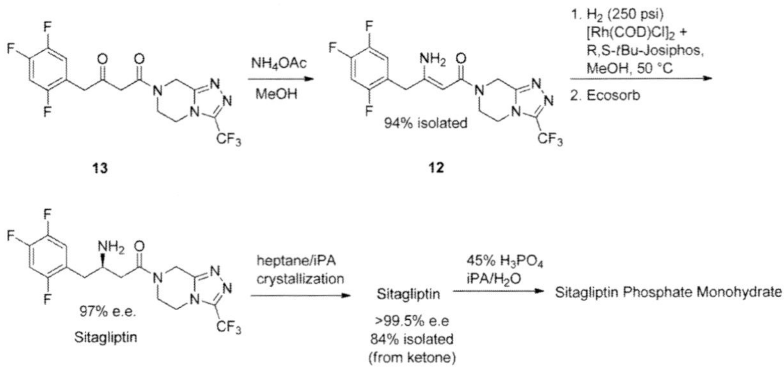

Figure 3.9 Synthesis of sitagliptin via asymmetric chemocatalytic hydrogenation.

manufacture of the drug and delivered a product with good chiral purity (97% e.e.), it was necessary to upgrade the product through crystallization to deliver the required enantiopurity of >99.5% e.e. In addition, the hydrogenation required very high pressures (250 psi); hence specialized manufacturing equipment was needed, and residual metal needed to be removed from the product so as to meet Food and Drug Administration (FDA) requirements. Nonetheless, this approach was a major improvement over earlier approaches and allowed Merck to win a 2006 Environmental Protection Agency (EPA) Presidential Green Chemistry Challenge Award.

Merck and Codexis investigated the potential of a TA-based approach, which would allow direct conversion of pro-sitagliptin ketone **13** into the required chiral amine with the desired enantioselectivity (i.e., >99.5% e.e. amine). Such a result would remove the need to upgrade the chiral amine, the need for specialized equipment, and the use of heavy metal catalysts (Fig. 3.10). In principle, this would be a highly attractive route to the amine product, but no known TA enzyme had any detectable activity on the target substrate.

Ketones with at least one small substituent (methyl or alternatively small cyclic ketones) are known to be accepted by natural *R*- and *S*-selective TAs, but ketones with two bulky substituents are not.[30–32] As such, initial screening was conducted on a truncated analogue of pro-sitagliptin ketone **13** that bore methyl ketone motif

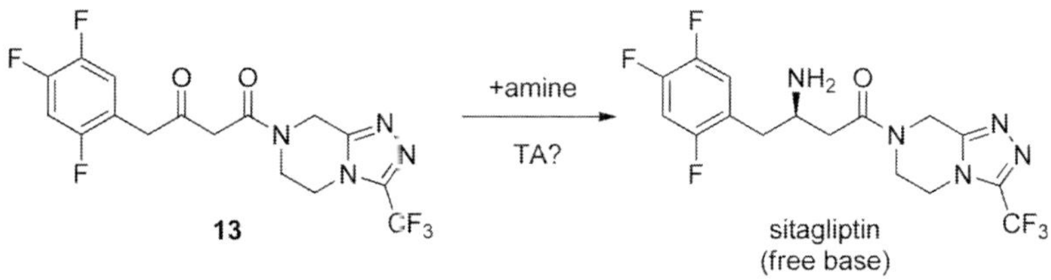

Figure 3.10 Aspired process for conversion of pro-sitagliptin ketone to sitagliptin.

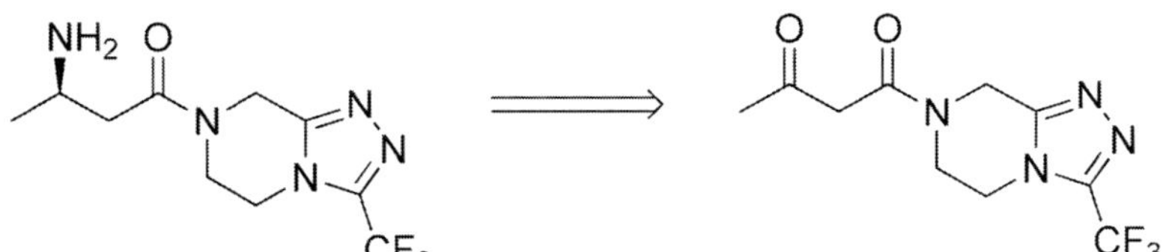

Figure 3.11 Initial development of TA for sitagliptin manufacture involved increasing the activity on a truncated analog.

14 (Fig. 3.11). This substrate was indeed accepted, and slight activity for the desired conversion was found. In silico studies suggested that the heterocyclic side of the ketone docked into the large binding pocket, whereas the methyl group fit into the small pocket. After optimization of the large binding pocket through evolution, enzymes were obtained with significantly improved activity for the truncated substrate **14**.

The next phase of the evolution program required accommodation of the full substrate **13**, which bears a bulky trifluorophenyl substituent on the methyl group. An in silico docking study was performed to identify residues that could potentially be mutated to allow expansion of the small binding pocket. Libraries containing combinations of small pocket mutations were synthesized and screened for activity on the full substrate. A variant containing three small pocket mutations provided the first detectable activity on pro-sitagliptin substrate **13**, giving an estimated turnover of $\sim$1 per day. Once initial—albeit extremely low—activity was found, this variant was optimized for commercial use over the course of 11 rounds of evolution.[33]

Evolution provided continuous improvements in activity, and once a reasonable level was reached (four rounds provided a 1,000-fold improvement over the first active variant), in-depth process development was initiated at Merck. Changes in the process were incorporated into the continuing evolution campaign to develop an enzyme capable of withstanding ever more challenging process conditions. By the end of the combined enzyme evolution and process development program, the reaction conditions had changed dramatically with a temperature increase from 25°C to 45°C and dimethyl sulfoxide (DMSO) co-solvent concentration from 5% to 50%. The equilibrium was shifted toward completion via a combination of increased concentration of isopropylamine and removal of acetone via mild vacuum and nitrogen sweep. The reaction could be run at substrate loadings as high as 275 g/L, using a 3 wt% catalyst loading, reaching completion in 24 hours. The overall yield of sitagliptin using the TA route was up to 13% higher than with the previous hydrogenation process. In addition, this biocatalytic process provided a 53% increase in productivity, a 19% decrease in total waste, elimination of all heavy metals, and a reduction in total manufacturing cost.[33]

Overall, the increase in biocatalyst activity was >25,000-fold from the first enzyme variant with detectable activity on pro-sitagliptin ketone **13**, and the product was obtained in high yield with perfect enantiopurity, that is, the *S*-enantiomer was never detected. The performance criteria for this process are summarized in Table 3.6. Merck and Codexis jointly received the 2010 EPA Presidential Green Chemistry Challenge Award for the sitagliptin transaminase process.

Table 3.6 Summary of the evolutionary trajectory of the biocatalytic process to sitaglipin

Parameter	Target	Initial variant	First active variant	Final catalyst
[Substrate] g/L	100	2	2	110
[Enzyme] g/L	≤5	10	10	5.5
Time (h)	≤24	72	24	24
% Conversion	>95	0	0.5	>95
Enantiopurity	>99	–	>99	>99.9

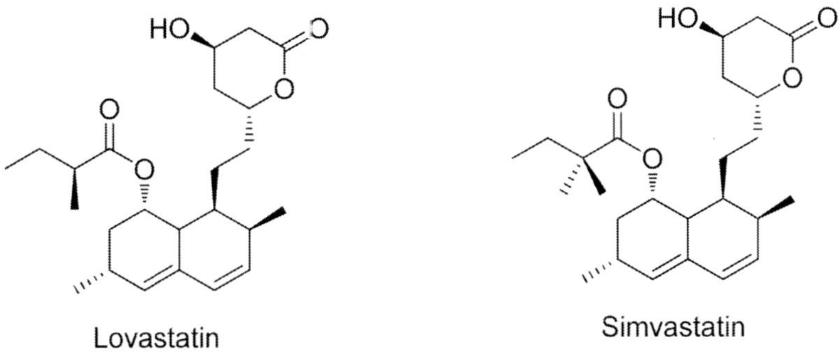

Figure 3.12 Structures of lovastatin and simvastatin.

3.6 Acyltransferase Approach to Simvastatin

Simvastatin is a hypolipidemic drug developed by Merck and marketed as Zocor®. It is a potent 3-hydroxy-3-methylglutaryl–coenzyme A (HMG-CoA) reductase inhibitor and blocks the biosynthetic pathway for cholesterol production. It is a synthetic derivative of lovastatin, a fermentation product from the fungus *Aspergillus terreus*, differing only by a single methyl group on the butyryl sidechain (Fig. 3.12). Zocor was the second-largest-selling cholesterol-lowering drug in 2005, with estimated sales of USD 4.3 billion and was later marketed as a combination therapy with ezetimibe (Vytorin®). In 2009, sales of generic simvastatin were approximately USD 558 million, while Vytorin had sales of USD 2.1 billion.

Simvastatin is generally manufactured by one of two main routes, both starting with lovastatin (Fig. 3.13). One approach involves deprotonation of the lovastatin sidechain, followed by direct methylation of the enolate. The second approach involves removal of the lovastatin side chain and replacement with the required dimethylbutyryl side chain. Both approaches require protecting group strategies, with the former also requiring cryogenic conditions and genotoxic alkylating agents and the latter requiring multiple steps.

A conceptually more efficient approach would be to achieve a regiospecific acylation without the need for protecting groups. Tang

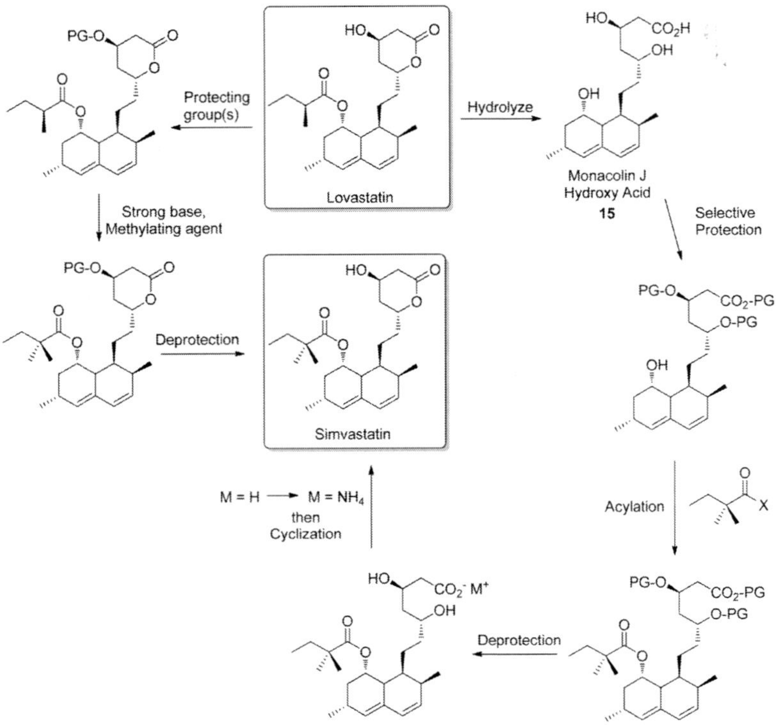

Figure 3.13 Chemical routes for conversion of lovastatin to simvastatin.

et al. at the University of California, Los Angeles (UCLA) isolated and cloned the *lovD* gene, which encodes the enzyme that catalyzes the final step of lovastatin biosynthesis, the installation of the methylbutyryl side chain onto the 8-hydroxy group of monacolin J **15**. They then demonstrated that this enzyme would accept a nonnatural dimethylbutyryl side chain donor to give simvastatin (Fig. 3.14).[34] Codexis licensed the technology with a view to developing an efficient biocatalytic process employing an isolated enzyme. The activity and stability of the initial enzyme were poor, and an extensive evolution campaign was used to improve enzyme performance with respect to activity, stability, and enzyme solubility (which was important for enzyme manufacture via overexpression in *E. coli*), and to overcome issues with substrate and product inhibition. In addition, the reaction was found to be an equilibrium

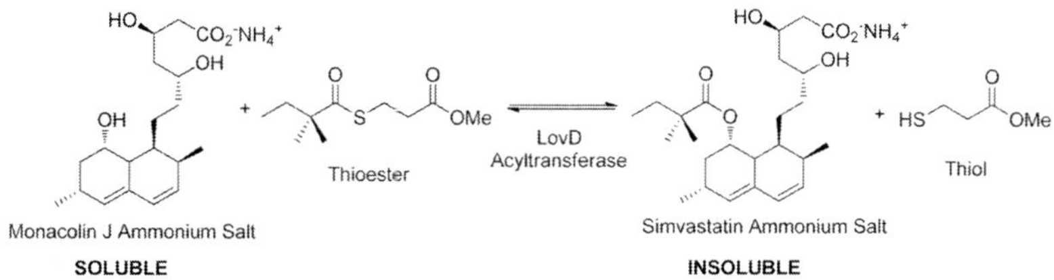

Figure 3.14 Biocatalytic approach for ammonium-simvastatin manufacture.

Table 3.7 Comparison of natural LovD and the final evolved variant in the biocatalytic process for simvastatin manufacture

Parameter	Wild type	Final catalyst/process
[Substrate] g/L	0.3	75
[Enzyme] g/L	0.9	0.75
[Thioester acyl donor]	4 equiv.	1.1 equiv.
% Conversion	50%	97%
Co-solvent	MeCN (1%)	None

reaction leading to incomplete conversion under typical conditions. To ensure that the minimum amount of thioester donor was used in the process while, at the same time, achieving high conversion, an innovative process was developed, in which the product was directly precipitated as it formed. Overall, the biocatalytic acylation was highly efficient, allowing the isolation of simvastatin ammonium salt in excellent purity via a simple direct-drop process that removed up to three steps from the traditional acylation approach (Table 3.7). The process is currently in scale-up for commercial manufacture.[35,36]

3.7 Integrated Biocatalytic Processes: Atorvastatin

Lipitor® is a blockbuster "statin" drug, developed by Pfizer and used to treat hypercholesterolemia. It is the largest-grossing therapeutic drug in history and was the first to exceed annual sales of $10 billion. Like simvastatin, atorvastatin, the API in Lipitor, is a competitive

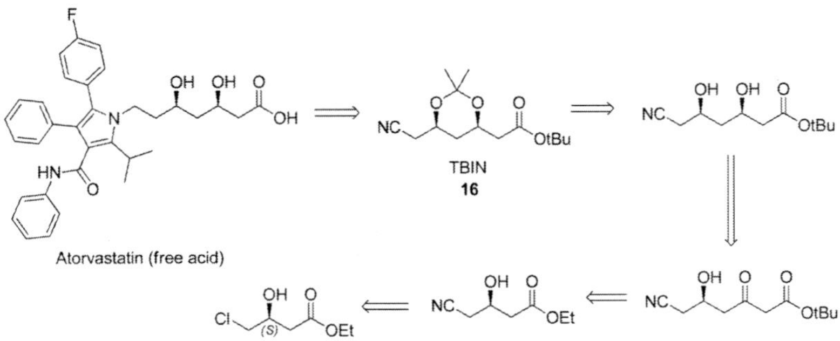

Figure 3.15 Retrosynthesis of atorvastatin.

inhibitor of HMG-CoA reductase and as such interferes with the rate-limiting step in hepatic cholesterol biosynthesis. The key pharmacophore of the molecule is a synthetically challenging 3,5-dihydroxycarboxylic acid moiety. The main advanced intermediate for atorvastatin, known as *tert*-butyl isopropylidene nitrile (TBIN) or ATS8 (**16**), bears all of the requisite chiral functionality and is an attractive target for a more efficient manufacturing approach. Codexis developed an integrated biocatalytic process to **16**, leveraging four enzymes over three crucial steps. The preferred synthetic approach to **16** started with ethyl (*S*)-4-chloro-3-hydroxybutyrate **17** and followed a sequence of cyanation, Claisen condensation, diastereoselective reduction, and protection (disconnections shown in Fig. 3.15). Codexis looked to integrate biocatalysis into this route in several places.

The starting chiral hydroxyester **17** was accessible through a number of approaches, including chiral pool methods (from (*S*)-3-hydroxybutyrolactone, itself obtained via decomposition of complex sugars or through high-temperature hydrogenation of malic acid), microbial reduction, or asymmetric hydrogenation of ethyl 4-chloroacetoacetate **18**. Codexis evolved a KRED enzyme (from *Candida magnolia*) to perform the asymmetric reduction of **18**. The NADP co-factor was regenerated using glucose/GDH, with the GDH being evolved for stability under the process conditions (Fig. 3.16).[37] This provided the key chiral hydroxyester in exquisite

Figure 3.16 Biocatalytic route to ethyl (*S*)-4-chloro-3-hydroxybutyrate **17**.

Table 3.8 Summary of the biocatalytic process for manufacture of ethyl (*S*)-4-chloro-3-hydroxybutyrate **17**

Parameter	Performance
[Substrate] g/L	180
[Enzyme] g/L	0.7
[NADP] g/L	0.13
Isolated yield	97%
Stereopurity	>99.9% e.e.
Comments	20% butyl acetate; pH 7.0 isolation via continuous countercurrent extraction

chiral purity in a highly productive process. The process conditions are listed in Table 3.8.

The second step of the reaction sequence required nucleophilic displacement of the chlorine substituent with a cyanide group to give hydroxynitrile **19** (Fig. 3.17). Typical methods of achieving this transformation required the use of high temperatures and basic conditions, which resulted in a range of undesired by-products. This issue was compounded by the fact that the product was not crystalline, and tedious, high-vacuum fractional distillation was required to obtain a product of sufficient quality. An innovative biocatalytic solution was developed that allowed the nucleophilic displacement to happen under neutral conditions using a highly regioselective, evolved halohydrin dehalogenase (HHDH) enzyme. This enzyme first catalyzed the closure of the chlorohydrin moiety of **17** to give an epoxide intermediate and then aided the epoxide ring opening with cyanide under neutral conditions, rendering the reaction irreversible. The performance of the wild-type enzyme was improved approximately 4,000-fold in terms of activity, productivity, stability, and inhibition to deliver a commercially attractive process,[12] with process details given in Table 3.9. The achievements

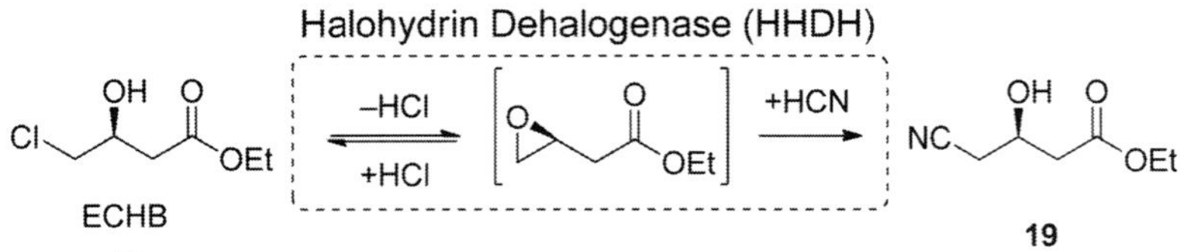

17

19

Figure 3.17 Biocatalytic route to hydroxynitrile **19**.

Table 3.9 Summary of the biocatalytic process for manufacture of hydroxynitrile **19**

Parameter	Performance
[Substrate] g/L	140
[Enzyme] g/L	1.2
Isolated yield	92%
Stereopurity	>99.9% e.e.
Comments	pH 7; isolation via continuous countercurrent extraction and wiped-file distillation

in the development of the enzyme-mediated process for the conversion of ethyl 4-chloroacetoacetate **18** into hydroxynitrile **19** were recognized through the award of the EPA Presidential Green Chemistry Challenge Award in 2006.[37]

The resulting hydroxynitrile **19** was subjected to Claisen condensation with *tert*-butyl acetate to give hydroxyketoester **20**. The key to the final part of the synthesis lay in the ability to reduce the keto function with near-perfect stereoselectivity. One of the reported approaches to this intermediate (utilized at multikilo scale)[38] involved a challenging diastereoselective reduction. The substrate solution needed to be cooled to −85°C and was treated first with diethylmethoxyborane to form a chelated complex, and the reduction of the ketone was achieved using $NaBH_4$ at an even lower temperature to give diol **21** (Fig. 3.18). Even under such stringent conditions, complete control of diastereoselectivity could not be achieved. After quenching, the borate waste was removed using multiple vacuum distillations with methanol. The dihydroxyester was not crystalline and could not be purified, so it had to be carried forward to the crystalline TBIN intermediate, which could

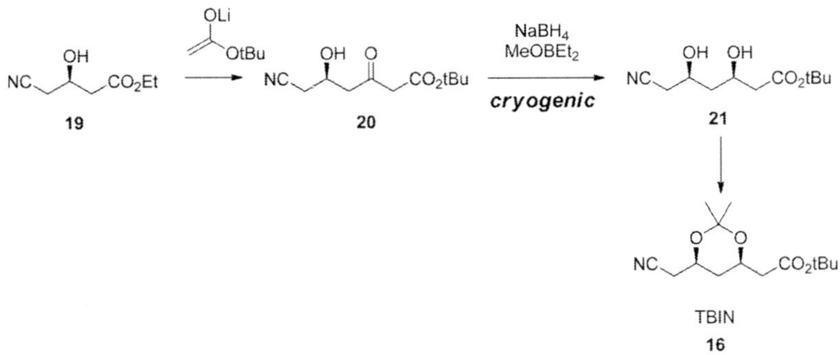

Figure 3.18 Chemical reduction approach to TBIN.

be upgraded. Thus, the diol was converted to an acetonide to purge the undesired diastereomer through crystallization. Although this valuable intermediate could be obtained in high purity, the route involved the use of hazardous stoichiometric borane reagents. This generated substantial amounts of waste, in turn, demanding large volumes of solvent for waste removal. In addition, the difficulty in achieving the directed diastereoselective reduction required that energy-intensive cryogenic conditions were needed. Clearly a greener and more economical approach to this reduction was desirable.

Again, engineering of a natural KRED provided an enzyme capable of delivering the desired dihydroxy ester **21** in excellent diastereopurity (Fig. 3.19). The activity of the wild-type KRED was 100 times lower than desired, and its stability was similarly challenged. Using a combination of directed evolution techniques provided a much-improved variant capable of operating at very high substrate loads. The reaction proceeded to >98% conversion, giving the desired product **21** with unprecedented diastereopurity (>99.9% d.e.) with a glucose/GDH-coupled co-factor-recycling system (Table 3.10).[39] Diol **21** could be taken forward to the key TBIN intermediate. The green metrics of the biocatalytic reaction compared very favorably with the original chemical reduction: the substrate load was tripled over that reported in the patented process;[38] hazardous boron reagents, concomitant removal of borate waste, and cryogenic conditions were avoided along with an

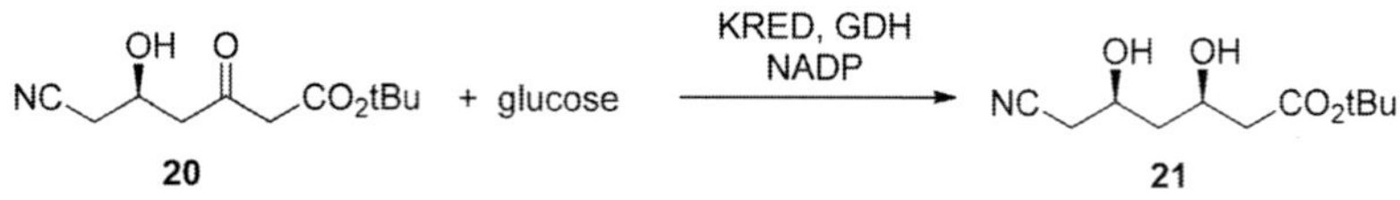

Figure 3.19 Biocatalytic route to diol **21**.

Table 3.10 Summary of the biocatalytic process for manufacture of diol **21**

Parameter	Performance
[Substrate] g/L	300
[Enzyme] g/L	1.5
Conversion	>98%
Stereopurity	>99.9% d.e.
Comments	The reaction substrate contains t-Bu-acetoacetate by-product from the Claisen condensation, complicating the HTS for KRED optimization.

85% reduction in organic solvent usage as it operated in a mainly aqueous medium. This process is run at a large scale, delivering >5 mT per month of high-quality intermediate.

3.8 Conclusions and Outlook

Biocatalysis can be a truly green technology, as discussed herein. The use of engineered biocatalysts has become an established strategy for the manufacture of a range of chiral raw materials, advanced intermediates, and active ingredients. The remarkable stereoselectivity of enzymes can be augmented through directed evolution to provide significant improvements in activity and process stability to deliver economic, green, and sustainable manufacturing processes. Many of the examples included have been demonstrated on a large scale, and intermediates made by these biocatalytic processes have found their way into millions of doses of important medicines all over the world.

Reflecting on the examples earlier, the links between biocatalysis and the principles of green chemistry, such as catalysis, design for degradation, and inherently safer chemistry for accident prevention, are evident. KREDs and TAs use renewable feedstocks as co-substrates, and product-specific examples of the principles include prevention (e.g., intermediates for atorvastatin and duloxetine), atom economy (e.g., simvastatin and the chiral alcohol for ezetimibe), less hazardous chemical synthesis (e.g., AOSA and sitagliptin), safer solvents and auxiliaries (e.g., intermediates for montelukast and atorvastatin), design for energy efficiency (e.g., sitagliptin and the montelukast chiral intermediate), and reduced derivatization (e.g., simvastatin and the intermediate for ezetimibe).

Many of the early successes of this approach leveraged the KRED enzyme class, delivering high-value chiral alcohols. More recent developments have broadened the availability of the biocatalytic solutions portfolio. TAs are becoming an established commercial technology and are seeing increasing application. The future for biocatalysis is bright, and different chemistries are becoming accessible with the development of oxidative enzymes such as amine oxidases[40,41] and monooxygenases.[42] The expansion in available enzyme classes has been complimented by significant developments in evolution technologies, allowing higher-quality libraries of enzymes to be developed and screened more rapidly than ever, further accelerating the evolutionary trajectory required to reach commercial targets.

Acknowledgments

We would like to acknowledge the diligence and creativity of the many scientists at Codexis in Redwood City and Singapore who were involved in the various programs described in this chapter.

References

1. Constable, D. J. C.; Dunn, P. J.; Hayler, J. D.; Humphrey, G. R.; Leazer, J. L., Jr.; Linderman, R. J.; Lorenz, K.; Manley, J.; Pearlman, B. A.; Wells, A.; Zaks, A.; Zhang, T. Y. (2007). *Green Chem.*, **9**, 411–420.
2. (a) Stemmer, W. P. C. (1994). *Nature*, **370**, 389–391; (b) Powell, K. A.; Ramer, S. W.; Del Cardayre, S. B.; Stemmer, W. P. C.; Tobin, M. B.;

Longchamp, P. F.; Huisman, G. W. (2001). *Angew. Chem., Int. Ed.*, **40**, 3948–3959.

3. Taylor, S. J. C.; Brown, R. C.; Keene, P. A.; Taylor, I. N. (1999). *Bioorg. Med. Chem.*, **7**, 2163–2168.
4. Serizawa, N.; Matsuoka, A. (1991). *Biochim. Biophys. Acta*, **1084**, 35–40.
5. Rozzell, J. D. (1999). *Bioorg. Med. Chem.*, **7**, 2253–2261.
6. Schoemaker, H. E.; Mink, D.; Wubbolts, M. G. (2003). *Science*, **299**, 1694–1697.
7. Huisman, G. W.; Lalonde, J. J. (2006). *Biocatalysis in the Pharmaceutical and Biotechnology Industries*, (R. Patel, Ed.). CRC Press, 717–742.
8. Böttcher, D.; Bornscheuer, T. (2010). *Curr. Opin. Microbiol.*, **13**, 274–282.
9. Reetz, M. T. (2011). *Angew. Chem., Int. Ed. Engl.*, **50**, 138–174.
10. Fox, R. J.; Clay, M. D. (2009). *Trends Biotechnol.*, **27**, 137–140.
11. Fox, R. J.; Huisman, G. W. (2008). *Trends Biotechnol.*, **26**, 132–138.
12. Fox, R. J.; Davis, S. C.; Mundorff, E. C.; Newman, L. M.; Gavrilovic, V.; Ma, S. K.; Chung, L. M.; Ching, C.; Tam, S.; Muley, S.; Grate, J.; Gruber, J.; Whitman, J. C.; Sheldon, R. A.; Huisman, G. W. (2007). *Nature Biotechnol.*, **25**, 338–344.
13. Rozzell, D. J.; Lato, S. (2008). *sp2 Mag.*, **Sept.**, 32–33.
14. Anastas, P.; Warner, J. (1998). *Green Chemistry: Theory and Practice*. New York, NY: Oxford University Press, 30.
15. Matsuda, T.; Yamankak, R.; Nakamura, K. (2009). *Tetrahedron Asym.*, **20**, 513–557.
16. Patel, R. (Ed.). (2006). *Biocatalysis in the Pharmaceutical and Biotechnology Industries*. CRC Press.
17. de Wildeman, S. M. A.; Sonke, T.; Schoemaker, H. E.; May, O. (2007). *Acc. Chem. Res.*, **40**, 1260–1266.
18. Moore, J. C.; Pollard, D. J.; Kosjek, B.; Devine, P. N. (2007). *Acc. Chem. Res.*, **40**, 1412–1419.
19. Shafiee, A.; Motamedi, H.; King, A. (1998). *Appl. Microbiol. Biotechnol.*, **49**, 709–717.
20. Liang, J.; Lalonde, J.; Borup, B.; Mitchell, V.; Mundorff, E.; Trinh, N.; Kochrekar, D. A.; Cherat, R. N.; Pai, G. G. (2010). *Org. Proc. Res. Dev.*, **14**, 193–198.
21. Savile, C.; Gruber, J. M.; Mundorff, E.; Huisman, G. W.; Collier, S. J. WO2010025287.
22. Savile, C.; Gruber, J. M.; Mundorff, E.; Huisman, G. W.; Collier, S. J. WO2010025238.

23. Fu, X.; McAllister, T. L.; Thiruvengadam, T. K.; Tann, C.-H.; Su, D. (2003). *Tetrahedron Lett.*, **44**, 801–804.
24. Mundorff, E.; de Vries, E. WO2010025085.
25. Noyori, R.; Ikeda, T.; Ohkuma, T.; Widhalm, M.; Kitamura, M.; Takaya, H.; Akutagawa, S.; Sayo, N.; Saito, T. (1989). *J. Am. Soc. Chem.*, **111**, 9134–9135.
26. Berks, A. H. (1996). *Tetrahedron*, **52**, 331–375.
27. Campapiano, O.; Mundorff, E.; Borup, B.; Voladri, R. WO2009046153.
28. Crump, S. P.; Rozzell J. D. (1992). *Biocatalytic Production of Amino Acids and Derivatives* (J. D. Rozzell; F. Wagner, Eds.). New York, NY: Wiley, 43–58.
29. Matcham, G. W.; Bowen, A. R. (1996). *Chim. Oggi/Chem. Today*, **June**, 20–24.
30. Höhne, M.; Kuhl, S.; Robins, K.; Bornscheuer, U. T. (2008). *ChemBioChem*, **9**, 363–365.
31. Koszelewski, D.; Lavandera, I.; Clay, D.; Guebitz, G. M.; Rozzell, D.; Kroutil, W. (2008). *Angew. Chem , Int. Ed.*, **47**, 9337–9340.
32. Koszelewski, D.; Lavandera, I.; Clay, D.; Rozzell, D.; Kroutil, W. (2008). *Adv. Synth. Catal.*, **350**, 2761–2766.
33. Savile, C. K.; Janey, J. M.; Mundorff, E.; Moore, J. C.; Tam, S.; Jarvis, W. R.; Colbeck, J. C.; Krebber, A.; Fleitz, F. J.; Brands, J.; Devine, P. N.; Huisman, G. W.; Hughes, G. J. (2010). *Science*, **329**, 305–311.
34. Xie, X.; Watanabe, K.; Wojcicki, W. A.; Wang, C. C. C.; Tang, Y. (2006). *Chem. Biol.*, **13**, 1161–1169.
35. Gilson, L.; Collier, S. J.; Sukumaran, J.; Yeo, W. L.; Alviso, O.; Teo, E. L.; Wilson, R. J.; Xu, J. WO2011041231.
36. Collier, S. J.; Teo, E. L.; Sukumaran, J.; Wilson, R. J.; Xu, J. WO2011041233.
37. Ma, S. K.; Gruber, J.; Davis, C.; Newman, L.; Gray, D.; Wang, A.; Grate, G.; Huisman, G. W.; Sheldon R. A. (2010). *Green Chem.*, **12**, 81–86.
38. Butler, D. E.; Le, T. V.; Millar, A.; Nanninga, T. N. (1992). US 5155251.
39. Giver, L. J.; Newman, L. M.; Mundorff, E.; Huisman G. W.; Jenne, S. J.; Zhu, J.; Behrouzian, B.; Grate, J. H.; Lalonde, J. WO2008042876.
40. Mijts, B.; Muley,S.; Liang, J.; Newman, L.; Zhang, X.; Lalonde, J.; Clay, M. D.; Zhu, J.; Gruber, J. M.; Colbeck, J. M.; Munger, J. D.; Mavinhalli, J.; Sheldon, R. WO201008828.
41. Lalonde, J.; Liang, J. (2010). *Asymmetric Catalysis on an Industrial Scale: Challenges, Approaches and Solutions* (2nd ed.). Wiley, 41–60.
42. Mayhew, M.; Kambourakis, S.; Rozzell, D. (2008). *sp2 Mag.*, **May**, 28–29.

Chapter 4

Application of Green Metrics to Scalable Industrial Synthesis Plans: Approaches to Oseltamivir Phosphate (Tamiflu®)

John Andraos
Department of Chemistry, York University, 4700 Keele Street, Toronto M3J 1P3, Canada
c1000@careerchem.com

4.1 Introduction

Why should industrial chemists consider thinking "green"? This is a question that continues to recur in the current chemistry literature and is almost unavoidable, given the modern concerns of scarcity and security of resources and good environmental stewardship. A noteworthy article answering this question from a philosophical point of view has recently appeared.[1] The joint American Chemical Society and Green Chemistry Institute Pharmaceutical Roundtable has recently published a wish list of what it considers as priorities for implementing green chemistry practices in the pharmaceutical industry.[2,3] A corollary practical question that arises is, Why should industrial chemists consider using metrics as a routine tool as

Scalable Green Chemistry: Case Studies from the Pharmaceutical Industry
Edited by Stefan G. Koenig

ISBN 978-981-4316-49-1 (Hardcover), 978-981-4364-81-2 (eBook)
www.panstanford.com

they brainstorm for ideas at the design stage of synthesizing a target molecule? The implementation of green chemistry practices in the industry will depend on not only innovation but also a sound and honest metrics assessment of the processes invented. Performance metrics are the only way that one process may be judged against another with respect to some given set of criteria. The aim of this chapter is to convince the industrial chemist of this fact and, by means of an example, to showcase key reaction and synthesis metrics, how they are calculated, and how they may be used in synthesis plan design at any scale of production. The target molecule selected is oseltamivir phosphate (Tamiflu®), which has been extensively reviewed[4–8] and its synthesis plans assessed from a green metrics perspective.[9] The present chapter updates the extensive green metrics analysis published in 2009 to include 19 new plans published since then, bringing the total number of plans examined for this compound to 34.

4.2 Green Metrics Essentials for Industrial Chemists

Before beginning to implement green chemistry practices, one needs to precisely enumerate what constitutes waste. The simplistic method of calculating waste for any process is to determine the mass difference between all input materials used and the final product obtained. While this yields a global determination of the waste produced, it gives no information as to its constitution. Knowledge of what waste is necessitates a full understanding of the chemistry involved in every reaction step in a synthesis plan. This baseline information is required before any attempts can be made to minimize waste materials. For any chemical reaction, waste is comprised of the following materials: unreacted reagents; excess reagents; by-products arising as a direct mechanistic consequence of producing the intended target product; side products arising from competing reactions occurring in parallel to the intended target reaction; the reaction solvent; auxiliary workup materials, including all aqueous washes; and auxiliary purification materials used in chromatographic and recrystallization operations. To become a green chemist, therefore, one needs to acquire and be adept in the following skills: balancing of chemical equations; basic numeracy, especially for the synthetic chemist who often is schooled to avoid it or downplay its importance; knowledge of reaction mechanisms; knowledge of reaction kinetics and thermodynamics; and knowledge of all chemical structures involved in a chemical transformation from reactants to products. In other words, to be a successful practicing green chemist one first needs to be a complete chemist, demonstrating proficiency in all these aspects.

A recent review nicely summarizing green chemistry and sustainability metrics has recently been published.[10] As a primer, the reader is directed to various publications providing derivations and illustrating the implementation of green metrics on various chemical reaction types and synthesis plans for important target molecules.[11–17] In the study of reaction metrics, there are always two perspectives in assessing performance. The first is from a positive point of view, that is, how much of the input materials end up in the target product. Metrics such as atom economy (AE) and reaction mass efficiency (RME) are key examples. The second is from the negative point of view, that is, how much unwanted waste material is produced in order to access the target product. Metrics based on E-factors, defined as the ratio of kilograms of waste per kilogram of product, are of this type. Clearly, since the law of conservation of mass applies to every chemical transformation, that is, chemical equations are balanced, these two groups of metrics are fundamentally related and therefore both positive and negative points of view are complementary to one another.

For a practicing industrial chemist, the fundamental metrics equations and parameters for individual reactions with respect to RME and E-factor breakdowns are as follows: for any chemical reaction, the global RME is given by

$$\text{RME} = \varepsilon(\text{AE})(1/\text{SF})(\text{MRP}) = \frac{\text{Mass of product}}{\text{Sum of masses of all input materials}} \tag{4.1}$$

where ε is the reaction yield with respect to the limiting reagent; AE is the atom economy; 1/SF is the inverse of the stoichiometric factor, taking into account excess reagent usage; and MRP is the material recovery parameter that takes into account all auxiliary material usage. The definitions of these parameters and their method of calculation have been described elsewhere.[11–13] Each of these variables is a fractional number between 0 and 1. The corresponding global E-factor expression is given by

$$E = (1/\mathrm{RME}) - 1 \tag{4.2}$$

The E-factor breakdowns that sum to E given in Eq. (4.2) are given by

$$E - \mathrm{Kernel} = \{1/\varepsilon\,(\mathrm{AE})\} - 1 \tag{4.3a}$$

$$E - \mathrm{Excess} = (\mathrm{SF} - 1)\,/\varepsilon\,(\mathrm{AE}) \tag{4.3b}$$

$$E - \mathrm{Aux} = \{\mathrm{SF}/\varepsilon\,(\mathrm{AE})\}\,\{(1/\mathrm{MRP}) - 1\} \tag{4.3c}$$

These latter equations refer to the core or kernel E-factor based on the reaction yield and AE, excess reagent contribution, and auxiliary material consumption, respectively. Equation (4.1) may be conveniently represented visually as a radial pentagon that provides an easy and rapid assessment of reaction performance. A simple Microsoft Excel spreadsheet with embedded formula cells was developed to easily calculate all individual reaction metrics once masses of all input materials are entered.[12]

For analyzing synthesis plans, synthesis tree diagrams have been demonstrated to handle linear and convergent plans, regardless of degree of complexity.[15] These diagrams provide an easy-to-read visual representation of a synthesis plan where the following basic parameters may be determined at once: the number of reaction steps; the number of branches (applicable to convergent plans); the number of input materials; the reaction yields for each step; the molecular weights of all input reagents, including corresponding stoichiometric coefficients; and the molecular weights of isolated intermediates and the target product. The mass amount of any material at any node in the tree diagram can be determined at once with respect to the mole scale of the final product by simply connecting the node in question to the target product node and tracing the contributing reaction yields along the way.

The fundamental metrics equations for synthesis plans with respect to E-factor breakdowns are necessarily more complicated

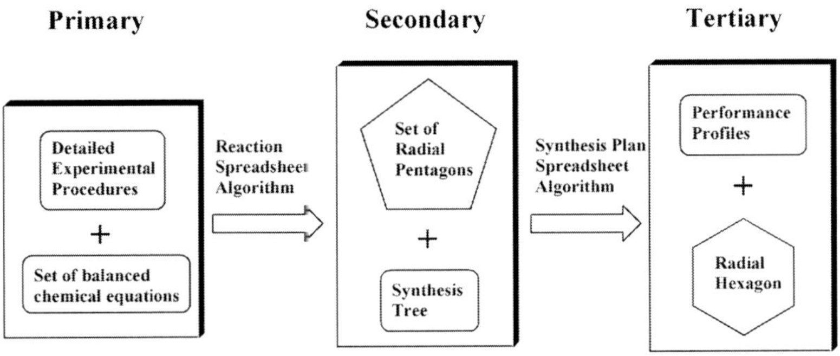

Figure 4.1 Paradigm chart linking experimental procedures to a radial pentagon, a radial hexagon, and synthesis tree analyses.

in form than Eqs. (4.3a–4.3c) and are given elsewhere.[9] Suffice it to say that Excel spreadsheets with embedded formulas were developed to easily calculate these parameters using the outputs of the radial pentagon analyses for each chemical reaction. In addition, synthesis strategy metrics such as degree of convergence, fraction of sacrificial reagents based on molecular weight, and fraction of the kernel waste produced due to target bond-forming reactions may also be determined. Again, by analogy with the concept of a radial pentagon for individual reactions, a radial hexagon may be drawn to depict visually the overall material and strategy performance of an entire synthesis plan. Figure 4.1 summarizes the overall paradigm of assessing synthesis plan performance.

Advantages of using spreadsheet algorithms and synthesis tree diagrams include:

(a) providing an intelligent in-depth critique of plans;
(b) allowing unbiased comparisons between plans;
(c) providing concrete proof that a newly reported plan is different and better by material consumption and strategic design than prior reported plans;
(d) providing a proofreading tool for patent examiners of patent applications and reviewers of manuscripts to screen for errors in experimental procedures;
(e) providing proof of novelty in improvements to process patents; and

(f) providing a powerful tool for scientists to do "what if" analyses and see the instant results of their choices during the design phase of a synthesis BEFORE going to the lab.

For every synthesis plan, the total number of steps may be subdivided into two broad categories, the number of target-bond-forming reactions and the number of non-target-bond-forming reactions, that is, sacrificial reactions. Target bond reactions are of the *aufbau* or construction type, including additions and couplings, as well as additive redox reactions in which atoms of the redox reagents end up in the target product. Sacrificial reactions are ones that involve protecting groups, covalently attached chiral auxiliaries to control stereochemistry, and redox adjustments using subtractive redox reagents whose atoms never get incorporated in the target structure. From a strategy perspective, the objective is to simultaneously minimize the total number of steps involved and to minimize the number of sacrificial reactions. From a waste management perspective, the prior objective translates to the simultaneous minimization of global waste and waste contributions from sacrificial reactions, so called "bad waste." The result of implementing both of these perspectives leads to material efficiency and therefore greener synthesis plans. In addition, telescoping or concatenation of steps to reduce the number of isolations of intermediates, reducing solvent demand per reaction step, reducing auxiliary material consumption, and choosing safe and benign materials and methodologies are further strategies to aim for overall greenness.

All industrial chemists and chemical engineers have learned that synthesis optimization is a large and complex max-min problem with respect to a given set of variables and applied constraints. The goal of true optimization is achieved when all performance variables gravitate in the positive direction together. This can be done only by trial and error and coming up with more than one way of making the target molecule. Since optimization is a comparative exercise nothing can be said if only one plan exists for a given target compound. Hence, when an optimal synthesis is announced it should mean that that plan has the best overall yield, the highest AE, the fewest number of steps, the fewest number of sacrificial reactions, and the least E-factor contributions in all three categories (kernel, excess reagent, and auxiliary material). If a set of plans to a target molecule shows a scattering in best-performance metrics, such as plan A has the best AE, plan B has the highest overall yield, and plan C has the least global E-factor, then optimization has not been achieved. The merits of each plan need to be brought together as far as possible into the same plan. The words most often used to advertise synthesis plans in the literature are "concise" and "efficient" as evidenced by their frequency even in the titles of the papers referenced in this chapter on the synthesis of oseltamivir phosphate. Such statements are never couched in a rigorous assessment of metrics beyond the number of steps and overall yield for the newly announced plan. Hence, there is no possibility of assessing fairly, objectively, and thoroughly how a new strategy ranks with prior published plans. The lack of synergistic optimization is a ubiquitous fault in all of the chemistry literature, and it is thus often disappointing to find out that a newly announced plan using the latest chemical transformations could be orders of magnitude worse in material performance than an earlier one published several decades prior at a time when the concept of "green chemistry" was not yet conceived. It is this void that the present chapter addresses using the synthesis of oseltamivir phosphate as an example.

4.3 Oseltamivir Approaches: Synthesis Strategies

Oseltamivir phosphate is a neuraminidase inhibitor that was originally developed by Gilead and is now manufactured and marketed as Tamiflu® by Hoffmann-LaRoche. Its structure, shown below in Fig. 4.2, is deceptively simple. It has a single six-membered ring with an endocyclic double bond in conjugation to an ester group. The feature that makes it a challenge to synthesize is that it has three contiguous stereogenic centers in a down-up-down arrangement around the ring. It was this structural feature, in addition to the compound's importance in combating the latest strains of influenza virus, particularly H5N1 (bird or avian flu) and H1N1 (swine flu), that drew the attention of several prominent synthesis labs around the world.

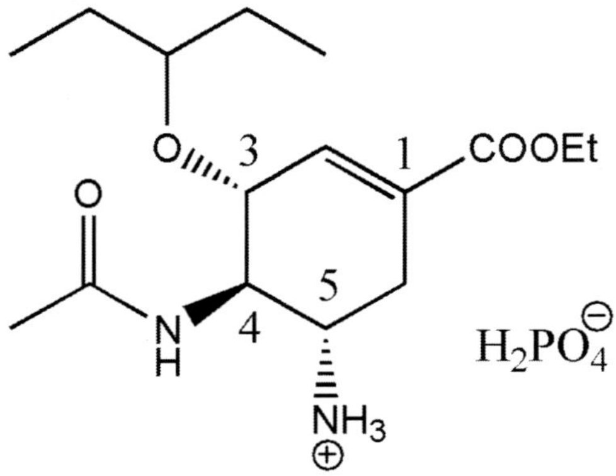

Figure 4.2 Structure of oseltamivir phosphate.

A previous publication examined 15 synthetic approaches to the target. This chapter inspects another 19 novel syntheses, bringing the total number of approaches to oseltamivir phosphate to 34.[19–52] The ring construction strategies for all plans are summarized in Table 4.1.

Table 4.1 Summary of ring construction strategies employed for the synthesis of oseltamivir phosphate by all plans to date

Ring construction strategy	Synthesis plan
[3 + 3]	Kann
[4 + 2]	Corey, Fukuyama, Hayashi G2, Hayashi G3, Kamimura, Lu, Okamura, Roche G4, Shibasaki G3, Shibasaki G4, Shibasaki G5, Trost
[6 + 0]	Fang G1, Ko, Mandai G1, Mandai G2, Shiniogi
[2 + 2 + 2]	Hayashi G1
Claisen rearrangement of six-membered *D*-glucal ring	Chen-Liu
Precursor with preformed ring: Bromobenzene	Banwell, Fang G2, Fang G3
Ethyl benzoate	Hudlicky G1, Hudlicky G2
Quinic acid	Gilead, Roche G1, Roche G2
Shikimic acid	Roche G3, Roche G6, Shi
1.4-Cyclohexadiene	Shibasaki G1, Shibasaki
2,6-Dimethoxyphenol	G2 Roche G5

4.4 Oseltamivir Syntheses: Material Efficiencies

In this section, various types of data are presented for all plans regarding the material efficiency performances. Table 4.2 summarizes the kernel metrics. The meritorious plans are the Shi (highest overall AE, least number of input materials, least number of reaction steps, lowest kernel mass of waste, highest kernel RME), Roche G3 (second-highest AE, second-lowest fraction of sacrificial reagents), Roche G6 (least number of reaction steps), Hayashi (least number of reaction stages), and Fang G2 (lowest fraction of sacrificial reagents). At the kernel level, there is a scattering of optimal parameters across plans and therefore there are further possibilities to achieve full optimization in both kernel material and synthesis strategy efficiency. Table 4.3 summarizes the E-factor breakdowns. This assessment shows a clear leader in the Roche G3 plan when solvent demand, excess reagent consumption, and auxiliary material consumption are taken into account. By taking the minimum values in each E-factor category, this data indicates that the target E-total for the next new plan is $7.4 + 24.6 + 198.6 = 230.6$ or 94 kg of waste per mole of the target product. This has essentially been achieved by the Roche G3 plan. Any further improvements would be further reductions in each E-factor category for the Roche G3 plan. Only minimum estimates for the Shi and Hayashi plans can be made because the full material consumption for auxiliary materials was not disclosed in these plans, particularly for solvents used in chromatographic purification. This is a typical oversight for all academic plans. Also, none of the academic plans proved their efficiencies upon scale-up to gram or kilogram quantities of the target product made.

Figures 4.3 and 4.4 show the radial hexagons for industrial and academic syntheses, respectively. Good-performing plans are characterized by pictures showing hexagons of maximum size, whereas poor performers are ones with distorted hexagons gravitating to the center of the diagram. Clearly, the Roche G3 plan has the largest-area radial hexagon among industrial plans consistent with its ranking in the preceding tables. Among academic plans, the Fang G2 and Fang G3 plans rank the highest, followed by the Hayashi

Table 4.2 Summary of kernel metrics for oseltamivir phosphate synthesis plans

Plan	Type[a]	N[b]	M[c]	I[d]	f(sac)	% yield	% AE	% Kernel RME	Kernel mass of waste[e]
Shi	L	8	8	14	0.661	47.6	21.7	11.9	3.0
Roche G3 (shikimic acid route)	L	13	13	19	0.457	39.0	21.0	11.5	3.1
Roche G2 (quinic acid route)	L	14	14	20	0.467	21.9	20.3	9.0	4.2
Fang G2	L	12	12	20	0.452	26.5	14.5	7.2	5.3
Fang G3	L	12	12	21	0.557	22.5	17.1	6.8	5.6
Lu	L	10	10	20	0.642	20.8	11.6	4.8	6.2
Trost (short)	L	9	9	17	0.630	29.9	16.1	5.6	6.9
Hayashi G3	C	11	13	20	0.717	24.2	9.5	4.3	6.9
Corey	L	11	11	17	0.743	22.4	17.2	5.5	7.2
Roche G5 (desymmetrization)	L	11	11	24	0.677	25.6	13.8	5.3	7.3
Hayashi G1	C	5	7	21	0.756	29.3	8.0	4.1	7.3
Hayashi G2	C	4	6	21	0.753	30.5	7.7	4.0	7.5
Roche G6	L	8	8	15	0.659	15.2	21.9	4.6	8.5
Trost (long)	L	12	12	21	0.690	16.2	13.4	4.0	9.8
Roche G1 (quinic acid route)	L	12	12	21	0.625	7.6	18.5	3.2	12.5
Fang G1	C	17	18	35	0.671	13.4	12.0	3.1	12.7
Hudlicky G2	L	9	9	21	0.705	5.3	19.9	3.1	12.7
Gilead	L	12	12	21	0.607	6.3	20.5	2.7	15.0
Hudlicky G1	C	12	13	23	0.741	3.5	20.6	2.4	16.5
Fukuyama	L	13	13	22	0.638	5.5	15.9	2.4	16.4
Mandai G2	C	16	18	26	0.493	8.2	12.6	1.8	17.1
Shiniogi	L	12	12	25	0.679	9	14.7	2.2	18.2
Mandai G1	L	18	18	30	0.714	7.0	7.8	1.5	20.4
Shibasaki G4	L	11	11	25	0.584	9.6	11.5	1.9	21.1
Banwell	L	14	14	29	0.656	4.6	13.1	1.9	21.3
Ko	L	16	16	32	0.798	6.7	7.4	1.4	21.5
Roche G4 (Diels–Alder)	L	9	9	17	0.505	1.1	23.9	1.5	27.4
Shibasaki G5	L	11	11	21	0.509	6.7	13.0	1.4	28.0
Okamura–Corey	L	13	13	25	0.754	2.6	16.8	1.3	32.0
Kamimura	L	14	14	26	0.690	2.1	15.9	1.2	32.7
Kann	L	15	15	25	0.788	3.4	11.8	0.9	47.4
Shibasaki G2	L	16	16	32	0.837	4.5	10.0	0.8	47.9
Shibasaki G3	L	11	11	23	0.481	1.4	16.1	0.6	73.6
Chen-Liu	L	21	21	33	0.651	1.7	9.4	0.5	86.8
Shibasaki G1	L	15	15	34	0.766	1.4	9.6	0.3	150.3

[a] L = linear; C = convergent.
[b] N = number of reaction stages.
[c] M = number of reaction steps.
[d] I = number of input materials.
[e] Kilograms per mole oseltamivir phosphate.

Table 4.3 Summary of E-factor breakdowns for oseltamivir phosphate synthesis plans

Plan	E_{kernel}	E_{excess}	E_{aux}	E_{total}	True mass of waste[h]
Roche G3 (shikimic acid route)	7.7	24.6	198.6	230.9	94.7
Roche G2 (quinic acid route)	10.1	30	267.7	307.9	126.2
Hayashi G1	23.4	177.8	>396	>597	>186
Shi	7.4	28.2	>429	>465	>191
Roche G1 (quinic acid route)	30.6	71.1	755.5	857.2	351.4
Roche G5 (desymmetrization)	17.8	68.4	847.4	933.6	382.8
Gilead	36.7	91.5	808.6	936.7	384.0
Fang G2	12.9	59.2	>1657	>1729	>709
Hayashi G2	24.1	92.1	>2329	>2445	>763
Roche G6	20.9	84.9	1784.1	1889.8	774.8
Fang G3	13.7	104.3	>1859	>1977	>810
Fang G1	31.0	274.8	>2275	>2581	>1058
Trost (short)[a]	16.8	141.5	>2527	>2685	>1101
Hudlicky G1[b]	40.2	133.4	>2593	>2767	>1134
Trost (long)[c]	23.8	144.4	>2691	>2859	>1172
Lu	19.9	321.2	>3717	>4058	>1266
Corey[d]	17.5	208.8	>3057	>3283	>1346
Hudlicky G2[1]	31.0	220.3	>3359	>3611	>1480
Fukuyama[e]	40.0	163.4	>3843	>4047	>1659
Ko	68.8	277.4	>5685	>6031	>1882
Shibasaki G4[f]	51.4	1162.5	>3741	>4955	>2031
Roche G4 (Diels-Alder)	66.8	181.2	>4856	>5104	>2092
Shibasaki G5[f]	68.3	1476.9	>4000	>5545	>2274
Hudlicky G1[g]	40.2	6174.1	>5792	>12007	>4923
Banwell	21.3	838.7	>11683	>12573	>5155
Shiniogi	44.4	453.0	>13041	>13539	>5551
Kann	115.5	285.9	>13238	>13640	>5592
Hayashi G3	22.0	200.2	>19742	>19964	>6229
Chen-Liu	211.6	337.9	>14945	>15494	>6353
Shibasaki G1[f]	366.6	3772.8	>12055	>16194	>6640
Kamimura	79.7	551.5	>18182	>18813	>7713
Shibasaki G2[f]	116.8	1279.9	>18818	>20215	>8288
Okumura-Corey	78	439.8	>21926	>22444	>9202
Shibasaki G3	179.5	1554.1	>24806	>26539	>10881

[a] Includes synthesis of Trost ligand and DuBois catalyst.
[b] Excludes excess water and fermentation medium in step 1.
[c] Includes synthesis of Trost ligand and DuBois catalyst.
[d] Includes synthesis of Corey-Shibata-Lee catalyst.
[e] Includes synthesis of MacMillan catalyst.
[f] Includes synthesis of Shibasaki ligand.
[g] Includes excess water and fermentation medium in step 1.
[h] Kilograms per mole oseltamivir phosphate.

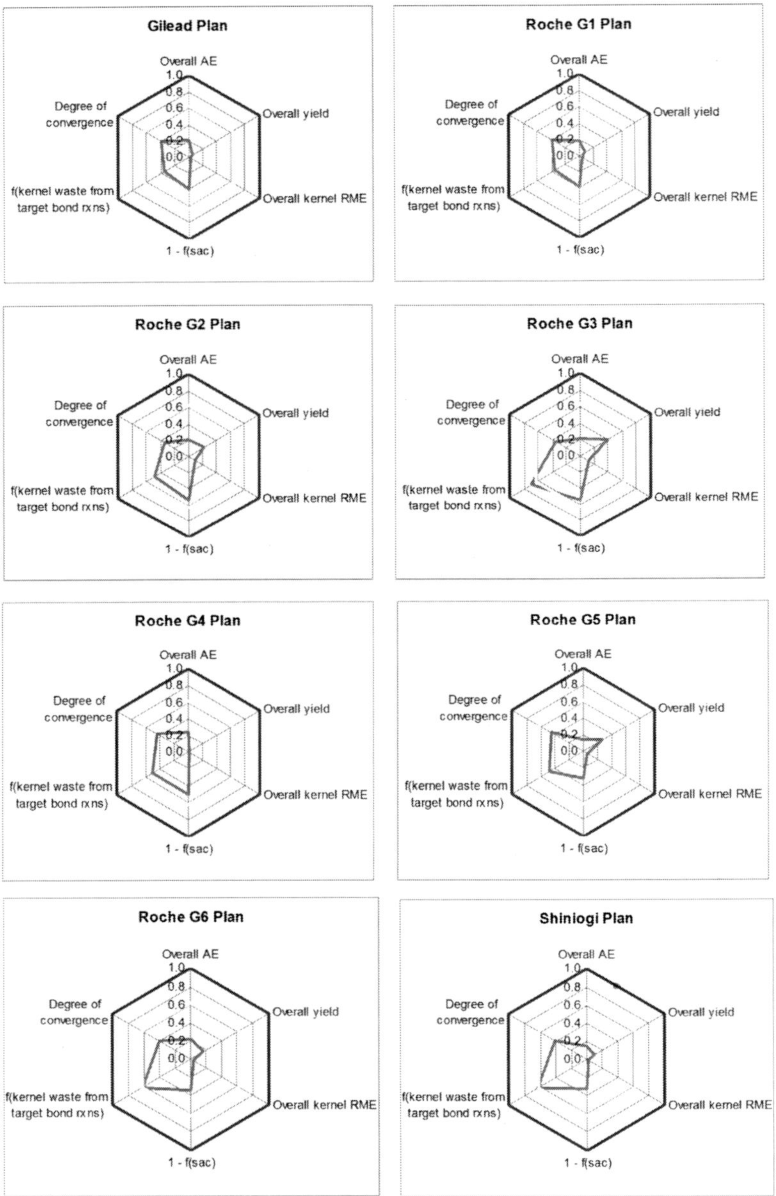

Figure 4.3 Radial hexagons for industrial synthesis plans for oseltamivir phosphate.

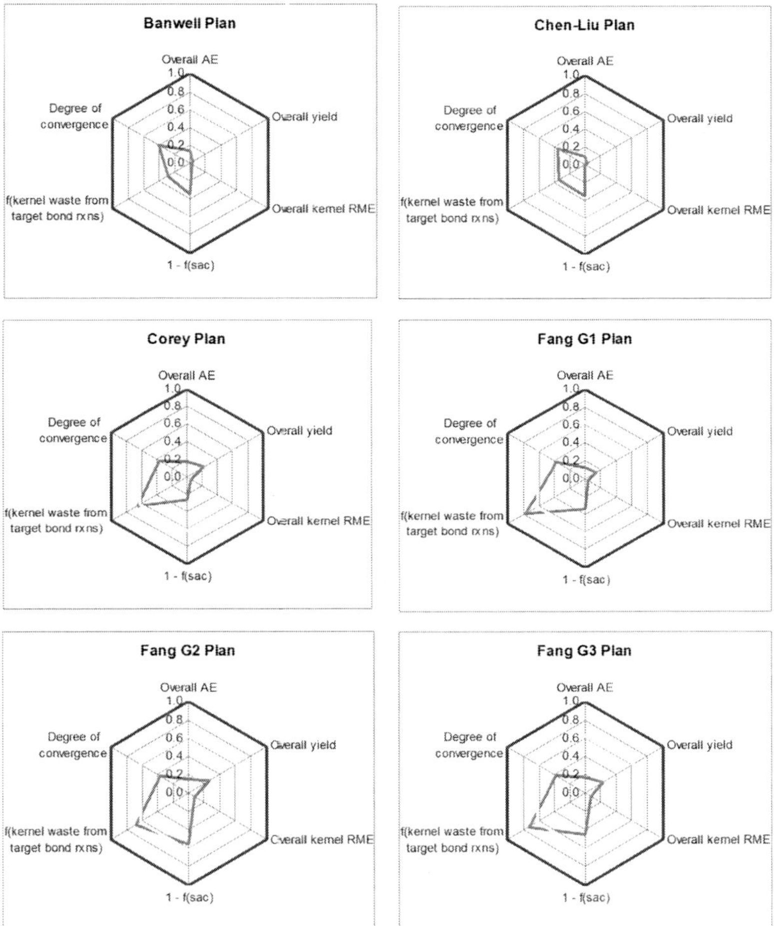

Figure 4.4 Radial hexagons for academic synthesis plans for oseltamivir phosphate.

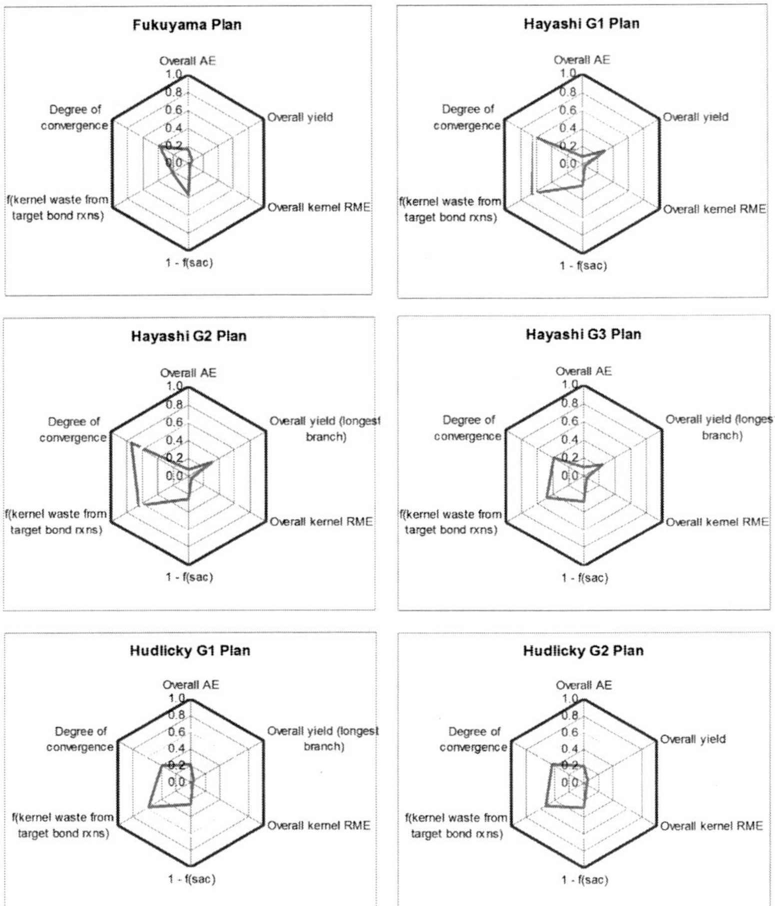

Figure 4.4 (*Continued*)

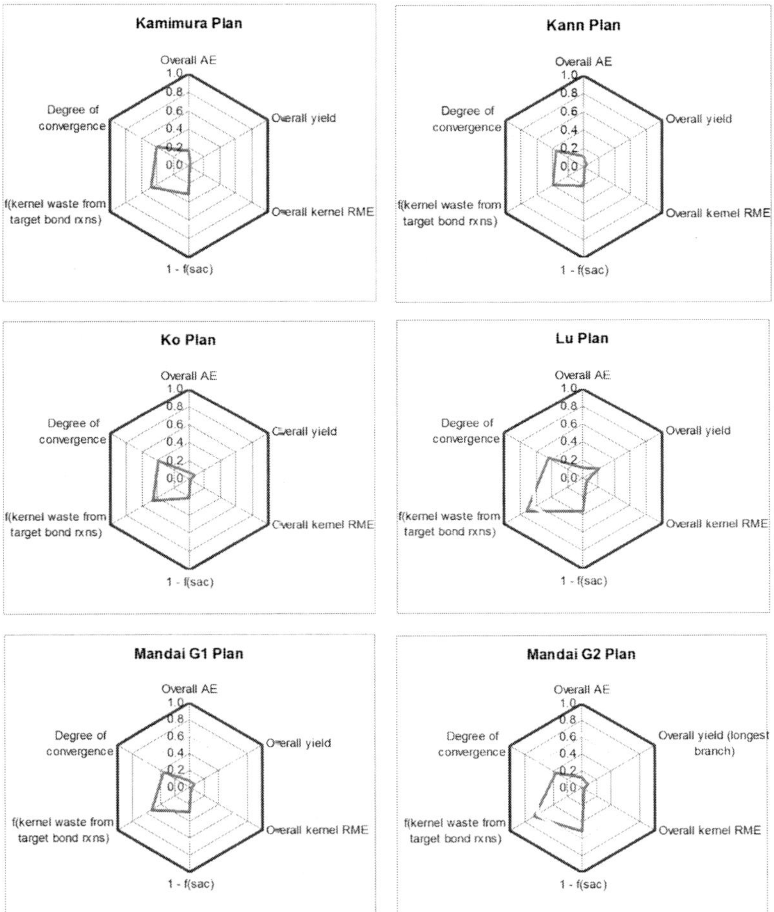

Figure 4.4 (*Continued*)

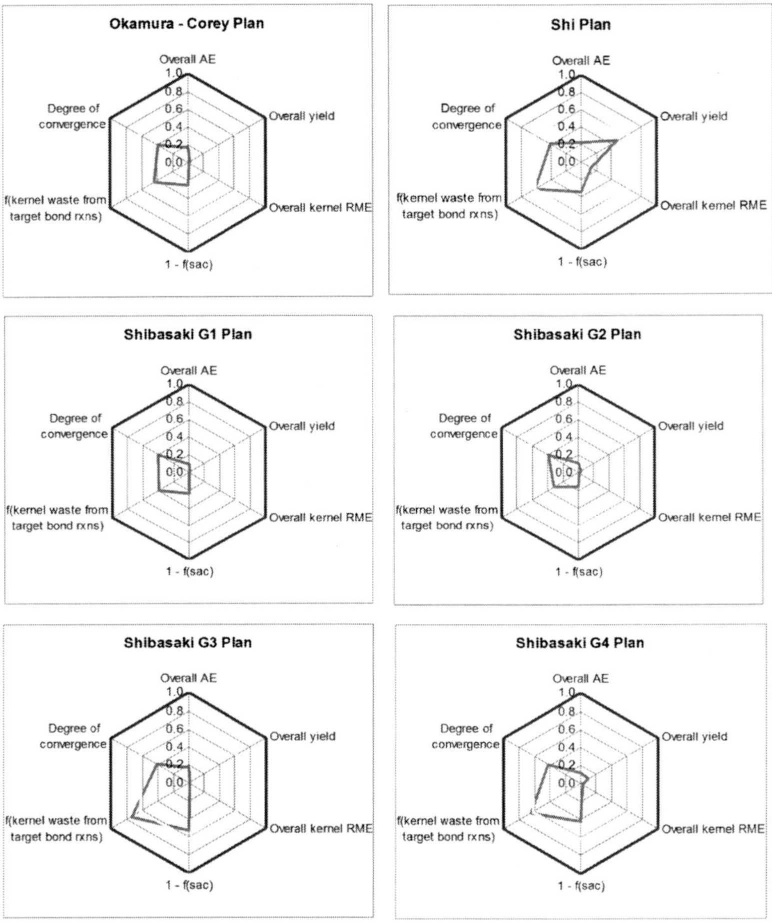

Figure 4.4 (*Continued*)

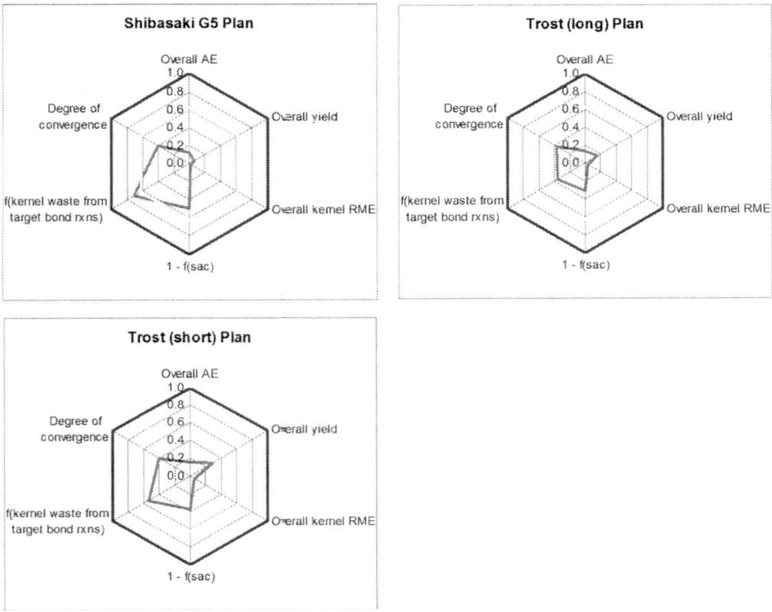

Figure 4.4 (*Continued*)

G1, Shi, Shibasaki G3, and Shibasaki G5 plans. Figure 4.5 shows a ranking of the kernel mass of waste produced per mole of the target product, as well as a breakdown of the waste between target bond–forming and sacrificial reactions. Here, the Shi and Roche G3 plans rank the highest with respect to the mass of kernel waste produced. With respect to the fraction of kernel waste produced from sacrificial reactions, the following plans have the greatest contributions: Trost (short), Roche G1, Gilead, Fukuyama, Banwell, Kann, Shibasaki G2, and Shibasaki G1. Figure 4.6 and Scheme 4.1 show the radial pentagons and the chemical route for the 13-step linear Roche G3 plan, respectively. From these series of pictures the reader may see at once where the bottlenecks are with respect to low atom-economical steps, low reaction yield steps, high excess reagent consumption, and high auxiliary material consumption. These are summarized in Table 4.4. All of these features manifest themselves as distorted radial pentagons toward the centers of the diagrams. When these bottlenecks are identified then further optimizations

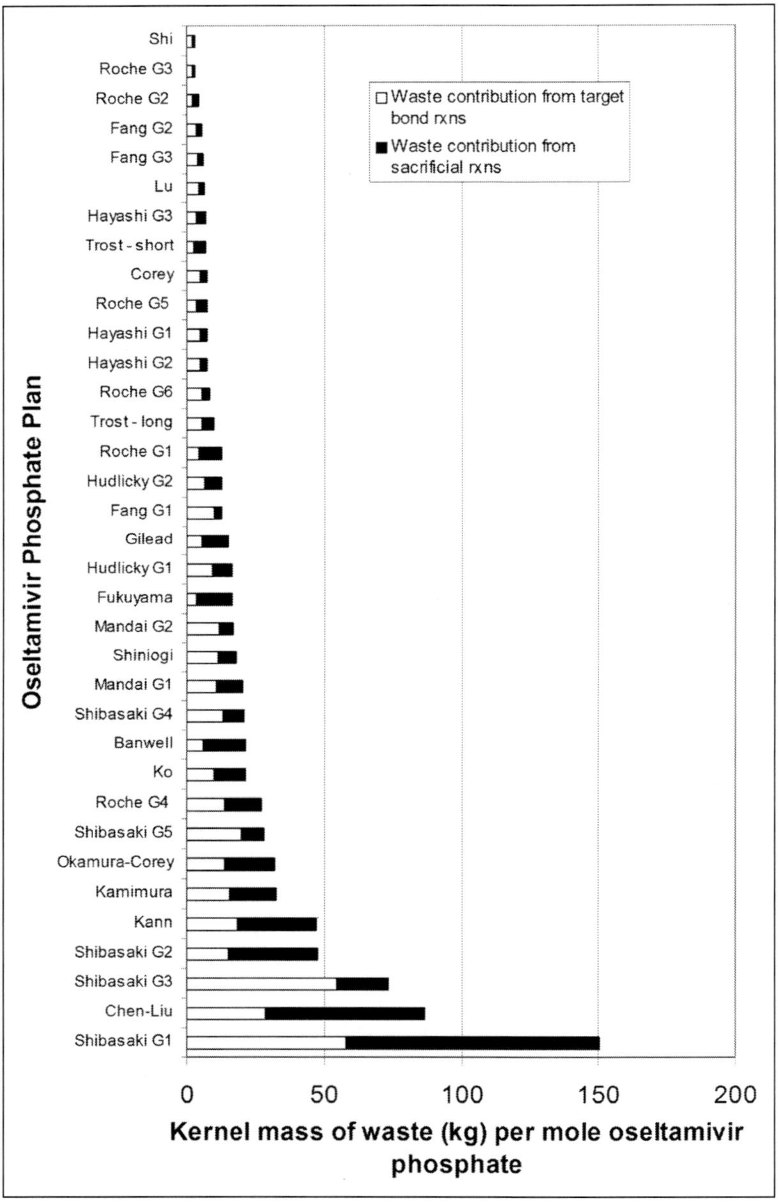

Figure 4.5 Bar graph showing sacrificial and building-up reaction proportions for each synthesis plan for oseltamivir phosphate. Plans are ranked according to kernel mass of waste produced per mole of oseltamivir phosphate.

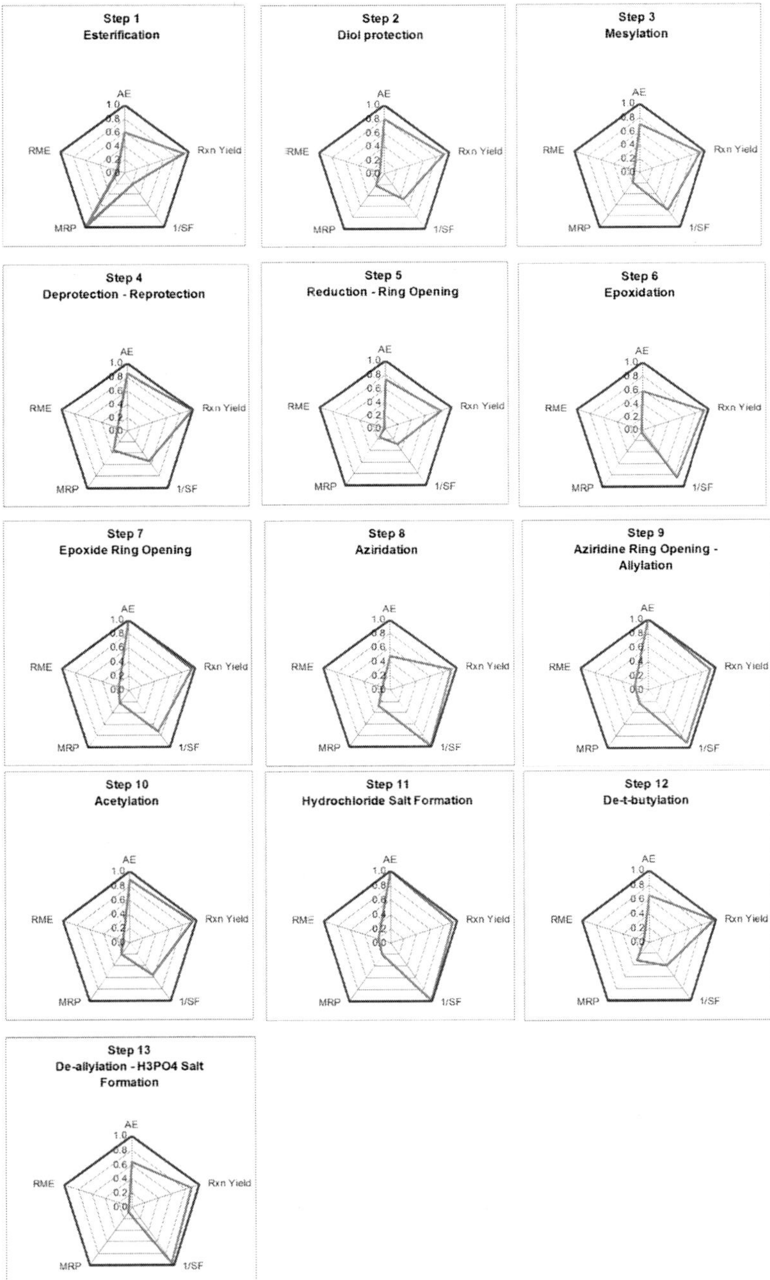

Figure 4.6 Radial pentagons for the Roche G3 synthesis plan.

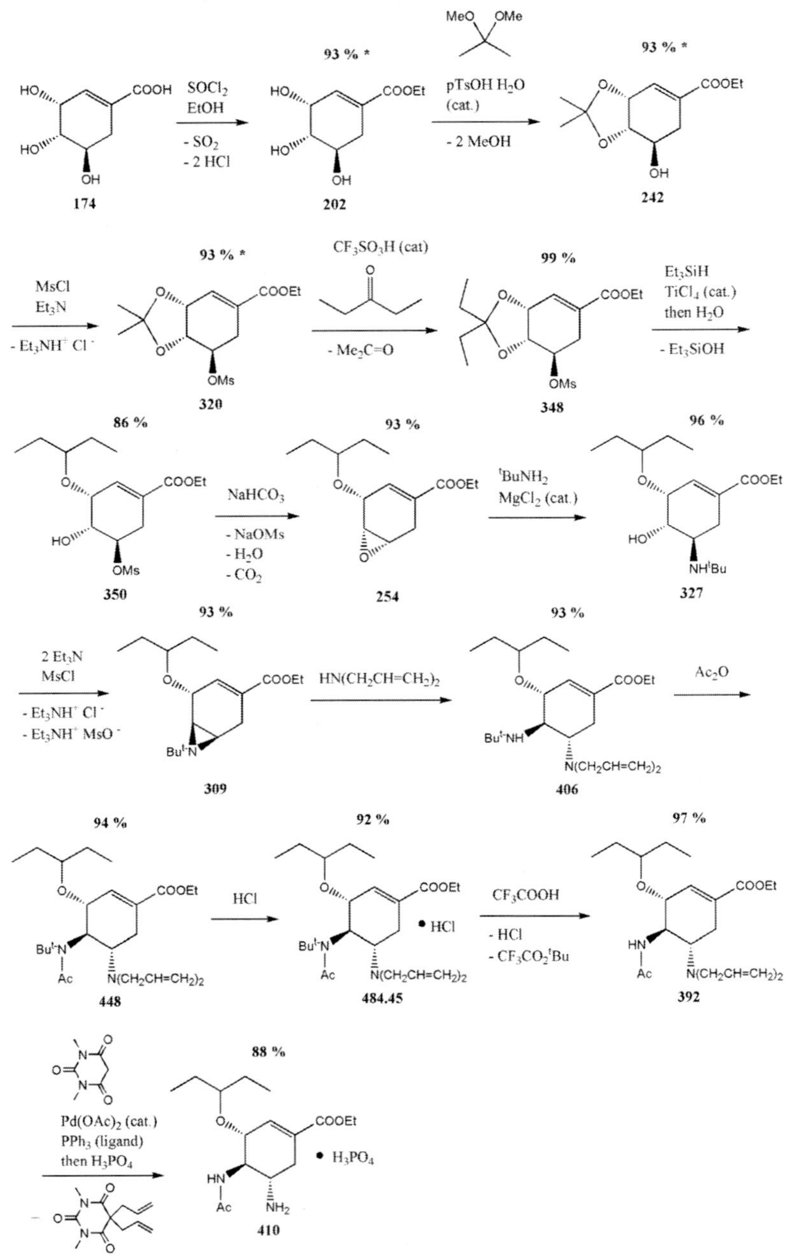

Scheme 4.1 Roche G3 synthesis plan.

Table 4.4 Summary of the bottleneck reaction steps in the Roche G3, Hayashi G1, and Shi plans for oseltamivir phosphate[a]

Plan	Reaction yield	AE	Excess reagent consumption	Auxiliary material consumption
Roche G3	None	Steps 1, **8**, 12, and 13	Steps **1**, 2, 5, and 12	Steps 2, 3, 5, **6**, and 13
Hayashi G1	Step **2**	Steps 1, 1*, and **5**	Steps 4 and **5**	Steps **1***, **2**, and 4
Shi	None	Steps 1, 2, and **4**	Steps **1**, 4, **6**, and 8	Steps 2, **3**, **4**, **5**, 6, **7**, and **8**

[a]The worst-performing step in each category is shown in bold.

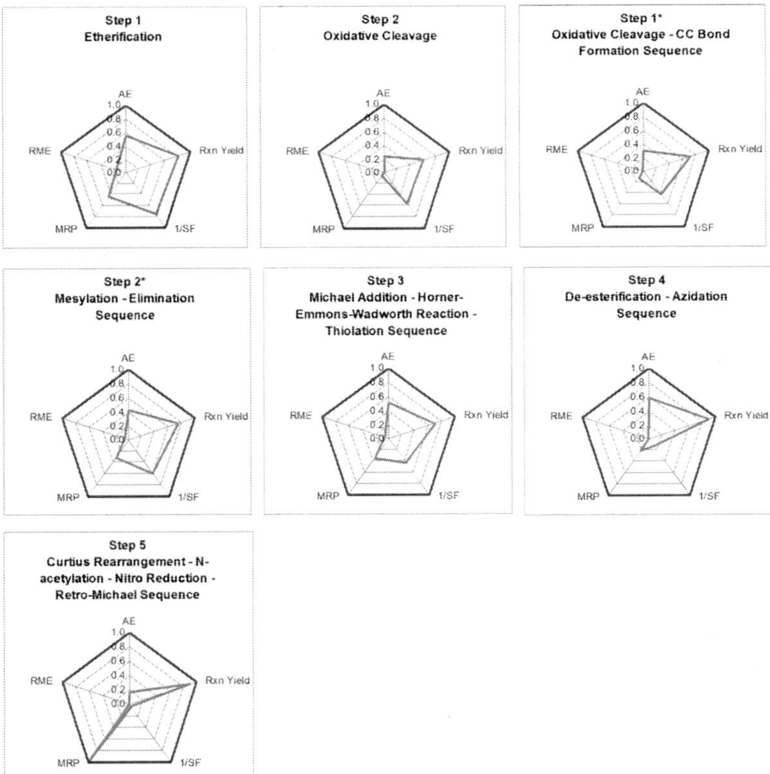

Figure 4.7 Radial pentagons for the Hayashi G1 synthesis plan.

may be made in a more directed fashion in order to address them. The outer perimeter pentagon is the "green" limit, where all five parameters have values of unity or 100%. Figure 4.7 and Scheme 4.2 show the results for the Hayashi plan, while Fig. 4.8 and Scheme 4.3 refer to the Shi plan.

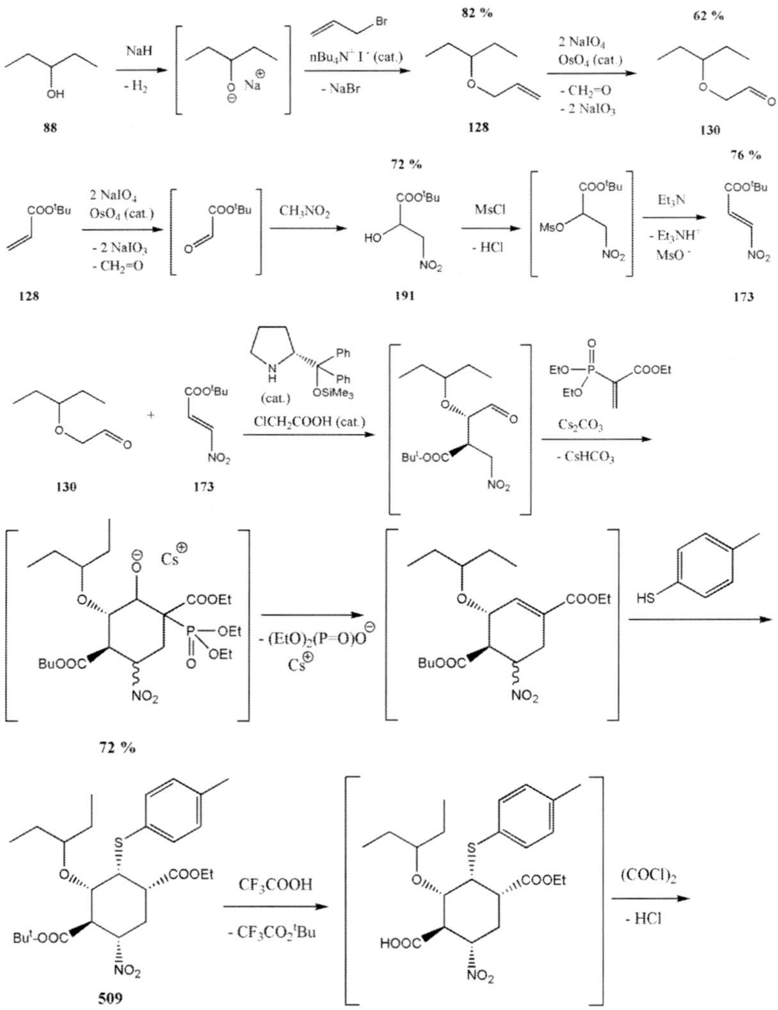

Scheme 4.2 Hayashi G1 synthesis plan.

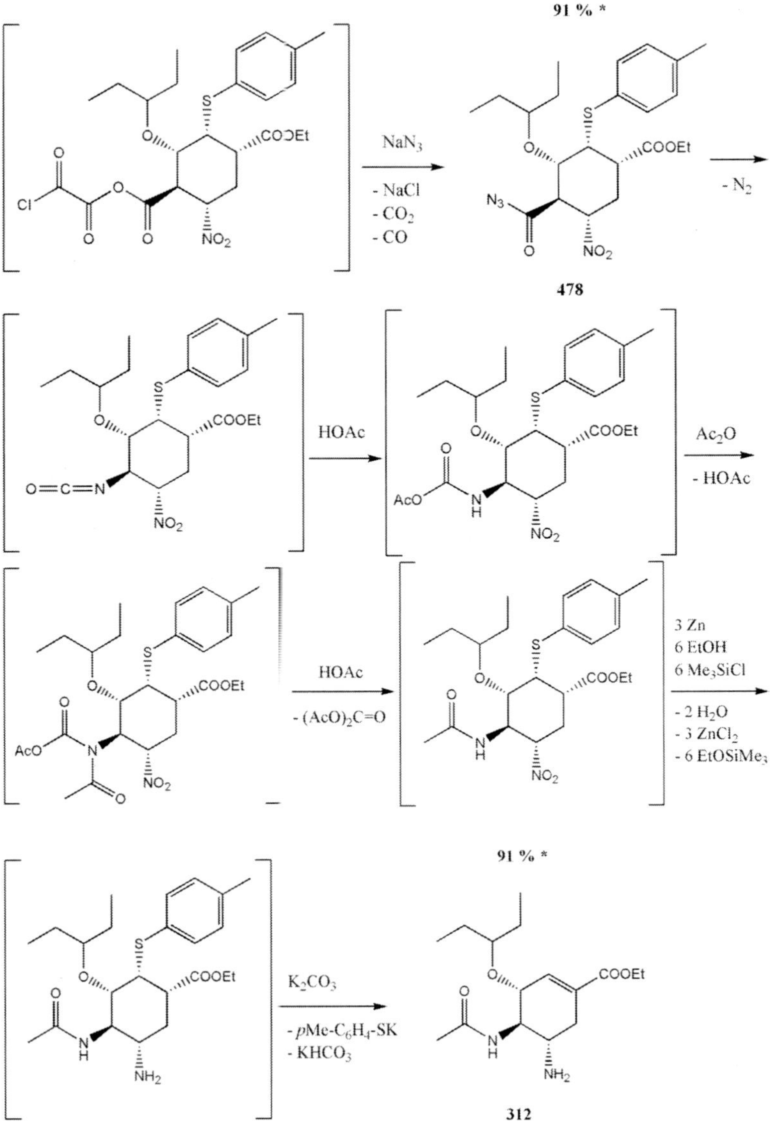

Scheme 4.2 (*Continued*)

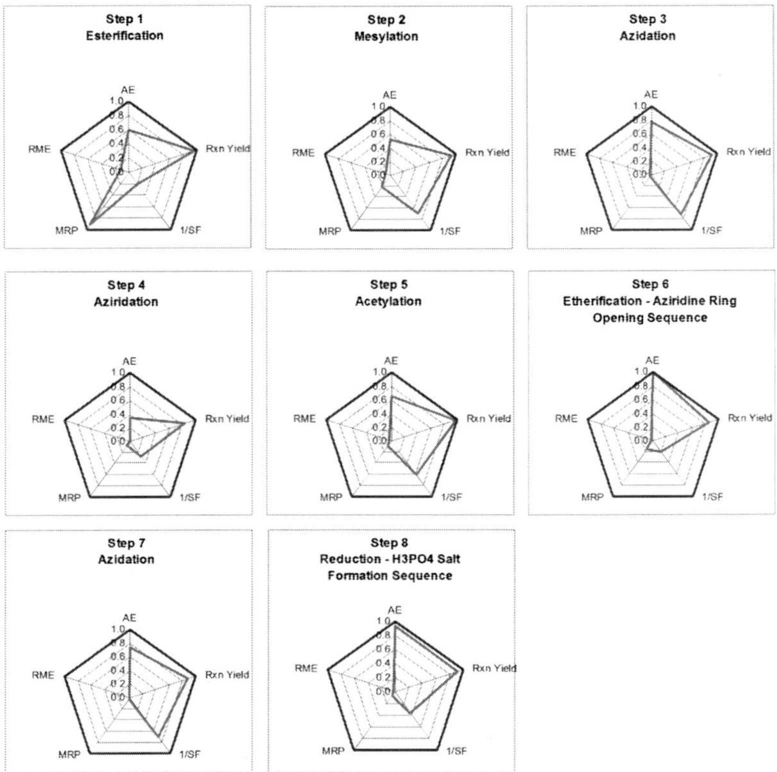

Figure 4.8 Radial pentagons for the Shi synthesis plan.

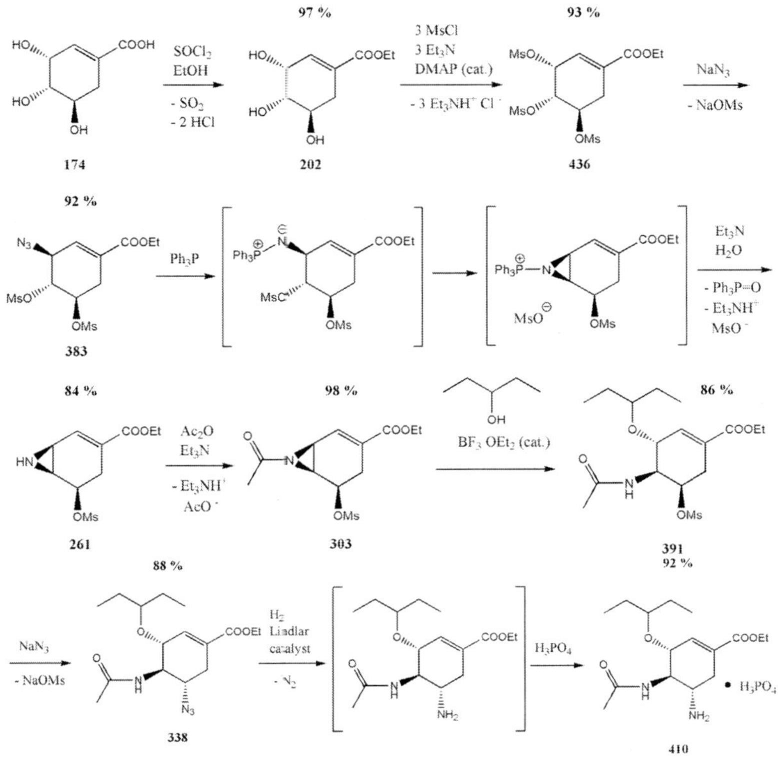

Scheme 4.3 Shi synthesis plan.

4.5 Concluding Remarks

The present updated analysis of 34 synthesis plans of oseltamivir phosphate clearly demonstrate that to date the use of shikimic acid as the starting material is an unbeatable strategy despite the efforts of several workers to come up with alternatives. The only conceivable precursor that could possibly compete with this is some yet to be found natural product that already has the central C4 amino group with the same stereo-orientation as the final product. Having this in hand at the outset would offer a unique advantage since the orientations of the neighboring groups at C3 and C5 could be controlled by the commonly employed successive aziridine ring formation/opening strategy (aziridine walk around the ring) already employed by the majority of existing plans.

In a general sense, "good" synthesis strategies have several attributes: (a) they involve the use of building-type reactions such as multicomponent, tandem, domino, cascade, and "one pot" addition reactions in synthesis planning, especially for the construction of ring systems; (b) they involve telescoping or concatenation of reaction steps, thus minimizing intermediate isolations; (c) they reduce reaction solvent demand; (d) they reduce auxiliary material demand; and (e) they use safe and benign methodologies as far as possible.

Efforts to reduce waste and optimize reaction performance have always been an ongoing activity in the chemical industry. The problem is that these efforts were conducted haphazardly, guided more by intuition and experience than by real quantitative measurements of waste reduction. As mentioned at the outset of this chapter, serious waste reduction can only be effectively achieved if one knows precisely the chemistry involved in each reaction step in a synthesis plan. First-generation waste reduction involves reducing consumption of reaction solvents and solvents used in workup and purification stages; replacing hazardous solvents with benign, environmentally friendly solvents; and recycling solvents in the same process or for use in another. When green chemistry was introduced as a guiding framework nearly 20 years ago, most of the original work was in these areas. This made sense since the total mass of all solvents used in any synthesis plan far exceeds that of input reagents used to make the final product. Evidence of this is readily apparent by looking at the proportion of E-aux to E-total in Table 4.3. Second-generation waste reduction involves optimizing reaction conditions to increase reaction yields. This is usually done by manipulating the reaction time, reaction temperature, reaction pressure, choice of solvent (solubility, heat capacity, compatibility with reagents), and use of catalysts or additives. Third-generation waste reduction involves reduction of by-products by redesigning reactions and restrategizing a synthesis plan to increase the number of target bond–forming reactions and additive redox-type reactions of high AE and to decrease the number of sacrificial reactions and subtractive redox-type reactions of low AE. When comparing these three generations of waste reduction, it is clear that efforts become progressively more challenging on going from solvent reduction to synthesis redesign. The desired strategy progression is that the nth-generation plan for a given target molecule reported in year X should aim to have a better set of metrics parameters than the $(n - 1)^{th}$ generation plan reported in year $(X - 1)$. The aim is to synergistically optimize all parameters in such a way that demonstration of novelty and substantial E-factor reductions are both satisfied. There is no doubt that this goal is hard to achieve, as is well demonstrated between the objectives of a medicinal chemistry route versus a process chemistry route to a pharmaceutical. In the former case, the objectives are speed and versatility in order for a large library or database of analog compounds to be constructed from a generalized synthesis plan. The latter objective is sought only after the key lead structure having the desired pharmaceutical properties has been identified in the compound library. Chemists can get a tremendous head start to achieving directed, informed, and intelligent optimization of a given target molecule by carrying out detailed analyses, as shown here for the case of oseltamivir phosphate, of prior published synthesis plans before embarking on a new plan. There are obvious advantages when there is a significant literature on a given compound. Synthesis optimization is a continuous iterative exercise based on a comparative analysis of a set of plans. One must always bear in mind that ranking is the inevitable consequence of metrics analysis and that the achievement of the "best" plan is only possible on the shoulders of prior poorer-performing ones. In this regard, all plans have value.

Green chemistry is a subject that is rooted in decision making. Moreover, informed decision making can only be made on the basis of thorough metrics analysis, such as that demonstrated here. The current problem in the literature is that there are many false claims of "greenness" based on one or two metrics or a small subset of the Anastas–Warner 12 green principles. The challenge is compounded because of a lack of consensus and agreement on standardization of metrics usage and the disparate reporting styles of authors, particularly in the disclosure of experimental details in procedures. The field is currently at a key crossroads. If these issues remain unaddressed, green chemistry could lose its credibility as a rigorous discipline as more "green washing" publications appear in the literature. The negative backlash is that skepticism, particularly from the chemical industry, will increase and slow down progress toward achieving sustainable chemical processes, as emphasized by Tucker.[1] A number of methods should help green chemistry to better take root in the industry: education, literature reporting, and standardization.

References

1. Tucker, J. L. (2010). *Org. Process Res. Dev.*, **14**, 328–331.
2. Constable, D. J. C.; Dunn, P. J.; Hayler, J. D.; Humphrey, G. R.; Leazer, J. L., Jr.; Linderman, R. J.; Lorenz, K.; Manley, J.; Pearlman, B. A.; Wells, A.; Zaks, A.; Zhang, T. Y. (2007). *Green Chem.*, **9**, 411–420.
3. Carey, J. S.; Laffan, D.; Thomson, C.; Williams, M. T. (2006). *Org. Biomol. Chem.*, **4**, 2337–2347.
4. Shibasaki, M.; Kanai, M. (2008). *Eur. J. Org. Chem.*, 1839–1850.
5. Graening, T. (2009). *Nachrichten Chem.*, **57**, 654–659.
6. Abrecht, S.; Harrington, P.; Iding, H.; Karpf, M.; Trussardi, R.; Wirz, B.; Zutter, U. (2004). *Chimia*, **58**, 621–629.
7. Farina, V.; Brown, J. D. (2006). *Angew. Chem., Int. Ed.*, **45**, 7330–7334.
8. Magano, J. (2009). *Chem. Rev.*, **109**, 4398–4438.
9. Andraos, J. (2009). *Org. Process Res. Dev.*, **13**, 161–185.
10. Calvo-Flores, F. G. (2009). *ChemSusChem*, **2**, 905–919 (and references therein).
11. Andraos, J. (2008). Application of green metrics analysis to chemical reactions and synthesis plans in *Green Chemistry Metrics: Measuring and Monitoring Sustainable Processes* (A. Lapkin; D. J. C. Constable, Eds.). Wiley, Oxford, 69–199.
12. Andraos, J.; Sayed, M. (2007). *J. Chem. Educ.*, **84**, 1004–1010.
13. Andraos, J. (2005). *Org. Process Res. Dev.*, **9**, 149–163.
14. Andraos, J. (2005). *Org. Process Res. Dev.*, **9**, 404–431.
15. Andraos, J. (2006). *Org. Process Res. Dev.*, **10**, 212–240.
16. Andraos, J.; Izhakova, J. (2006). *Chim. Oggi/Chem. Today*, **24**(6 suppl.), 31–36.
17. Andraos, J. (2007). *Can. Chem. News*, **59**(4), 14–17.

18. Matveenko, M.; Willis, A. C.; Banwell, M. G. (2008). *Tetrahedron Lett.*, **49**, 7018–7020.

19. Yeung, Y. Y.; Hong, S.; Corey, E. J. (2006). *J. Am. Chem. Soc.*, **128**, 6310–6311.

20. Shie, J. J.; Fang, J. M.; Wang, S. Y.; Tsai, K. C.; Cheng, Y. S. E.; Yang, A. S.; Hsaio, S. C.; Su, C. Y.; Wong, C. H. (2007). *J. Am. Chem. Soc.*, **129**, 11892–11893.

21. Shie, J. J.; Fang, J. M.; Wong, C. H. (2008). *Angew. Chem., Int. Ed.*, **47**, 5788–5791.

22. Satoh, N.; Akiba, T.; Yokoshima, S.; Fukuyama, T. (2007). *Angew. Chem., Int. Ed.*, **46**, 5734–5736.

23. Satoh, N.; Akiba, T.; Yokoshima, S.; Fukuyama, T. (2009). *Tetrahedron*, **65**, 3239–3245.

24. Rohloff, J. C.; Kent, K. M.; Postich, M. J.; Becker, M. W.; Chapman, H. H.; Kelly, D. E.; Lew, W.; Louie, M. S.; McGee, L. R.; Prisbe, E. J.; Schultze, L. M.; Yu, R. H.; Zhang, L. (1998). *J. Org. Chem.*, **63**, 4545–4550.

25. Ishikawa, H.; Suzuki, T.; Hayashi, Y. (2009). *Angew. Chem., Int. Eng. Ed.*, **48**, 1304–1307.

26. Sullivan, B.; Carrera, I.; Drouin, M.; Hudlicky, T. (2009). *Angew. Chem., Int. Ed.*, **48**, 4229–4231.

27. Werner, L.; Machara, A.; Hudlicky, T. (2010). *Adv. Synth. Catal.*, **352**, 195–200.

28. Bromfield, K. M.; Gradén, H.; Hagberg, D. P.; Olsson, T.; Kann, N. (2007). *Chem. Commun.*, 3183–3185.

29. Mandai, T.; Oshitari, T. (2009). *Synlett*, 783–786.

30. Oshitari, T.; Mandai, T. (2009). *Synlett*, 787–789.

31. Kipassa, N. T.; Okamura, H.; Kina, K.; Hamada, T.; Iwagawa, T. (2008). *Org. Lett.*, **10**, 815–816.

32. Karpf, M.; Trussardi, R. (2001). *J. Org. Chem.*, **66**, 2044–2051.

33. Federspiel, M.; Fischer, R.: Hennig, M.; Mair, H. J.; Oberhauser, T.; Rimmler, G.; Albiez, T.; Bruhin, J.; Estermann, H.; Gandert, C.; Göckel, V.; Götzö, S.; Hoffmann, U.; Huber, G.; Janatsch, G.; Lauper, S.; Röckel-Stäbler, O.; Trussardi, R.; Zwahlen, A. G. (1999). *Org. Process Res. Dev.*, **3**, 266–274.

34. Harrington, P. J.; Brown, J. D.; Foderaro, T.; Hughes, R. C. (2004). *Org. Process Res. Dev.*, **8**, 86–91.

35. Abrecht, S.; Karpf, M.; Trussardi, R.; Wirz, B. (2000). EP 1127872 Tamiflu via Diels-Alder (F. Hoffmann-La Roche AG).

36. Zutter, U.; Iding, H.; Wirz (2000). EP 1146036 Tamiflu, gallocarboxylic acid approach (F. Hoffmann-La Roche AG).
37. Zutter, U.; Iding, H.; Spurr, P.; Wirz, B. (2008). *J. Org. Chem.*, **73**, 4895–4902.
38. Karpf, M.; Trussardi, R. (2009). *Angew. Chem., Int. Ed.*, **48**, 5760–5762.
39. Nie, L. D.; Shi, X. X.; Ko, K. H.; Lu, W. D. (2009). *J. Org. Chem.*, **74**, 3970–3973.
40. Nie, L. D.; Shi, X. X. (2009). *Tetrahedron Asymm.*, **20**, 124–129.
41. Fukuta, Y.; Mita, T.; Fukuda, N.; Kanai, M.; Shibasaki, M. (2006). *J. Am. Chem. Soc.*, **128**, 6312–6313.
42. Mita, T.; Fukuda, N.; Roca, F. X.; Kanai, M.; Shibasaki, M. (2007). *Org. Lett.*, **9**, 259–262.
43. Yamatsugu, K.; Kamijo, S.; Suto, Y.; Kanai, M.; Shibasaki, M. (2007). *Tetrahedron Lett.*, **48**, 1403–1406.
44. Yamatsugu, K.; Yin, L.; Kamijo, S.; Kimura, Y.; Kanai, M.; Shibasaki, M. (2009). *Angew. Chem., Int. Ed.*, **48**, 1070–1076.
45. Yamatsugu, K.; Kanai, M.; Shibasaki, M. (2009). *Tetrahedron*, **65**, 6017–6024.
46. Trost, B. M.; Zhang, T. (2008). *Angew. Chem., Int. Ed.*, **47**, 3759–3761.
47. Ma, J.; Zhao, Y.; Ng, S.; Zhang, J.; Zeng, J.; Than, A.; Chen, P.; Liu, X. W. (2010). *Chem. Eur. J.*, **16**, 4533–4540.
48. Ishikawa, H.; Suzuki, T.; Orita, H.; Uchimaru, T.; Hayashi, Y. (2010). *Chem. Eur. J.*, **16**, 12616–12626.
49. Kamimura, A.; Nakano, T. (2010). *J. Org. Chem.*, **75**, 3133–3136.
50. Ko, J. S.; Keum, J. E.; Ko, S. Y. (2010). *J. Org. Chem.*, **75**, 7006–7009.
51. Weng, J.; Li, Y. B.; Wang, R. B.; Li, F. Q.; Liu, C.; Chan, A. S. C.; Lu, G. (2010). *J. Org. Chem.*, **75**, 3125–3128.
52. Osato, H.; Jones, I. L.; Chen, A.; Chai, C. L. L. (2010). *Org. Lett.*, **12**, 60–63.

Chapter 5

The Road to Becoming Green: Process Development of AR-A2, an Active Pharmaceutical Ingredient with Antidepressant Activity

Hans-Jürgen Federsel, Martin Hedberg, Fredrik R. Qvarnström, and Wei Tian

Pharmaceutical Development, AstraZeneca, 151 85 Södertälje, Sweden

hans-jurgen.federsel@astrazeneca.com, martin.hedberg@sp.se, fqv@hotmail.com, and tianwei58@gmail.com

5.1 The Green Agenda

In the first decade of the new millennium we have seen the road map driving toward a sustainable society being widely addressed by politicians, scientists, the media, environmentalists, and laypersons. Unless you have a clearly expressed view on matters related to sustainability in the widest sense, you run the obvious risk of being seen as hopelessly obsolete. The concept is complex and far reaching and encompasses subareas such as lean production, clean technologies, environmentally considerate processes, recycling and

Scalable Green Chemistry: Case Studies from the Pharmaceutical Industry
Edited by Stefan G. Koenig

ISBN 978-981-4316-49-1 (Hardcover), 978-981-4364-81-2 (eBook)
www.panstanford.com

reuse of spent material, energy efficiency, waste reduction, and minimization of raw material utilization. Expressed in this way, it is bound to have a deep impact on individuals' day-to-day lives. This is further emphasized by a fundamental paradigm upon which the vision for a sustainable world rests (as seen from a chemistry perspective): "meeting the needs of the current generation while preserving the ability of future generations to meet their needs."[1]

The chemical industry, both with its reputation of being a major polluter and a documented attitude of being forward looking and innovative when it comes to adopting novel ideas to improve the business, has fully embraced the green chemistry initiative as a subset of corporate sustainability targets. Thus, first launched in the late 1990s,[2] the goal of designing and developing benign manufacturing procedures, captured under the banner of green chemistry, has now been widely accepted across a whole range of industries—bulk, fine chemicals, and pharmaceutical producers alike. A set of 12 principles[3] underpin green chemistry, and these are now used to provide guidance when setting out to construct new processes, as well as evaluate the performance of existing ones. Focusing on the pharmaceutical industry and, more specifically, the chemistry conducted therein,[4] it is important to realize that the demands on medicinal chemistry are quite different from what applies in a process research and development (PR&D) organization. In the former, making large numbers of potentially interesting molecules in small amounts, many of which will be discarded after testing, is a key driver for which virtually any synthetic methodology will suffice. For PR&D, however, concentrating on relatively few selected compounds, there will be an expectation that the best synthetic routes will be delivered to meet a number of tough criteria—for instance, from environmental and safety points of view, allowing operation on a large scale, offering cost competitiveness, avoiding patent infringements, or sustainability for long-term production, etc.[5–8] A key aspect of the novel, environmentally concerned processing paradigm is to minimize the use of solvents,[9] both for manufacturing purposes and, not least, when conducting cleaning procedures in production equipment, the latter being highly significant in a business operating under the strictest good manufacturing practice (GMP) legislation.

The intention of this chapter is to focus on issues that have been addressed during this transition from an early laboratory-based synthesis to a commercially viable process by focusing on a specific drug project, AR-A2.[10–12] Examples of changes that have been put in place will be provided to highlight the gradual improvements and changes implemented along the timeline and their consequences from the environmental point of view. It is seldom the case that this endeavor, as an outcome, sees entirely novel synthetic transformations emerging and more often results in fine-tuning and optimization of prior art methods, albeit with the addition of a considerable degree of innovation, smartness and practicality that makes the chemistry work at high efficiency in real life. In other words, it delivers a good process that, by definition, fulfills the criteria of being green!

5.2 Starting the New Project

In an effort to identify new compounds capable of allowing severely debilitating central nervous system (CNS) diseases, notably depression and anxiety, to be better addressed compared to current treatment regimes—a stream of work that has been ongoing for decades in our company (or its legacy parts)—a chiral aminotetralin, code name AR-A2 (**1a**), was identified and appointed as a candidate drug (CD) toward the end of the 1990s (Fig 5.1). In fact it was later proven that HBr salt **1b** was best suited for development as the drug product. The plan was to document this rather complex molecule, acting specifically as an antagonist on the 5-HT_{1B} receptor, by showing superior efficacy and reduced side effects.[13,14] As is often the case in the drug-developing industry, most projects face discontinuation before reaching marketing approval, especially in the CNS domain and its associated diseases, where the average attrition, globally speaking, is on the high end of the range (>90%). And in this respect AR-A2 does not constitute an exception, as the readout of clinical phase II studies (a relatively limited number of patients affected by the disease in question) clearly showed that the compound was lacking efficacy. It should be noted that finding relevant animal models on which to base forecasts on the

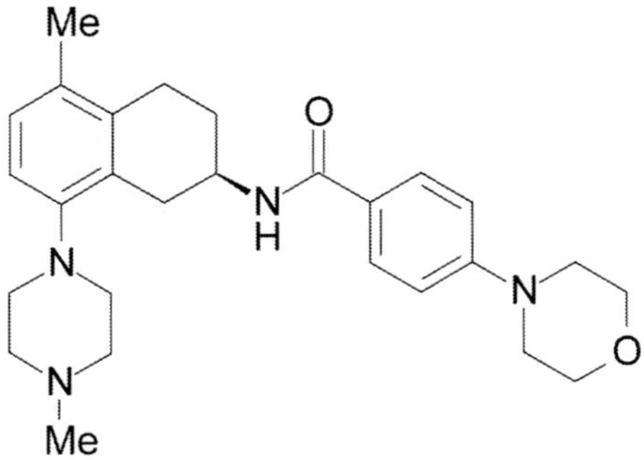

AR-A2
1a= free base
1b= mono-HBr salt

Figure 5.1 The CNS-active drug molecule AR-A2 was first developed as the free base **1a** but later replaced by mono-HBr salt **1b**.

performance of a given drug in man is extremely difficult in the CNS field, which to some degree explains the exceptionally high failure rate. Nonetheless, a whole raft of excellent chemistry was performed during the years that AR-A2 had the status of a CD, and this will be the main focus of this chapter.

When aiming to design novel molecular entities in medicinal chemistry, the widely accepted approach demands that a fairly wide structural space be covered and thoroughly investigated. Either this could mean starting from scratch by conducting a random screening exercise of a given compound collection, for example, those obtained from natural sources or a library from synthetic efforts. Or it could be initiated on the basis of the knowledge that some sort of chemical architecture (ß-lactams in antibiotics belonging to the penicillin and cephalosporin families or the [1,4]-benzodiazepine moiety as a general and significant core motif in many products used as tranquilizers) is abundantly present in certain classes of drug molecules. Thus, with this backdrop the prime focus when initiating this particular project was to build on previous and well-documented experience that 2-aminotetralins exert clinically relevant and useful effects on human receptor targets in the CNS, with a proven track record of beneficial treatments for various mental disorders.[15–17] When setting out to find novel molecules that constitute improvements to existing therapies, it goes without saying that the search has to follow a broad scope even if the compound family is predefined (tetralins in this case). This approach is also reflected in the way that synthetic pathways are designed inasmuch as obtaining a maximum structural diversity is normally seen as the highest priority. Therefore, many of the routes devised and used in medicinal chemistry are not tailor-made for a particular molecule but rather enable smooth synthesis of a whole range of analogs and derivatives. A good testimony to that effect is provided by the AR-A2 case and the first synthetic method[10–12,18] to be operated on a pilot plant scale (Scheme 5.1).

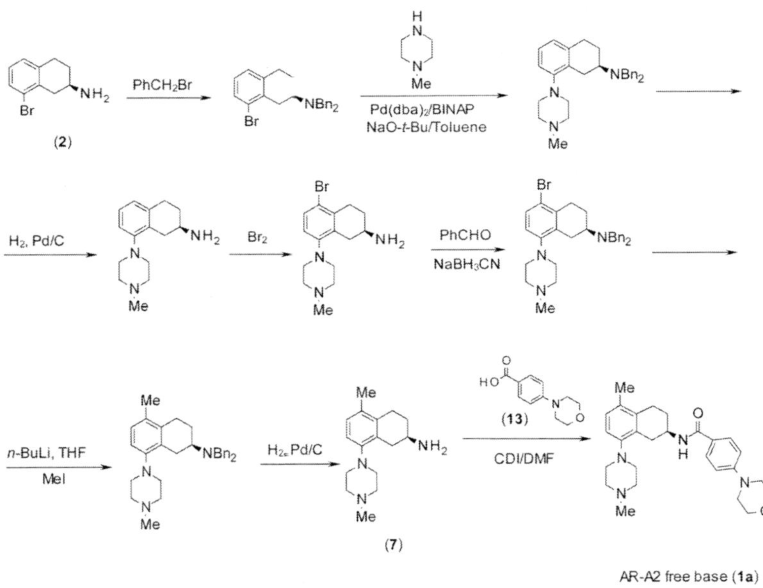

Scheme 5.1 The synthesis of AR-A2 (**1a**) following the procedure devised by medicinal chemistry. *Abbreviation*: CDI, *N,N'*-carbonyl-diimidazole.

Before discussing the features of this route and its environmental profile, it is worth pointing out that there were actually two preceding syntheses in place, which were both judged as unsuitable for scale-up beyond the laboratory. The first of these was constructed around the Smiles rearrangement[19,20]—subjecting acetamidoalkyl-substituted phenolic moieties to harsh conditions (NaH at elevated temperatures >100°C)—as the core constituent to create the appropriate aniline derivative followed by ring closure using an aza-diacid and reduction of the imide formed to generate the piperazine motif (Scheme 5.2). In the other approach, the ring formation was, instead, conducted using *bis*-2-chloroethyl-*N*-methylamine[21,22]—a notorious carcinogen and highly unpleasant reactant—that was

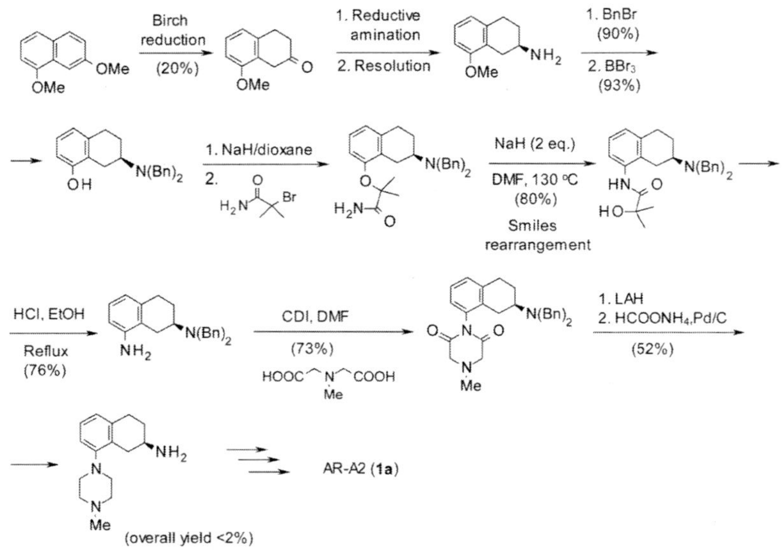

Scheme 5.2 A literature-based method involving a key Smiles rearrangement that eventually leads to a close analog of the target molecule AR-A2 (**1a**), where the C-5 methyl (Me) group is missing. This and other options were explored in the earliest project stages but abandoned due to unsuitability for scale-up.

somewhat more straightforward but nonetheless discarded on the basis of the extremely unfavorable risk profile that use in multipurpose pilot equipment would have rendered. Both of these approaches were, however, soon to be replaced by an elegant catalytic methodology, devised by Buchwald[23] and Hartwig,[24] respectively, that became available around the mid-1990s.

Scrutinizing the first synthesis to be taken into the pilot plant (Scheme 5.1) shows a main linear stream with a short arm to attach the benzoic acid side chain onto the NH_2 functionality. So, formally speaking, it can be characterized as convergent, but as the preparation of the morpholinobenzoic acid moiety proceeded at an extremely high efficiency[25] (yield on pilot scale ≥96%), it is more relevant to focus on the main synthetic branch. Some highlights of the work leading to the manufacturing method for the side chain will be shared later. What stands out in this route and what distinguishes it from the others is the starting material itself—(*R*)-2-amino-8-bromotetralin (**2**). At the time, this advanced chiral building block had become available in stereochemically defined form in small kilogram quantities from a single supplier,[26] who, using proprietary biocatalytic technology, managed to convert the pro-chiral ketone precursor using a specific transaminase enzyme.[27] This seemingly ideal entry point for direct synthesis of the target compound had two drawbacks. First, the cost of the material surpassed USD 100,000/kg due to the small production volume, and second, the methyl substituent was absent at the C-5 position. The knock-on effect of this structural shortcoming is the need to incorporate a Me-adding step into the synthetic strategy. Every attempt to avoid this stage-adding protocol by including the methyl in the starting material failed, as the transamination could not be made operable with the corresponding C-5 methyl tetralone. A further consequence is that a fairly sophisticated protecting group strategy had to be followed—first to ensure clean attachment of the piperazine ring onto C-8 in the Buchwald–Hartwig step and again to avoid side reactions of the C-2 amine before the Me insertion. Irrespective of this being a somewhat circumventing scheme with a lot to be desired in terms of atom economy, many of the eight total steps offered yields into the 90% range, giving an impressive overall outcome of 21–38%. This was seen as fully acceptable for an early-stage delivery and was shown to suffice for the production of a few hundred grams of AR-A2.

5.3 Time for Change: The Road Map to Better Synthesis

5.3.1 *A Novel Synthetic Approach*

As the AR-A2 project proceeded and the demand on larger quantities of the active pharmaceutical ingredient (API) became more pronounced, it became obvious that a better and more cost-effective process had to be designed. When approaching this kind of task it is often imperative to think out of the box in order to identify something sufficiently novel to stand a chance of becoming successful. In the present case the problem did not reside in the yield, which instead was seen as the prime strength of the synthesis. Rather, it was the extremely high cost of the starting material (aminotetralin **2**) and the associated risk with just one supplier who could only manufacture small amounts in a laboratory-like, batch process. The first important step was to disconnect from the prejudiced thinking that we could not initiate the synthesis from anything other than an intact tetralin. Second, a key target was incorporation of the eventual C-5 Me group in the final molecule. It goes without saying that achieving this methyl inclusion without adding synthetic steps would benefit any new route.

The route scouting thus initiated—essentially a paper exercise—relatively quickly homed in on compounds where only the aromatic ring had been retained, that is, various benzene derivatives. Foremost amongst these, 3-methylphenylacetic acid (**3a**) stood out by virtue of fulfilling the crucial demands outlined before. In addition, its availability in bulk quantities at a rather attractive cost helped the succinct second-generation manufacturing procedure fall into place (Scheme 5.3). A key transformation in the overall synthesis, after conducting the somewhat trivial but not very selective or high-yielding bromination[28] to afford 2-bromo-phenylacetic acid **4a** prior to acid chloride formation, is the cyclization to create the tetralone ring. Here ethene was used as an inexpensive C_2 equivalent in a Friedel–Crafts-type reaction to afford the properly substituted tetralone **5** in acceptable yields of 60–70%, an elegant way of creating this motif in just three steps from a commercially available starting material.[29,30] In examining the bromine addition step, however, it was apparent that with a delivery of only 30–35% of the desired product **4a** (albeit of high purity at 98%), the process economy would turn out as unfavorable. The two regioisomeric by-products—the 4-bromo-3-methyl- and 2-bromo-3-methyl analogs, respectively—generated in a combined ratio against the desired product of 0.75:1 could, fortuitously, be recirculated from the mother liquor after workup and subsequently subjected to reducing conditions by virtue of a catalytic hydrogenolysis to regenerate the

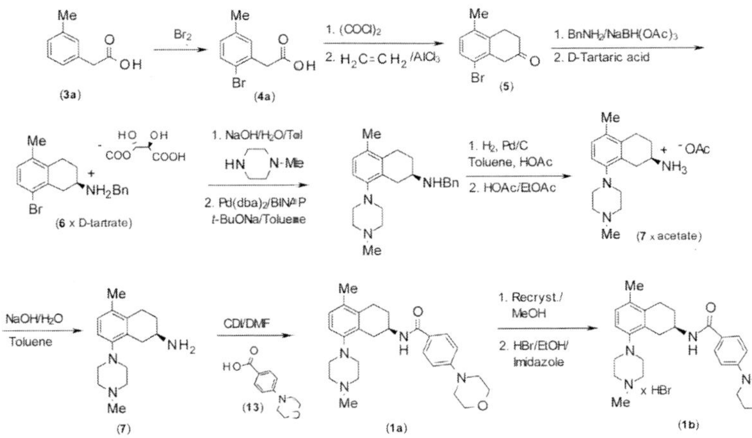

Scheme 5.3 The newly designed second-generation synthesis of AR-A2 × HBr (**1b**) provided a step change in achieving a complex molecular architecture. Starting with a compound (**3a**) where the key C-5 methyl group in **5** was already present eliminated the need to add this substituent using a multistep procedure.

starting material. Conducting the reaction in this manner offered a considerable gain in yield and a much better overall environmental performance.

Transforming pro-chiral ketone intermediate **5** thus prepared into the amine congener was conducted in a one-pot procedure by combining two stages. In the first of these reductive amination[31] was performed using benzylamine as the NH_2 donor, followed by in situ diastereomeric resolution with *D*-tartaric acid as the resolving agent. Still, this rather straightforward method had a significant drawback, namely an irritatingly low yield of 25%, corresponding to the isolation of only 50% of the theoretical content of (*R*)-enantiomer **6** present in the process stream. With the loss of three-fourths of the total material content in this step alone and with a combined yield of barely 4–6% midway to the final product without access to a technology that would allow recirculation and reprocessing, both the financial prospects and the sustainability aspects of this method were made plain. In contrast to the rather poor performance of the initial stages, the remaining steps, including coupling and deprotection, were conducted relatively smoothly with an overall yield of >60% leading to (*R*)-2-amino-8-piperazinyltetralin **7**. Formation of this intermediate—common to all three in-house routes applied on the pilot plant scale—was followed by amide formation to generate AR-A2 as the free base (**1a**) and precipitation of the end product as HBr salt (**1b**). Our conclusions at this point were that though the yield for the whole sequence was 3–4%, considerably lower than for the first synthesis (Scheme 5.1), it still represented a big step forward in terms of cost and suitability for large-scale production. The flaws expressed above, however, were serious enough to maintain focus on finding a better route.

5.3.2 *Refinements Leading to the Final Route*

Mindful of the disappointingly poor yield achieved in the resolution, a key question posed was whether the reductive amination of tetralone **5** could preferably generate the desired stereoisomer (*R*)-aminotetralin **6**. As a matter of fact this was entirely feasible and the simple trick was to replace the achiral NH_2 donor benzylamine with a chiral one, namely (*S*)-phenylethylamine (**8**). This methodology (Scheme 5.4) of utilizing an amine as a chiral auxiliary in a diastereoselective reductive amination protocol has been applied in the preparation of various compound classes such as amino acids,[32,33] cyclohexylamines,[34] phenylalkylamines,[35] and tetralins.[36] Imine formation between tetralone **5** and amine **8** proceeded smoothly in toluene at 40°C by azeotropic evaporation under vacuum, either in the presence or in the absence of *p*-TSA, a quantitative conversion according to ^{1}H nuclear magnetic resonance (NMR) analysis. Water produced during the reaction was completely removed when half the amount of toluene had been distilled off. The resulting imine product **9** was subjected to reduction without isolation, performed in a mixture of *iso*-propanol and MeOH (entry 1, Table 5.1), with the former ensuring dissolution of the reducing agent ($NaBH_4$) and the latter the substrate. Pleasingly, the outcome was successful with the desired (*R*, *S*)-isomer **10** predominating over (*S*, *S*)-isomer **11** in a 4:1 diastereomeric ratio. In the absence

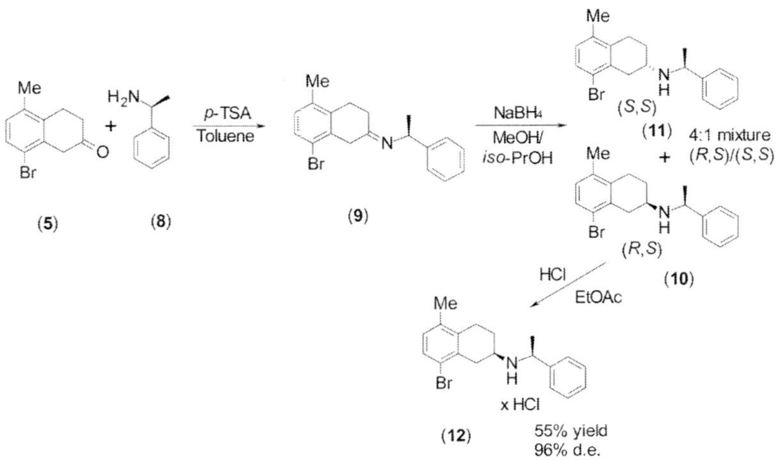

Scheme 5.4 With the steric induction provided by (*S*)-phenylethylamine (**8**), a high degree of stereoselectivity was obtained. *Abbreviation*: *p*-TSA, *para*-toluenesulfonic acid.

Table 5.1 The diastereoselective reductive amination of tetralone **5** with (*S*)-phenylethylamine (**8**) under various conditions

Entry	Reducing agent	Solvent	Selectivity (*R*,*S*):(*S*,*S*)	Reaction conditions
1	$NaBH_4$	*iso*-PrOH/MeOH	4:1	5–15°C
2	NaBH3CN	MeOH	4:3	pH = 5, AcOH, 23°C
3	Zn/AcOH	Toluene	No reaction	65°C
4	BH3pyridine	MeOH	3:2	23°C
5	Pt/C-H_2	MeOH	1:1	30°C
6	$NiCl_2$-$NaBH_4$	*iso*-PrOH/MeOH	2:1	23–65°C

of *iso*-propanol, the reduction of imine to amine was slow and accompanied by a concomitant loss in yield, though the selectivity was not impacted. Alternative reducing agents/conditions tested (entries 2–6, Table 5.1) showed stereoselectivity to be dependent on the reducing agent chosen, for example, with $NaBH_3CN$ the (*R,S*)/(*S,S*)-ratio dropped to 4:3 (entry 2). Moreover, catalytic hydrogenation (H_2) with Pd in MeOH gave no selectivity whatsoever (entry 5).

(R, S)-diastereomer **10** was conveniently isolated directly from the reaction mixture by salt formation with HCl, resulting in a product with an optical purity of 94% d.e. One reslurrying of the crude salt in a mixture of EtOH and EtOAc (2.3:1 v/v) brought the chiral purity of the final product (**12**) to 97% d.e., in line with the requirements defined in the specification. This third-generation process (Scheme 5.5) was scaled up in the pilot plant to manufacture

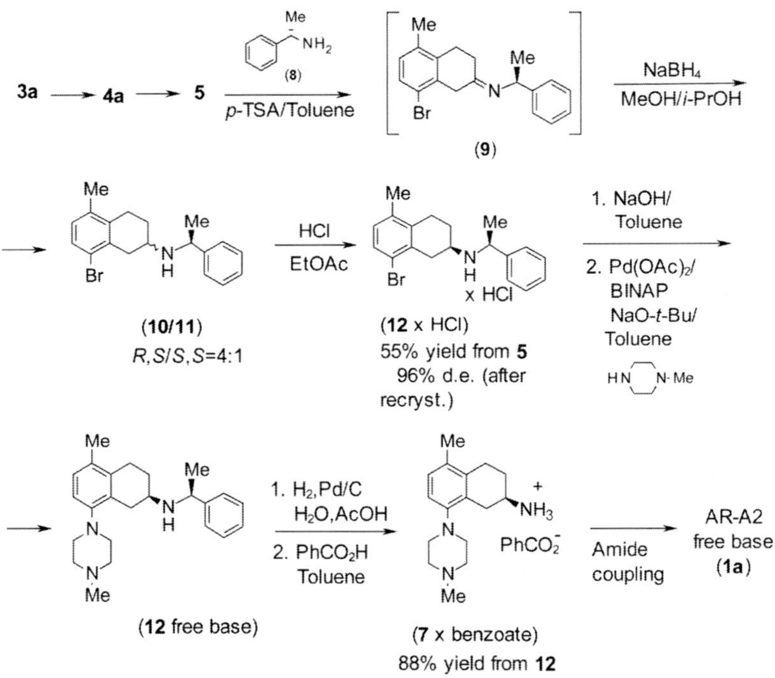

Scheme 5.5 The final route to AR-A2 × HBr (**1b**): a straightforward way of synthesizing a fairly complicated molecule in 10 steps from a simple, commercially available starting material.

216 kg of compound **12** in two successful batches with a total yield of 55% over three stages calculated from tetralone **5**. In conclusion, the third-generation process doubled the yield compared to the previous second-generation yield of just 25%.

5.4 Homing in on Safety, Health, and Environment (SHE) Improvements

5.4.1 *The Steps Leading up to the Resolved Aminotetralin*

Since none of the brominated species **4a–c** was commercially available at the time, the new synthetic strategy required back integration to 3-methylphenylacetic acid (**3a**) or to the corresponding amide **3b** or nitrile **3c** moieties (Scheme 5.6). Subjecting these

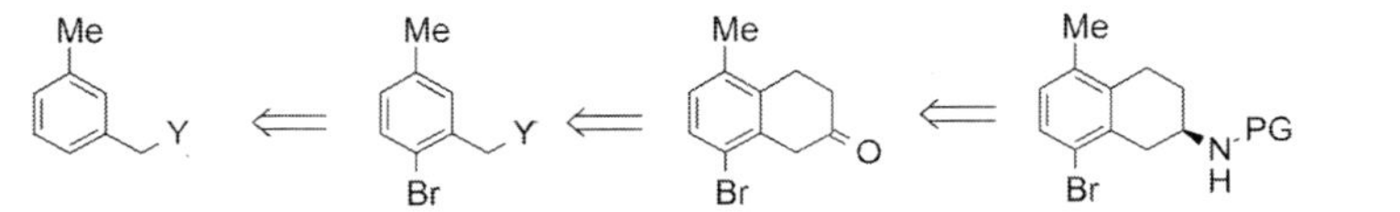

Scheme 5.6 Retrosynthetic analysis leading from the aminotetralin motif via the appropriate tetralone to substituted phenylacetic acid and its derivatives as ultimate starting materials.

starting materials to a bromination screening, using Br_2 under either acidic or mildly alkaline conditions, led to the selection of **3a** in the presence of a base as the ideal reaction system. Though an improvement, these conditions still only provided around 50% of the desired product and included the aforementioned regioisomeric mono-bromo analogs as main by-products alongside different dibrominated species.

Despite large efforts to optimize the reaction regarding formation of by-products, the best conditions still only gave 60 Area-% of 2-bromo-5-methylphenylacetic acid (**4a**). The optimized procedure (tested on a 5 L scale) was to dissolve the starting material in 20% K_2CO_3 solution (3.2 L/kg starting material) and add 1.01 equiv. of Br_2 slowly, keeping the temperature <20°C. Quenching of the reaction mixture was achieved by addition of sodium sulfite. Treatment with HCl precipitated the product and enabled subsequent isolation by filtration. The crude product thus obtained contained a large amount of the unwanted Br isomers. Fortunately, these by-products could be removed rather easily using two consecutive recrystallizations—first in an *iso*-propanol/H_2O/AcOH mixture and then in EtOH—to give a pure product (>98.6% purity) in 34.5% overall yield.

At the time of this study, 3-methylphenylacetic acid (**3a**) was commercially available from two different bulk producers, yet the price was high enough to motivate further work to increase the efficiency of the transformation. Since our efforts to reduce the formation of regioisomers and di-brominated by-products and to improve the total yield did not give the desired result, we decided to develop a method of recirculation of the undesired isomers. De-halogenation with H_2 gas and heterogeneous catalysis is well known for large-scale applications. And though consumption of the brominated components was quick when the hydrogenation was applied directly to the recrystallization mother liquor, that is, *iso*-propanol/H_2O/AcOH under acidic conditions, all experiments led to the formation of significant amounts of the isopropyl ester of **3a**. When alkaline reaction conditions were used instead, by adding NaOH or KOH to reach a pH >12, the reaction was equally rapid and no isopropyl ester was formed. Addition of HCl allowed **3a** precipitation and isolation in an impressive 90% yield. Sadly, while these results looked promising, the drug project was discontinued before any further optimization or scaling-up of the process could be conducted.

For the second-generation process, the stereochemistry was introduced in a telescoped, two-staged procedure comprised of a reductive amination of tetralone **5** followed by a resolution. The first part was nonselective and provided a racemic mixture of 2-benzylaminotetralin, which was subsequently resolved with *D*-tartaric acid to afford the desired diastereomeric salt **6** × tartrate. This process was applied to the manufacture of 108 kg of the tartrate in two batches with a best chemical yield of 25%. The optical purity was increased to 95% e.e. by applying a threefold reslurry of the precipitated salt in EtOH, though isolation using filtration took days or even weeks to complete due to the small particle size of the crystals. This process (reductive amination and resolution) generated large quantities of chemical waste and a rather unfavorable E-factor of 267.[37,38] Switching to the third-generation method offered a clear benefit in terms of environmental characteristics with a significant reduction in the amount of chemical waste (E-factor reduced to 116) for the corresponding transformation, now only involving reductive amination.

5.4.2 *The Pd-Catalyzed Coupling*

Much effort went into investigating the introduction of the *N*-methyl-piperazine moiety, an activity initiated in the second-generation synthesis but that received even more attention in the development of the third-generation route. The overall changes, however, mainly concerned the robustness and operability of the reaction rather than the environmental impact of the manufacturing process.

The first route to AR-A2 already profited from the then recently developed Buchwald–Hartwig reaction[23,24] to couple nitrogen-containing substituents onto aryl bromides to create C–N bonds. Our reaction used $Pd(dba)_2$ and (*R*)-BINAP as the catalyst and the ligand, respectively; NaO^tBu (dissolved in *tert*-butyl methyl ether) as the base; and toluene as the solvent, giving a conversion close to quantitative. To prevent the primary NH_2 group in the C-2 position from reacting in this amination, it was protected with two benzyl substituents. By the time the second route was developed, we discovered that this double protection was unnecessary. Even with the amine protected as the mono-benzyl derivative, the resulting secondary amine, with potential to react with another equivalent of aryl bromide during the Buchwald–Hartwig step, fortuitously did not show any such behavior. Only the desired coupling product, where the methyl-piperazine moiety had been attached to the aromatic ring, was found in the reaction mixture. This change in the protective group strategy led to increased atom efficiency (reduction in atomic "ballast" by about 21%) and also opened the door for the new approach with a stereo-preferred formation of the tetralin via chiral induction. Apart from these modifications, the reaction conditions were maintained largely unchanged in the second-generation process.

For the third-generation route, we exchanged the $Pd(dba)_2$ for $Pd(OAc)_2$, leading to both a faster reaction as well as lower cost, and avoided the repeated charging of catalyst. Since the amination reaction in route 2 was nonstereoselective, one could expect that enantiomerically pure (*R*)-BINAP would not be necessary. Indeed, we showed that the reaction worked with racemic BINAP but were surprised to find that in bulk quantities, the (*R*)-enantiomer was cheaper and more readily available than the racemic counterpart. In the third-generation route NaO^tBu was added in slurry form in toluene instead of in solution (*tert*-butyl methyl ether), thus removing one solvent and leading to easier handling.

5.4.3 *The End Game*

De-protection of the amino functionality was done through "standard" hydrogenolysis using Pd/C, a method frequently used on a large scale.[39] Thus, in the first-generation synthesis the hydrogenation was performed in glacial AcOH, while in the second generation the solvent was swapped for a toluene/AcOH mixture. The main drawback of either method was that the product, constituting a primary amine, formed a salt in the presence of AcOH. This salt formed an oily layer at the bottom of the vessel, making sampling and workup very difficult and lowering the efficiency of the heterogeneous catalyst. Despite using 30% w/w of 10% Pd/C and charging the catalyst in two portions, the reaction still took 36 hours to complete. For the third-generation route, the group to be removed was no longer benzyl but instead 1-methylbenzyl, and it was suspected that this would further decrease the reaction rate. By instead running the reaction in H_2O and AcOH, the problem of forming a separate phase was eliminated. Coupled with some optimization, using only 7.7% w/w of 5% Pd/C (0.59 mol %) allowed the reaction to reach full conversion in four hours and a one-time catalyst charge instead of the previous two additions. As an extra bonus, the in-process control samples were more homogeneous and therefore more reliable.

This "H_2O-based" method also had benefits regarding the use of solvents and, therefore, the amount of waste produced. Earlier, three extractions were necessary after the Buchwald–Hartwig reaction to remove the catalyst, reagents, and by-products that might interfere with the subsequent hydrogenation and transfer of the substrate into the toluene solution. One of these extractions could now be omitted and the hydrogenation run in a more concentrated fashion. In the earlier routes, NaOH and H_2O were added to the mixture after conversion to the primary amine and before filtering off the Pd/C. In the third-generation synthesis, Pd was removed first, followed by the addition of NaOH and toluene. In the end, all routes led to the free base of (*R*)-8-(*N*-piperazinyl)-2-aminotetralin **7** as a solution in toluene. Since **7** represented the penultimate intermediate in all synthetic pathways investigated and, hence, constituted an important control point to guarantee product quality in a GMP regime, isolation of this product was seen as a necessity. Though this was initially achieved by chromatography, the isolation and purification of free base **7** on silica gel has limitations when applied to large-scale manufacturing.

An isolation method based upon precipitating **7** as a salt was highly desirable. The di-HCl salt was not entirely suitable, so a method was developed where the mono-acetate salt was isolated instead for use in the second-generation process. However, as mentioned earlier, one major problem with the acetate salt of **7** was the oiling out in toluene. To produce a solid crystalline material, a solvent swap to EtOAc had to be performed to precipitate the salt in a controlled manner. Although the yield and purification effect in this isolation were good, the solid-phase properties of the salt were not satisfactory. In an early campaign where 26 kg of the final product AR-A2 (**1a**) were manufactured, the filtration of **7** × acetate took three weeks!

Development of new crystallization conditions was highly prioritized, and after some screening work it was decided to change the counterion to benzoate. It turned out that **7** in the form of the mono-benzoate formed very nice crystals in toluene, so the solvent change (to EtOAc) became superfluous. The only problem experienced was the poor solubility of solid benzoic acid in toluene, but this was tackled by conducting the dissolution in *iso*-propanol before mixing with **7** dissolved in toluene. The benzoate salt of **7** really stood up to the expectations regarding filterability, rendering filtration times less than one hour even on a 50 kg scale.

5.5 Focusing the Morpholinobenzoic Acid Side Chain: Design of the Ultimate Process

The first synthesis of 4-(*N*-morpholino)benzoic acid (**13**) to be reported (1970s) started from the highly toxic methyl bis-(2-chloroethyl)aminobenzoate (**14**).[40,41] Later, a more production-friendly and less hazardous, hydrolysis-based method from morpholinobenzonitrile (**15b**) was devised by using HCl.[42] We considered nitrile **15b** as well as the related analogs **15c-e** as convenient precursors to product **13** and, therefore, decided to enter into the development of a robust process for preparing these compounds on a large scale (Scheme 5.7). Aromatic morpholinylation had been

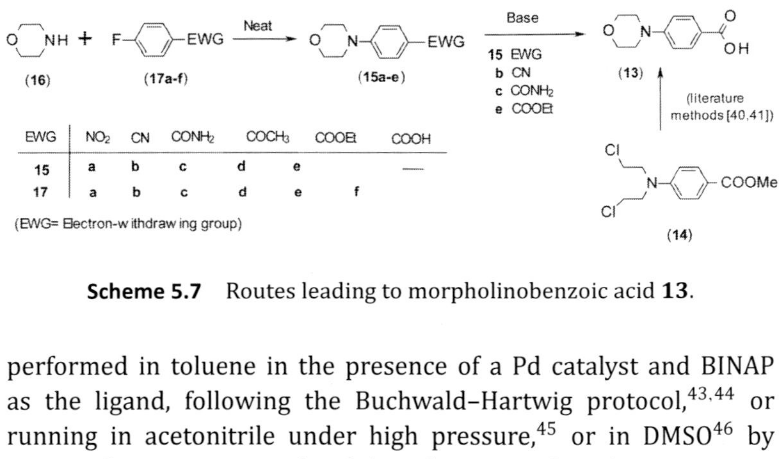

Scheme 5.7 Routes leading to morpholinobenzoic acid **13**.

performed in toluene in the presence of a Pd catalyst and BINAP as the ligand, following the Buchwald–Hartwig protocol,[43,44] or running in acetonitrile under high pressure,[45] or in DMSO[46] by means of an aromatic nucleophilic substitution (S_NAr). Our aim was to develop a process for morpholinylation without employing metal catalysts, avoiding demanding reaction conditions (high pressure), or utilizing any solvent in order to minimize chemical waste and reduce cost.

Compounds **15a–e** were obtained in high yield by simply heating a mixture of morpholine (**16**) and the respective substrate **17a–e** under neat conditions. Both reaction time and temperature had to be adjusted according to the nature of the electron-withdrawing group (EWG). As expected, the stronger the effect exerted by the EWG, the faster the reaction and, consequently, the lower the temperature required (compare entry 1 with entries 2–5 in Table 5.2). The

Table 5.2 Synthesis of morpholinobenzenes **15a–e** from a neat reaction of morpholine (**16**) with various fluorobenzene derivatives **17a–e**

Entry	Substrate	EWG	Product	Yield (%)	Reaction time (h)	Temperature (°C)	Melting point (°C)
1	**17a**	NO_2	**15a**	96	0.5	40	152–153
2	**17b**	CN	**15b**	95	5	120	82–83
3	**17c**	$CONH_2$	**15c**	94	10	120	220–221
4	**17d**	$COCH_3$	**15d**	93	10	120	95–96
5	**17e**	COOEt	**15e**	89	24	120	82–83

less expensive reactant **16** was employed in excess to maintain a homogeneous solution during the reaction. Products **15a–e** were conveniently isolated by precipitation (from H_2O), followed by filtration. Replacement of the fluorine atom in **17b** by Br, Cl, and F_3CSO_3 (triflate), as a means to avoid excessive corrosion due to F-induced pitting, unfortunately resulted in slow turnover and poor yields (52–65%) of **15b**. Attempts to prepare **13** directly following the same procedure by reacting **16** with **17f** failed due to salt formation. It was found, however, that the desired product (**13**) could be obtained in nearly quantitative yield from **17b**, **c**, or **e** on basic hydrolysis (1 N NaOH) followed by acidification (1 N HCl). As a matter of fact, the coupling reaction was so clean that the product could be directly hydrolyzed without first having to be isolated and purified. Thus, an efficient and scalable one-pot process for the synthesis of **13** starting from nitrile **17b**, the amide **17c**, and the ester **17e** was successfully developed.[25] It should be mentioned that methyl ketone **15d** is also a potential precursor to **13** by virtue of a NaOCl (hypochlorite)-mediated oxidation, although the yields obtained with this reaction were rather poor.

The substitution in neat morpholine followed by basic hydrolysis was used for the preparation of 80 kg of the side chain (**13**) in a single batch starting from **16** and **17b**. The E-factor for the synthesis of **15b** by this one-pot procedure is only 1.9, which compares extremely favorably with the 137 achieved when applying the Buchwald–Hartwig method described in the literature.[44] In other words, from a production point of view, the "solvent-less" method generated only 1.4% of the chemical waste generated by the Pd-catalyzed aromatic amination method.

5.6 Reaching the End Product: SHE Perspective on the Final Stages

Succeeding in developing a synthesis (Schemes 5.4 and 5.5) from tetralone **5** to the (*R*)-2-aminotetralin motif (**12**) that outperformed previous-generation methods (Schemes 5.1 and 5.3) by at least a factor of two, there still remained some steps to complete the sequence. This entailed the formation of an amide bond and subsequent precipitation of the target molecule in its final form, which after some development work was found to be a mono-HBr salt. As will be evident later, the efforts invested in these two last stages improved the process dramatically, both from an economy and a capacity point of view as well as in the SHE impact.

5.6.1 *Amide Coupling: An Example of Continuous Improvement of Raw Material and Solvent Utilization*

5.6.1.1 Initial approach

Initially, the method applied for the amide coupling originated from medicinal chemistry, where dimethylformamide (DMF) had been used as a solvent in combination with *N,N'*-carbonyl diimidazole (CDI) as a coupling agent. This method (Scheme 5.8; Flowchart 5.1) was scaled up largely unchanged to produce the first two larger batches of product **1a** in 3 kg and 26 kg amounts, respectively. Of course, while other coupling reagents were considered and tested in laboratory scale, CDI was selected for scale-up and further development because it provided very clean coupling conditions with few by-products in small amounts. Furthermore, CDI is an economically attractive reagent featuring relatively harmless by-products and SHE-friendly properties.[45] In contrast, making the acid chloride of the side chain moiety **13** by reaction with thionyl chloride ($SOCl_2$) and then subjecting the crude material thus obtained to the coupling with aminotetralin **7** afforded lower yields and a messy process stream. This was, at least in part, due to acid- and heat-induced morpholine ring scission during the chlorination step. Later, imidazole was found to play a crucial role in selective

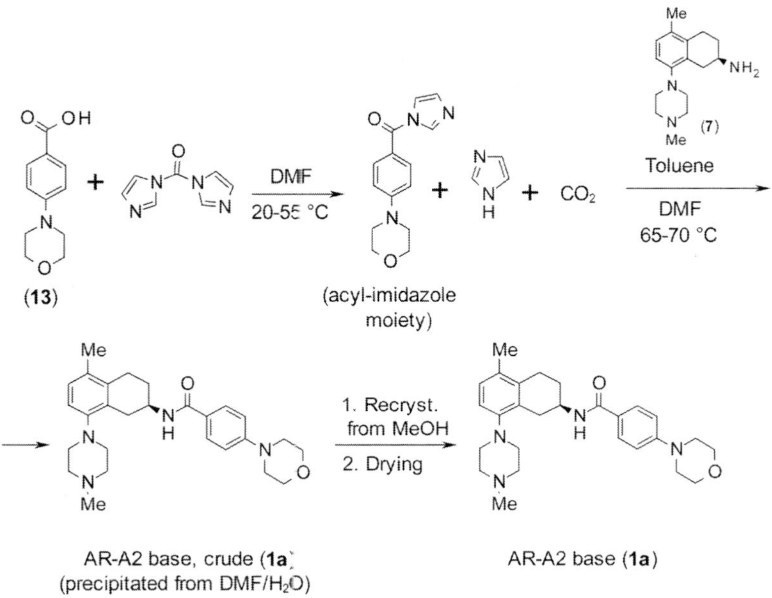

Scheme 5.8 Amide coupling: side chain **13** and intermediate **7** are connected using CDI as a condensing agent, leading to AR-A2 free base (**1a**), including isolation and drying of MeOH solvate (second-generation synthesis, Scheme 5.3).

formation of the desired mono-HBr salt (**1b**), so the choice to keep CDI as a coupling agent was rather straightforward.

As is evident from Flowchart 5.1, the amide coupling started with a release of the free aminotetralin base (**7**) by extraction between toluene and NaOH (aq). The volume of the resulting solution of **7** in toluene was reduced by vacuum distillation to give a completely dry solution by the efficient azeotrope between toluene and H_2O. This solution was then added to the acyl-imidazole-activated form of **13** with agitation at room temperature, followed by heating to 65–70°C. Under these conditions, this very mildly exothermic reaction[48] was fairly rapid and reached completion after four hours. Most of the DMF and all of the remaining toluene were removed by vacuum distillation, after which crude AR-A2 (**1a**) was precipitated from DMF/H_2O. Subsequently, the free base of **1** was isolated by recrystallization from MeOH. The crystal-bound MeOH present in

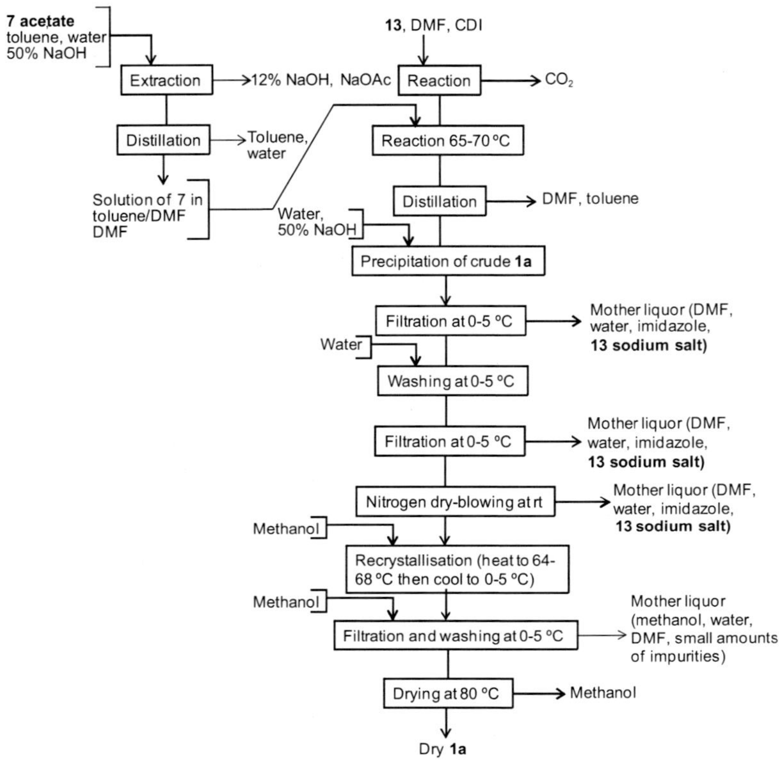

Flowchart 5.1 The first large-scale coupling process to AR-A2 free base (**1a**); see Scheme 5.8.

the first MeOH solvate of **1a** made it necessary to increase the drying temperature to 80°C to achieve solvent-free material. The **1a** manufactured with this method was used in toxicological studies prior to the start of clinical phase I trials.

5.6.1.2 Optimized approach

The first-generation amide coupling process (Flowchart 5.1) was rather complex and utilized three different solvents and two isolations of solid materials to generate **1a**, creating large amounts of different and rather complex mixtures of waste. The choice of solvents was clearly not optimal from a chemical point of view, as the sequence resulted in only moderate isolated yields (70%) based on the amount of aminotetralin (7 × acetate) charged. This was not primarily a result of side reactions but instead due to losses in the three mother liquors. Another significant weakness is that the salt formation was not conducted in an integrated fashion, without prior isolation of the pure AR-A2 free base. A key feature of the free base was its strong MeOH-binding properties, generating a solvate that required high-temperature vacuum drying to transition into a solvent-free form. Mindful that crystal-bound MeOH disturbs the crystallization of the final mono-HBr salt **1b** and, potentially more seriously, causes residual MeOH levels to exceed International Conference on Harmonisation (ICH) guideline recommendations[49] for the isolated crystalline free base, at the time of the first scale-up campaigns still considered as a possible final form of the API, some countermeasures were necessary.

Evaluation of different options showed that shifting to toluene as the lone solvent for the amide coupling was a significant improvement. The isolated yield in the amide-forming step was not only significantly better (from 70% to 85% yield), but this change also allowed for a much more facile crystallization of the mono-HBr form of AR-A2 (**1b**) as is evident from an illustration of the more mature second-generation procedure (Flowcharts 5.2 and 5.4).

CDI activation of **13** at 80°C (offering an acyl-imidazole moiety; see Scheme 5.8) was followed by addition of an adduct to a solution of **7** in toluene at 60°C. This was followed by heating to 90°C to give a reaction as clean and fast as when DMF was used as the solvent. As the reaction proceeded, a viscous off-white to yellow slurry formed and efficient agitation became important to minimize precipitation of material on the agitator and reactor walls. The reaction lent itself to interruption (by cooling) in the event of a technical failure and could easily be restarted by simple resumption of heating.

Reversing the order of addition compared to the regime operated in the previous method (Flowchart 5.1) offered no significant change in behavior of the reaction. Importantly, however, it provided a facile way of introducing necessary polishing filtrations of the two process streams **7** and activated **13**, respectively. In the latter case, this operation was required to prevent the nonactivated **13** from being transferred into the coupling vessel. Performing this unit operation before the amide coupling made it possible to avoid the

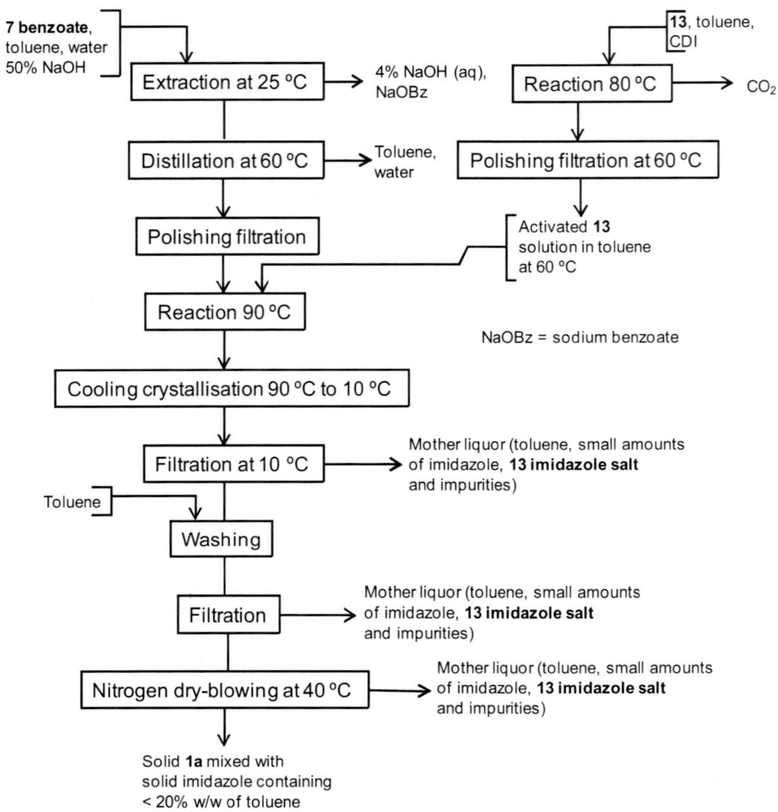

Flowchart 5.2 More process-friendly final coupling method leading to AR-A2 free base (**1a**).

operationally tricky polishing filtration during the subsequent salt formation stage and consequently provided a much simpler overall process.

After the reaction, the desired product was crystallized together with solid crystalline imidazole—a by-product from the CDI reagent—in the filter cake, in an approximate molar ratio AR-A2/imidazole of 1:2, by simple cooling of the reaction mixture from 90°C to 10°C. Aging at 10°C was followed by filtration and washing with fresh toluene. Filtration of the combined washing liquors and nitrogen-mediated dry-blowing drove the content of residual toluene to <20% w/w in the filter cake to give an AR-A2 free base/imidazole mixture with the right composition for direct use in the mono-HBr salt–forming step. As a spin-off, a relevant lesson from this work was that imidazole sublimes during vacuum drying. To keep most of the 2 mol equiv. of imidazole in the filter cake and prevent it from undergoing sublimation and crystallization elsewhere in the pressure filter, the centrifuge, and/or in the tubing, the method of drying was switched from vacuum drying to use of a warm stream of N_2.

Yields of AR-A2 free base (**1a**) produced as described (Flowchart 5.2) were consistently maintained at around 85% without sacrificing the quality. This constituted a dramatic improvement compared with the level of 70% reached for the first-generation process. The new procedure was successfully scaled up to 110 kg for delivery of material intended for clinical phase II and III trials as well as long-term formulation development. Importantly, the capacity and E-factor for the mature process improved dramatically by about four- and threefold, respectively, even without considering solvent recovery of toluene. Capacity and E-factor data for the last stages of the production process are summarized later (Table 5.5).

5.6.2 *Selective Precipitation of AR-A2 Mono-HBr Salt: A SHE-Friendly Process*

As mentioned before, the first form of AR-A2 to be considered for further development was the free base (**1a**). However, shortly after the first large-scale delivery (3 kg) for safety assessment studies, it became evident that a more soluble form of AR-A2 was needed as otherwise the development of the formulated product was seen as potentially problematic. Initial salt screening performed at the pharmacy department unfortunately did not result in many promising candidates. A rather hygroscopic lactate was considered together with a di-HBr salt, accidentally crystallized in medicinal chemistry. Further progression of the lactate salt was terminated at an early stage as the product sometimes dissociated to generate a precipitate of AR-A2 free base and lactic acid in solution. Consequently, there was some discussion as to whether the lactate complex was a true salt or a co-crystal in the solid state. In short, there was an urgent need for a robust procedure that would reproducibly generate a well-defined crystalline salt suitable for development to the final dosage form.

Initially, there was some hesitation around the choice of an HBr salt for further development, partly because of the initially uncontrolled nature of the process for making the di-HBr analog coupled with the high risk of corrosion in standard processing equipment on exposure to HBr. More importantly, though, there was some fear that known pharmacological effects[50,51] caused by the bromide ion, for example, the documented use in treating epilepsy and for the relief of restlessness and anxiety, could eventually prevent this option from being pursued. Lacking alternative salts, however, the project team agreed to try and develop a reliable process for preparing the di- or, preferably, the pharmacologically more attractive mono-HBr compounds. The PR&D team decided to investigate the mono-HBr track, since the differences in pK_a values between the three basic nitrogen atoms of AR-A2 are fairly large. Thus, with five to six units between the highest pK_a (8.3) and the next highest (2–3), as estimated from in silico data, and with the least basic position being about one further pK_a unit lower, this was seen as a worthwhile approach.[10] And indeed, by a careful design of the experimental procedure, it was entirely feasible to achieve a successful preparation of the desired mono-HBr moiety **1b** (Scheme 5.9).

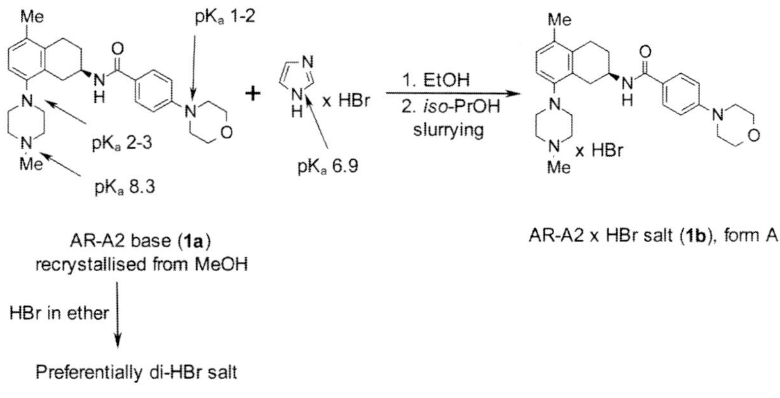

Scheme 5.9 The first HBr salt formation process where fresh imidazole × HBr was prepared prior to mixing with purified AR-A2 free base (**1a**); see Scheme 5.3.

Realizing that an un-buffered solution of HBr in solvent would kinetically protonate any nitrogen present in AR-A2 (**1a**), potentially resulting in statistical mixtures of solid mono-, di- and even tri-HBr salts as the product, it was clear that a "weaker/milder" form of HBr was needed to guarantee better control. With the knowledge that imidazole was a by-product formed in stoichiometric amounts in the amide coupling and had the intrinsic property of a pK_a value of 6.9,[45] imidazole × HBr was an obvious choice to investigate as a selective HBr salt–forming agent by virtue of the approximately 1.4 pK_a units difference (Scheme 5.9). Some other buffering candidates (*N,N*-dimethylaniline, *N*-methylmorpholine) were identified, but these were not pursued any further. Instead, focusing on imidazole and its unique fit for purpose, intensified investigational studies rapidly led to the design of the first-generation manufacturing process (Flowchart 5.3).

In the first-generation salt formation process, solid **1a** was mixed with EtOH and the slurry obtained heated to 65°C. A slight molar excess of imidazole was added followed by almost 1.5 mol equiv. of HBr (33% in AcOH). Upon continued agitation at 65°C, the slurry was gradually transformed into a solution, subjected to polishing filtration and subsequently heated to 80°C for two hours to secure full conversion to the salt. The mixture was cooled to 65°C to enable seeding by the addition of seed crystals, followed by a slow cooling ramp to −10°C. Filtration, washing and drying gave the pure crystalline product **1b**, occasionally consisting of the single-crystal form chosen for formulation development, "form A."

Solid-state analysis (powder X-ray diffraction [XRD]) of isolated material not only showed that there was a batch-to-batch variation but also revealed the presence of a second type of polymorph, consequently given the name "form B." One rationale for this observation was to assume a batch-dependent difference in the degree of the Fischer esterification of the solvent (EtOH + AcOH → EtOAc + H_2O). At any rate, this finding triggered an in-depth investigation focused on producing the desired pure form A alone. This work resulted in a reslurrying procedure (Flowchart 5.3), where *iso*-propanol became a necessary "additive" to guarantee a reliable and robust formation of crystal form A. In trial runs up to a 14 kg scale, the combined salt formation/reslurry process

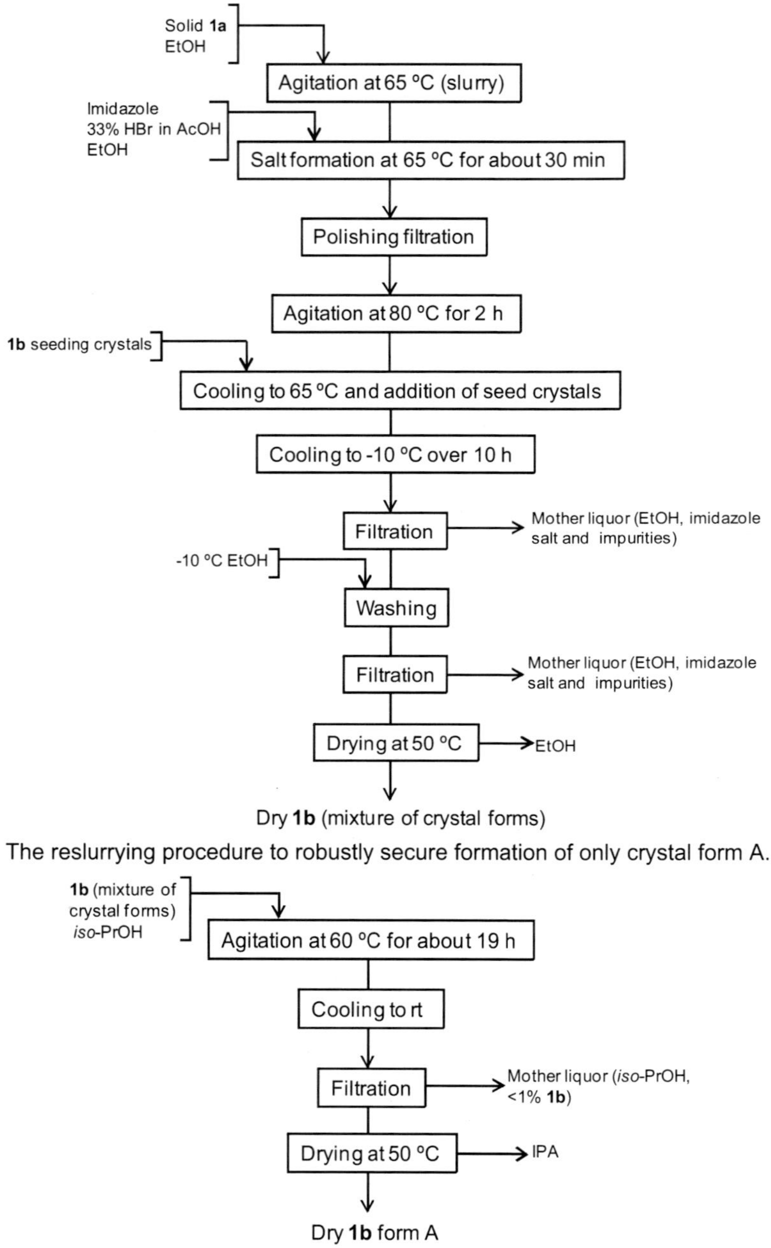

Flowchart 5.3 The first-generation salt formation scale-up: manufacture of 14 kg AR-A2 × HBr (**1b**) with associated slurry-based reprocessing to ensure exclusive formation of polymorph A.

Table 5.3 Selected data for an early 14 kg batch AR-A2 × HBr (**1b**) for use in early clinical trials

Analysis method	HPLC-purity (%)	Optical purity (%)	Total solvents (% w/w)	Assay by titration (% w/w)	Crystal form (XRD)	Pd (μg/g)
AR-A2 × HBr (batch 700/01)	99.7	99.6	0.7	97.5	Form A	<0.4

Abbreviation: HPLC, high-performance liquid chromatography.

consistently gave the desired product form in yields around 90%. This amounted to an overall yield of about 60%, calculated from the aminotetralin base (**7**). Clearly, this was not satisfactory for the final stages of a process aspiring to reach commercial standards but still considered to be good enough to deliver the requested amounts of clinical material matching the necessary high-quality requirements (Table 5.3).

Form A of product **1b** was found to be an anhydrate by thermal gravimetric analysis (TGA), and furthermore, it did not absorb water as measured by dynamic vapor sorption (DVS). On the other hand, form B was found to be a hydrate using a combination of TGA, DVS, and Karl Fischer titration. Although form B released water on drying, leading to an unstable crystal modification, water was readily reabsorbed to return to form B.

The conclusion from stability studies was that **1b** form A has greater or equal chemical stability to crystalline AR-A2 as the free base (**1a**) in all cases except for the solution in H_2O at 60°C, where slight by-product formation occurred due to morpholine ring scission with the acidic HBr salt. Various tests showed that the stability of bulk material toward temperature and light was very high and virtually identical for both **1a** and **1b** in crystal modification A.

Having identified **1b** as a stable, suitable form for formulation development, it was clear that the process for making this salt had to be further developed and integrated with the formation of the free base to improve capacity and reduce environmental impact. After further optimization, a final process was designed utilizing the AR-A2 free base (**1a**) work described before, which gave a mixture of

1a and imidazole (from the coupling reaction with CDI) as a solid containing small amounts of residual toluene (Flowchart 5.4).

The toluene-containing filter cake thus obtained, consisting of AR-A2/imidazole in a ratio of 1:2, was reslurried directly in the filter with pure *iso*-PrOH and transferred to another reactor. Stirring in *iso*-PrOH was continued for 30 minutes to avoid lumps that dissolve more slowly on addition of HBr/AcOH and cause a slurry-to-slurry transformation, an undesired process found to negatively impact purity and crystal form homogeneity. In another vessel, HBr (33%) in AcOH was mixed with *iso*-PrOH and the resulting solution added to the slurry of **1a** at 20°C, giving a clear solution due to toluene residue in the filter cake. On charging HBr/AcOH to *iso*-PrOH, some of the AcOH underwent Fischer esterification, forming isopropyl acetate and H_2O. These events did not pose any problem as the by-products were washed away with the mother liquors. Given the high corrosiveness of HBr, the mixing of its solution in AcOH with *iso*-PrOH was performed in a glass-lined reactor and thereafter transferred to the AR-A2 slurry as soon as possible. At this point, the pH reached 5.5–6.5, rendering the mother liquor essentially noncorrosive.

The temperature was increased to 60–65°C, resulting in crystallization of **1b** within roughly one hour. No seed crystals were needed to kick-start crystallization in this process, simplifying things operationally. Lowering the inner temperature to 10°C as fast as possible, limited largely by batch reactor cooling capacity, allowed the salt to be filtered off directly. In case the isolated crystals were not of 100% pure form A or of satisfactory purity, crude **1b** was reslurried in pure *iso*-PrOH at 70°C, followed by cooling to 10°C, filtration, and vacuum drying at 50°C, to give pure AR-A2 × HBr (**1b**) in an overall yield of 72–75% from **7**.

Unfortunately, the material has a strong affinity for *iso*-PrOH in the solid state and at the time of project termination (2004), the problem of residual levels above the recommended ICH guideline[49] of <0.5% had not been resolved. Since this was not a concern from a toxicological point of view, the product was approved for use in clinical trials without the need to further reduce the content of *iso*-PrOH. Analytical data for batches manufactured from 37 to 70 kg of **7** benzoate salt, giving 39 to 72 kg (74–76% isolated yield) of

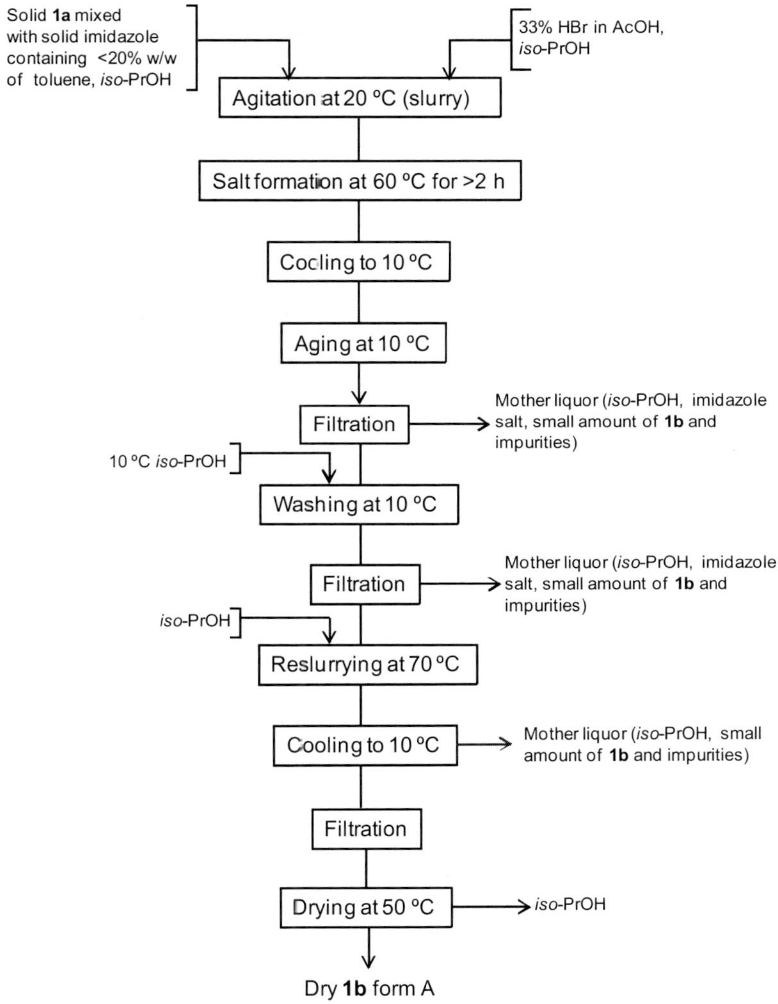

Flowchart 5.4 Optimized process for manufacture of **1b** using excess imidazole from amide coupling as a basic buffer. Having a toluene residue of <20% is important to make the AR-A2 × HBr salt solution clear and meta-stable at 20°C. If no toluene remains, the salt has been shown to precipitate within minutes of charging HBr, creating a less optimal particle size distribution.

Table 5.4 Selected analytical data for two batches of AR-A2 × HBr (**1b**) intended for clinical phase II trials

Batch	HPLC-	Imidazole	Optical	Total solvents	Assay	Crystal
Batch	purity (%)	(% w/w)	purity (% e.e.)	(% w/w)	(% w/w)	form (XRD)
103/03	99.5	n.d.	>99.8	0.8	98	A
SD/0054	99.8	<0.2	>99.8	1.0 (*iso*-PrOH)	98	A

Abbreviation: n.d. not detected.

1b (Table 5.4), show the largest organic impurity in both batches, apart from residual imidazole in batch SD/0054, to be unreacted side chain **13** at around 0.1% by HPLC.

Interestingly, it was found that if HBr (33% in AcOH) was added to **1a** (in the form of an *iso*-PrOH slurry) prior to dilution with *iso*-PrOH, an impurity was formed, most likely resulting from benzylic cation formation and quenched with residual H_2O and/or acetate nucleophiles (Scheme 5.10). Considering this behavior and the

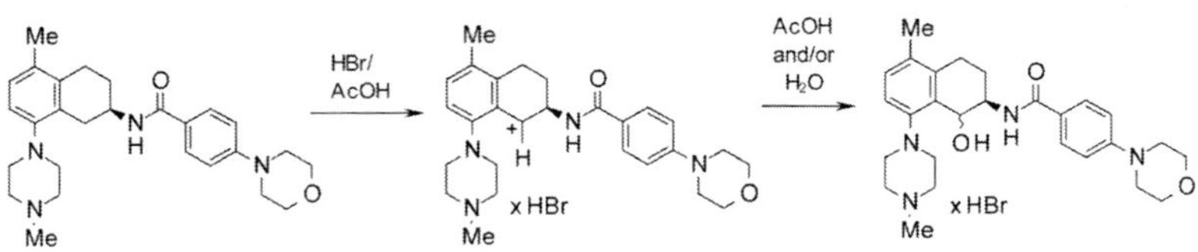

(1a) Benzylic carbocation Characterised impurity

Scheme 5.10 The proposed formation of a characterized alcohol impurity with HBr/AcOH applied without *iso*-PrOH predilution in the salt-forming step.

associated risk of product contamination, it was decided to keep the procedure (Flowchart 5.4) unchanged for manufacture and no further development was conducted.

5.7 Summary: Improvements in Environmental Impact of the AR-A2 x HBr Process

In the beginning, the environmental impact of the AR-A2 × HBr (**1b**) process was huge, with an overall yield of only about 3% for the second-generation route (Scheme 5.3). The initial stages up

Table 5.5 Key process parameters showing (1) increased capacity and (2) decreased environmental impact in the final stages of the process from **6** × tartrate and **12**, respectively, leading to the final product AR-A2 × HBr (**1b**)

Parameter	Second-generation process from 6 × tartrate	Third-generation process from 12	Optimized process (laboratory scale)
E-factor	148 kg/kg	103 kg/kg	62 kg/kg
API batch size	137 kg	168 kg	300 kg
Production capacity	0.34 kg/h	0.90 kg/h	1.36 kg/h
Process complexity	170 operations	150 operations	145 operations

to the point of the enantiopure key intermediates **6** × tartrate or **12** (Schemes 5.3–5.5, respectively) were never assessed in greater detail concerning their consequences on the environment (mainly because of resource constraints) before termination of the project. Lack of feasible alternatives for the synthesis of stereochemically defined and enriched aminotetralin derivatives, for example, **12** and other suitable analogs, was, however, a convincing factor that the drastic improvements in yield achieved by switching to the third-generation route provided at least an embryo of the ultimate process. Moreover, the incorporation of the recycling method of the wrong regioisomeric bromo-phenylacetic acids (analogs to **4a**) was, in fact, never introduced into any manufacturing campaign before the project ended, but would no doubt have resulted in at least a doubling of the overall yield up to intermediate **12**, dependent on the number of repetitive loops.

In contrast, the final stages—from the Buchwald–Hartwig amination onward (Scheme 5.5)—have been systematically evaluated and assessed concerning significant large-scale production aspects (Table 5.5). It is clear that the development work described herein for the final three stages, including the salt formation, has gradually taken us to a simpler process using fewer raw materials and solvents. This in turn has resulted in a dramatic improvement of both the capacity of the process (increase by a factor of 4 calculated as the amount of product output in kilograms manufactured in a given set of equipment per time unit) and the environmental impact (reduction in E-factor[37,38] by a factor of 2.4). It is important to mention that the E-factor, as per definition, has been calculated without consideration of aqueous waste, the reason being that this would be accommodated by an in-house wastewater treatment system. Furthermore, the figures presented (Table 5.5) do not include the potential reuse of any solvent stream to be recovered within the respective process and, consequently, are far higher than one would expect to achieve under a routine commercial production scenario.

References

1. Mours, M. (2008). *ChemSusChem*, **1**, 59–62.
2. Anastas, P. T.; Warner, J. C. (1998). *Green Chemistry: Theory and Practice* Oxford University Press.
3. Anastas, P. T.; Kirchhoff, M. M (2002). *Acc. Chem. Res.*, **35**, 686–694.
4. Tucker, J. L. (2006). *Org. Process Res. Dev.*, **10**, 315–319.
5. Federsel, H.-J. (2003). *Nat. Rev. Drug Discov.*, **2**, 654–664.
6. Federsel, H.-J. (2003). *Curr. Opin. Drug Discov. Dev.*, **6**, 838–847.
7. Federsel, H.-J. (2006). *Drug Discov. Today*, **11**, 966–974.
8. Federsel, H.-J. (2009). *Acc. Chem. Res.*, **42**, 671–680.
9. Clark, J. H.; Tavener, S. J. (2007). *Org. Process Res. Dev.*, **11**, 149–155.
10. Federsel, H.-J.; Hedberg, M.; Qvarnström, F. R.; Sjögren, M. P. T.; Tian, W. (2007). *Acc. Chem. Res.*, **40**, 1377–1384.
11. Federsel, H.-J.; Hedberg, M.; Qvarnström, F. R.; Tian W. (2008). *Org. Process Res. Dev.*, **12**, 512–521.
12. Federsel, H.-J. (2010). *Bioorg. Med. Chem.*, **18**, 5775–5794.
13. Stenfors, C.; Ahlgren, C.; Yu, H.; Larsson, L.-G.; Ross, S. B. (2004). *Psychopharmacology*, **172**, 333–340.
14. Stenfors, C.; Hallerbäck, T.; Larsson, L.-G.; Wallsten, C.; Ross, S. B. (2004). *Naunyn-Schmiedeberg's Arch. Pharmacol.*, **369**, 330–337.
15. Arvidsson, L.-E.; Hacksell, U.; Nilsson, J. L. G.; Hjorth, S.; Carlsson, A.; Lindberg, P.; Sanchez, D.; Wikström, H. (1981). *J. Med. Chem.*, **24**, 921–923.
16. Mason, J. P.; Dring, L. G.; Caldwell, J. (1995). *Xenobiotica*, **25**, 1371–1380.
17. Sari, Y. (2004). *Neurosci. Biobehav. Rev.*, **28**, 565–582.
18. Berg, S.; Linderberg, M.; Ross, S. B.; Thorberg, S.-O; Ulff, B. (1999). *PCT Int. Appl.* WO 9905,134.

19. Levy, A. A.; Rains H. C.; Smiles, S. (1931). *J. Chem. Soc.*, 3264–3269.
20. Evans, W. J.; Smiles, S. (1935). *J. Chem. Soc.*, 181–188.
21. Sommers, A. H.; Barnes, J. D. (1955). *J. Am. Chem. Soc.*, **77**, 4171–4172.
22. Süsse, M.; Skubatz, R.; Demus, D.; Zaschke, H. (1986). *J. Prakt. Chem.*, **328**, 349–358.
23. Guram, A. S.; Rennels, R. A.; Buchwald, S. L. (1995). *Angew. Chem. Int. Ed. Engl.*, **34**, 1348–1350.
24. Louie, J.; Hartwig, J. F. (1995). *Tetrahedron Lett.*, **36**, 3609–3612.
25. Tian, W. (2001). *PCT Int. Appl.* WO 01/87865.
26. Celgene Corp., P. O. Box 4914, 7 Powder Horn Drive, Warren, NJ 07059, USA.
27. Stirling, D. I.; Zeitlin, A. L.; Matcham, G. W. (1990). *Enantiomeric Enrichment and Stereoselective synthesis of Chiral Amines*, US Patent 4,950,606.
28. Rajšner, M.; Svátek, E.; Metyšová, J.; Bartošová, M.; Mikšik, F.; Protiva, M. (1977). *Collect. Czech. Chem. Commun.*, **42**, 3079–3093.
29. Sims, J. J.; Selman, L. H.; Cadogan, M. (1971). *Org. Synth.*, **51**, 109–112.
30. Hunden, D. C. (1984). *Org. Prep. Proc. Int.*, **16**, 294–297.
31. Abdel-Magid, A. F.; Carson, K. G.; Harris, B. D.; Maryanoff, C. A.; Shah, R. D. (1996). *J. Org. Chem.*, **61**, 3849–3862.
32. Soloshonok, V. A.; Ono, T.; Soloshonok, I. V. (1997). *J. Org. Chem.*, **62**, 7538–7539.
33. Zhong, H. M.; Cohen, J. H.; Abdel-Magid, A. F.; Kenney, B. D.; Maryanoff, C. A.; Shah, R. D.; Villani, F. J., Jr.; Zhang, F.; Zhang, X. (1999). *Tetrahedron Lett.*, **40**, 7721–7725.
34. Knupp, G.; Frahm, A. W. (1984). *Chem. Ber.*, **117**, 2076–2098.
35. Eleveld, M. B.; Hogeveen, H. (1984). *Tetrahedron Lett.*, **25**, 5187–5190.
36. Kanai, Y.; Saitou, K.; Koki, U. (1998). *Jpn. Kokai Koho* 10 072 411.
37. Constable, D. J. C.; Curzons, A. D.; Cunningham, V. L. (2002). *Green Chem.*, **4**, 521–527.
38. Sheldon, R. A. (2007). *Green Chem.*, **9**, 1273–1283.
39. Greene, T. W.; Wuts, P. G. M. (1999) *Protective Groups in Organic Synthesis* (3rd ed.). New York, NY: Wiley, 579–580.
40. Degutiene, A.; Rasteikiene, L.; Degutis, J. (1976). *4-Morpholinophenylalkanoic Acids*, Soviet Union Patent 510475 (in Russian).
41. Degutis, J.; Rasteikiene, L.; Degutiene, A. (1978). *Z. Org. Khim.* (*J. Org. Chem. USSR*), **14**, 2060–2064.

42. Setti, E. L.; Davis, D.; Janc, J. W.; Jeffery, D. A.; Cheung, H.; Yu, W. (2005). *Bioorg. Med. Chem. Lett.*, **15**, 1529–1534.
43. Wolfe, J. P.; Buchwald, S. L. (1997). *J. Org. Chem.*, **62**, 1264–1267.
44. Louie, J.; Driver, M. S.; Hamann, B. C.; Hartwig, J. F. (1997). *J. Org. Chem.*, **62**, 1268–1273.
45. Kotsuki, H.; Kobayashi, S.; Suenaga, H.; Nishizawa, H. (1990). *Synthesis*, 1145–1148.
46. Brown, G. R.; Foubister, A. J.; Ratcliffe, P. D. (1999). *Tetrahedron Lett.*, **40**, 1219–1222.
47. Merck & Co. (1996). Entry 4948 "Imidazole," in *The Merck Index* (12th ed.). Merck & Co., 844.
48. Allain Lorié, V. (2002). Unpublished results (AstraZeneca).
49. International Conference on Harmonisation of Technical Requirements for Registration of Pharmaceuticals for Human Use. (2009) *Impurities: Guideline for Residual Solvents Q3C(R4)*. http://www.ich.org/cache/compo/276-254-1.html (accessed August 16, 2010).
50. Merck & Co. (1996). Entry 7780 "Potassium Bromide," in *The Merck Index* (12th ed.). Merck & Co., 1313.
51. Ban, T. A. (2001). *Prog. Neuro-Psychopharmacol. Biol. Psychiat.*, **25**, 709–727.

Chapter 6

Improved and Greener Process for Pioglitazone and Its Pharmaceutically Acceptable Salts

Apurba Bhattacharya,[a] Rakeshwar Bandichhor,[b] and Lokeswara Rao Madivada[b]

[a]*Department of Chemistry, Texas A&M Kingsville, Kingsville, Texas 78363, USA*
[b]*Center of Excellence, Research & Development, Integrated Product Development, Dr. Reddy's Laboratories Ltd., Survey Nos. 42, 45, 46, & 54 Bachupally, Qutubullapur, Ranga Reddy Dist 500072, Andhra Pradesh, India*
kfab002@tamu.edu and rakeshwarb@drreddys.com

6.1 Introduction

6.1.1 *Green Chemistry in the Generic Pharmaceutical Industry*

As part of the desire to increase access to affordable medicines, the generic pharmaceutical industry has experienced explosive growth. Predictably, the emergence of such large-scale production has come with a significant environmental cost, a result of the industry's wasteful practices and highly unfavorable E-factors.[1–5]

Scalable Green Chemistry: Case Studies from the Pharmaceutical Industry
Edited by Stefan G. Koenig

ISBN 978-981-4316-49-1 (Hardcover), 978-981-4364-81-2 (eBook)
www.panstanford.com

Due to a lack of innovation in the form of green chemistry, the multifold production of generic versions of active pharmaceutical ingredients (APIs) typically yields exorbitant amounts of waste and contributes to an unbearable environmental burden. Considering the existing environmental scenario, it is clear that most of the generic manufacturers may not have taken green chemistry metrics into consideration when designing their chemistry or processes. This unfortunate situation, largely grounded in ignorance, deprived an exercise in the best possible chemistry and led to the creation of untold amounts of waste. Developing green chemistry–driven syntheses and processes is extremely challenging and time consuming. Frequently, scientists working in the generic industry generally adopt the easier, but more wasteful and less time-consuming, ways to deliver the product to be first in the market or enjoy the 180 days exclusivity afforded to the first-filing generic company. In the end, sustainability and the green chemistry principles are often neglected, creating an imbalance between the environment and the world economy.

Simultaneously, new chemical entity (NCE) innovators have been the key drivers for developing practicable, first-run chemistry. However, to get a new drug approved and patented in a competitive space, quite often, the synthetic route developed turned out to be suboptimal and nongreen. Since there was no competition to improve the patented synthesis before genericization, the nongreen legacy was carried through till the end of a pharmaceutical's lifecycle whereby a myriad of opportunities to reduce E-factors and develop cost-effective, green synthetic routes to APIs were largely ignored. In such a demanding situation, the philosophy of green chemistry, which is best defined as *the utilization of a set of principles that reduces or eliminates the use or generation of hazardous substances in the design, manufacture, and applications of chemical products,*[4] can still readily be adopted.

Over the past few years, the importance of green chemistry has gained momentum and significant amounts of research have been directed toward the development of new technologies and methodologies for environmentally benign processes at the innovator companies. Apart from a technological green paradigm shift, improvements in process conditions, economics, ever-increasing environmental controls and social pressure helped incorporate green chemistry into the synthesis of APIs where they have steadily been gaining priority in the pharmaceutical industry.

6.2 A Green Attitude

Thinking green in every aspect of our lives can do wonders for society. The generic pharmaceutical industry should adopt its own unique green strategy as a potential business driver. There are two aspects to a green attitude, and each can help make a business model more effective: 1) when green chemistry principles are implemented in the design phase of robust and sustainable processes, the positive results carry forward, and 2) products developed using green technologies can now often be sold at higher prices since this development requires simplification or finding alternatives for complicated chemistries and a high level of synthetic organic chemistry and engineering expertise, both of which can act as differentiators in the marketplace.

6.3 Synthesis of Pioglitazone

Pioglitazone **1** (Fig. 6.1) is a derivative of benzylthiazolidinedione, approved for the management of type II diabetes. Pioglitazone **1** is found to stimulate the peroxisome proliferator-activated receptor gamma (PPARγ) in order to modulate the transcription of insulin-sensitive genes that are involved in glucose and lipid metabolism.[6–8]

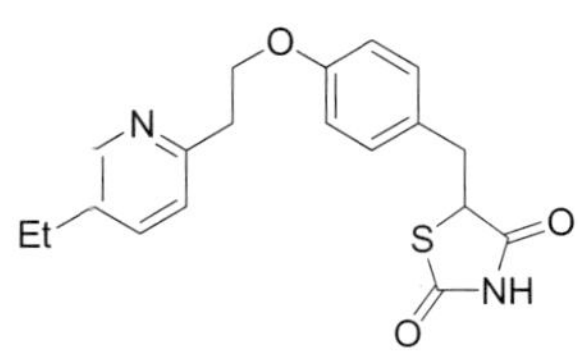

1

Figure 6.1 Structure of pioglitazone **1**.

6.3.1 Activation Group-Based Synthesis

The synthetic routes for pioglitazone **1**, as described in previous patents,[9,10] are shown in Scheme 6.1 and involve activation of 5-ethyl-2-pyridyl ethanol **2** with a *p*-toluene sulfonyl group to obtain intermediate **3**. This intermediate was subsequently subjected to nucleophilic substitution. In particular, the reaction between intermediate **3** and *p*-hydroxy benzaldehyde **4** in the presence of sodium hydroxide afforded intermediate **5**. The reaction between intermediate **5** and thiazolidinedione **6,** employing Knoevenagel conditions, afforded the penultimate intermediate **7**. The reduction of **7** in the presence of $Pd/C/H_2$ at 25°C yielded the desired compound **1**. In a separate disclosure,[11] the $CoCl_2{\cdot}6H_2O/NaBH_4$/dimethyl glyoxime system was also used for such a reduction to obtain **1**.

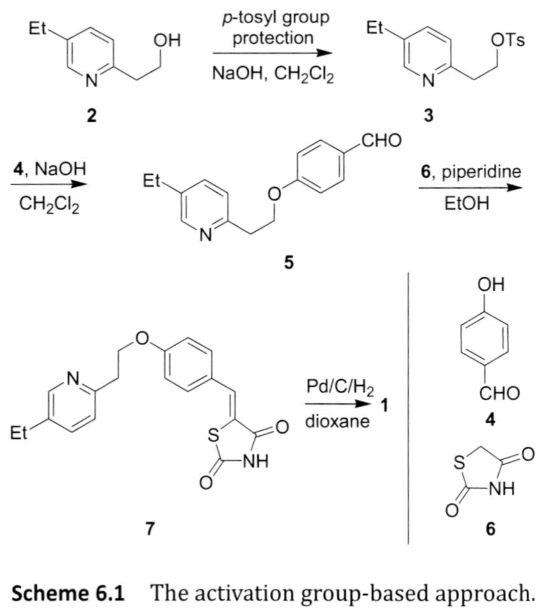

Scheme 6.1 The activation group-based approach.

Though activation of intermediate **2** with the tosyl group in **3** was high yielding, the production of an E2-elimination impurity during the reaction caused a cumbersome isolation and necessitated purification of intermediate **5**. Moreover, expensive Pd metal and partially recoverable solvents, for example, dioxane, were used in the reduction of **7** to **1.**

6.3.2 *The S_NAr-Based Approach*

Another approach (Scheme 6.2)[12] was reported, where starting material **2** was reacted with *p*-substituted fluorobenzene in an S_NAr approach to obtain intermediates **8** and **9**.

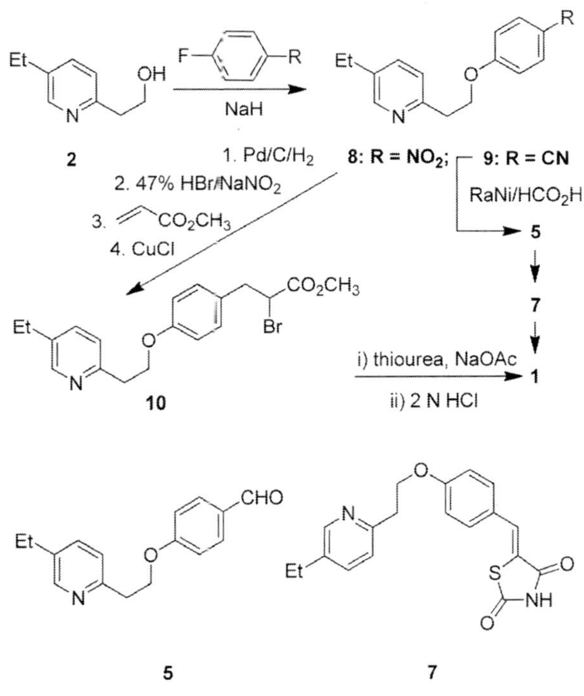

Scheme 6.2 The S_NAr-based approach.

Raney Ni/HCO_2H-mediated reductive hydrolysis of cyano derivative **9** afforded aldehyde intermediate **5**, which was converted to title compound **1** as per the procedure described in Scheme 6.1. Intermediate **8** was subjected to a cascade of reactions to obtain bromo derivative **10**. In this particular strategy, catalytic hydrogenation of the nitro group, diazotization of the aniline, and bromination of the diazonium salt were followed by Cu metal insertion across the aromatic C-Br bond and reaction with methyl acrylate to afford **10**. Subsequently, intermediate **10** was further utilized in the alkylation with thiourea and acid-catalyzed cyclization to render the desired species **1**. A convergent approach involving condensation of **4** and **6** to obtain **1** is also precedented.[13]

Apart from the cumbersome isolation and purification of intermediate **5**, there are some additional demerits associated with the strategy disclosed in Scheme 6.2, including the use of a plant-unfriendly, pyrophoric and non-nucleophilic base, NaH, in the transformation of **8** or **9** from **2**. Additionally, the cascade of reactions involving expensive metal (Pd), toxic acid (HBr), and the diazonium chloride intermediate, performed to yield intermediate **10**, were unfavorable, as were the use of expensive and partially recoverable solvents dioxane, tetrahydrofuran (THF) and CH_2Cl_2.

6.4 Development of a Greener Process: Improved Process for Pioglitazone

Considering the negatives associated with the processes described in Schemes 6.1 and 6.2, to achieve cost-effective and high-yielding transformations, an improved greener process for pioglitazone **1** and its pharmaceutically acceptable HCl salt was desired.

To improve the process, the synthesis of pioglitazone **1**, as shown in Scheme 6.3, started with the activation of 5-ethyl-2-pyridyl ethanol **2** with a methyl sulfonyl group to obtain intermediate **11**. Subsequently, this intermediate was subjected in situ to nucleophilic substitution. In particular, the reaction between intermediate **11** and *p*-hydroxy benzaldehyde **4** in the presence of potassium carbonate (K_2CO_3) afforded intermediate **5**. The reaction between intermediate **5** (not isolated) and thiazolidinedione **6**, employing Knoevenagel conditions, afforded penultimate intermediate **7** in 50% yield and 99% purity over the three steps. Moreover, the reduction of intermediate **7** in the presence of Co (II) salts as a catalyst at 30°C yielded the desired compound **1** in 90–98% yield and 98% purity.

In one of the reactions, following the literature procedure,[11] 15 mol % $CoCl_2 \cdot 6H_2O/NaBH_4$/dimethyl glyoxime system afforded **1** in

Route 3A

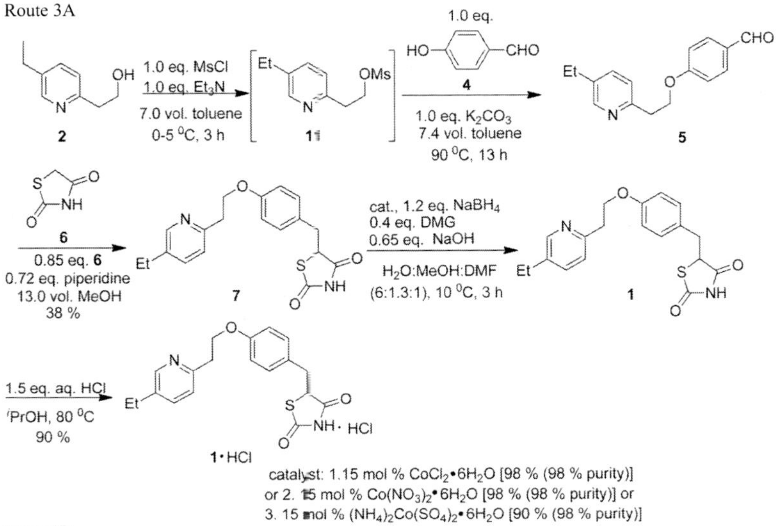

Route 3B

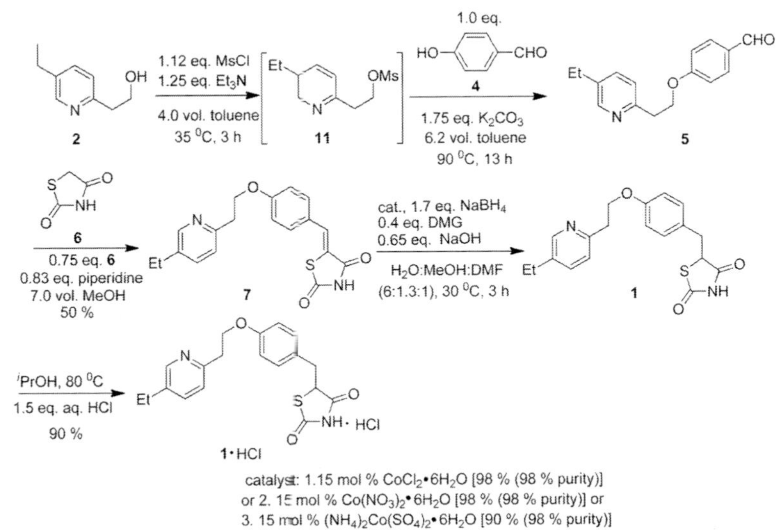

Scheme 6.3 Improved synthetic approach: Routes 3A (feasibility) and 3B (after optimization).

98% yield and 98% purity. This reduction was also performed in the presence of 15 mol % $Co(NO_3)_2{\cdot}6H_2O/NaBH_4$/dimethyl glyoxime as a catalytic system, giving rise to product **1** in the same yield and purity. Meanwhile, using cobalt ammonium sulfate as a catalyst, instead of cobalt chloride or nitrate, did not offer better yield, although the purity was found to be same. A pharmaceutically acceptable salt of **1** was prepared by employing 1.5 equiv. of aq. HCl and isopropanol in 90% yield and >99.7% purity (ICH grade).

6.5 Process Optimization

To minimize waste generation and develop an industrially viable and greener synthetic process, all the manipulations in the synthesis were optimized. Preparation of **11** involved mesylation of 2-(5-ethyl-2-pyridyl) ethanol **2** employing methanesulfonyl chloride (MsCl) and triethylamine (NEt_3) in toluene. During feasibility studies, 7.0 volumes of toluene were used, as shown in Scheme 6.3 (route 3A). We anticipated that the volume of the solvent could be further optimized to a lower quantity. With 3.0 volumes of solvent, a thick reaction mass was observed; however, with 4.0 or 5.0 volumes of solvent the stirring of the reaction mass was found to be efficient (Table 6.1, entry 2). In this study, all parameters except volume of solvent remained the same, as in Scheme 6.3 (route 3A). The process was run with 4.0 volumes of toluene henceforth.

As mentioned before, preparation of **11** involved use of NEt_3 in a toluene medium. During feasibility studies, 1.00 equiv. of NEt_3 was used, as shown in Scheme 6.3 (route 3A). To improve the yield, we reasoned that excess base could be detrimental and should be

Table 6.1 Optimization of solvent quantity used in the synthesis of intermediate **11**

		Toluene quantity		Yield of 11	
Entry	Quantity 2(g)	(mL)	vol.	(g)	(%)
1	25	75	3.0	33.7	88.8
2	**25**	**100**	**4.0**	**34.9**	**92.0**
3	25	125	5.0	33.8	89.1

Table 6.2 Optimization of NEt_3 equivalents in the synthesis of **11**

Entry	Quantity 2 (g)	NEt_3 quantity (mL)	equiv.	Yield of 11 (g)	(%)	% HPLC purity
1	25	23.0	1.00	32.9	87	61
2	25	25.8	1.12	33.1	87	70
3	**25**	**28.8**	**1.25**	**34.6**	**91**	**80**
4	25	31.2	1.35	34.9	92	81
5	25	34.6	1.50	34.4	91	82

optimized to the adequate level. With 1.12 equiv. of NEt_3, lesser purity of the product was observed. Additionally, different quantities of the base were investigated, but the best combination of purity and yield, considering the amount of base, was observed with 1.25 equiv. of NEt_3, as shown in Table 6.2 (entry 3). In this study, all parameters except NEt_3 were maintained the same, as in the Scheme 6.3 (route 3A).

As part of our strategy, the number of equivalents of MsCl in the preparation of **11** was also optimized. During feasibility studies 1.00 equiv. of MsCl was used, as shown in Scheme 6.3 (route 3A). In efforts to improve the yield, we anticipated that the quantity of mesylating reagent could also be detrimental. Employing 1.12 equiv. of MsCl, as shown in Table 6.3 (entry 2), we obtained the product with the best yield and purity. In the different reactions with the increased amount of MsCl, we isolated the product with lower yield and purity, as shown in Table 6.3. In this study, except MsCl, other reaction conditions remained the same, as in Scheme 6.3 (route 3A).

Table 6.3 Optimization of MsCl equivalents in the synthesis of **11**

Entry	Quantity 2(g)	MsCl quantity (mL)	equiv.	Yield of 11 (g)	(%)	% HPLC purity
1	25	19.0	1.00	30.2	80	78
2	**25**	**21.2**	**1.12**	**32.9**	**87**	**84**
3	25	23.7	1.25	33.1	87	55
4	25	25.6	1.35	32.4	85	75

Table 6.4 Optimization of temperature in the synthesis of **11**

Entry	Quantity 2 (g)	Temperature (°C)	Yield of 11 (g)	%	% HPLC purity
1	25	−5–0	33.0	87	85
2	25	0–5	32.9	87	85
3	**25**	**25–35**	**33.3**	**89**	**88**
4	25	45–50	24.2	64	83

Temperature always plays a pivotal role in chemical synthesis; therefore, as part of our strategy the preparation of **11** was studied at different temperatures. During feasibility studies, the reaction was conducted at 0 to 5°C, as shown in Scheme 6.3 (route 3A). To improve the yield, during optimization, we performed the reaction at 25–35°C and isolated the product with the best purity and yield, as shown in Table 6.4 (entry 3). In the different reactions at other temperatures, we were unable to isolate the product with better yield and purity. In this way we optimized the reaction conditions involved in the preparation of **11,** as shown in Scheme 6.3, route 3B.

Preparation of **5** involved alkylation of 4-hydroxy benzaldehyde **4** with 2-(5-ethyl-2-pyridyl) ethyl methanesulfonate **11** using K_2CO_3 in toluene. During feasibility studies 7.4 volumes of toluene were used, as shown in Scheme 6.3, route 3A. As per our optimization strategy, we realized that the volume of the solvent could be minimized. With 4.0 volumes of solvent, a thick reaction mass was observed; however, with 6.2 volumes of solvent, as shown in Table 6.5 (entry 2), the stirring of the reaction mass was found to be efficient. In this study, except the solvent, all other parameters were kept the same as in Scheme 6.3 (route 3A).

Table 6.5 Optimization of solvent quantity used in the synthesis of intermediate **5**

Entry	Quantity organic layer containing 11 (mL)	Toluene quantity (mL)	vol	Yield of 5 (g)	(%)	% HPLC purity
1	150	185	7.4	31.3	74.1	72.8
2	**150**	**155**	**6.2**	**32.4**	**76.7**	**80.8**
3	150	100	4.0	—	—	—

Table 6.6 Optimization of equivalents of 4-hydroxy benzaldehyde **4** in the preparation of **5**

Entry	Quantity organic layer containing 11 (mL)	Quantity 4 (g)	equiv.	Yield of 5 (g)	(%)	% HPLC purity
1	150	18.2	0.90	33.5	79.3	76.1
2	150	20.2	1.00	33.1	78.4	80.9
3	**150**	**21.6**	**1.07**	**33.8**	**80.0**	**85.4**
4	150	24.2	1.20	32.6	77.2	84.2

Alkylation of 4-hydroxy benzaldehyde **4** with 2-(5-ethyl-2-pyridyl) ethyl methanesulfonate **11** led to **5**. During feasibility studies, 1.00 equiv. of **4** was used; however, after optimization, the yield and purity in the preparation of **5** were found to be optimal with 1.07 equiv. of **4** (Table 6.6, entry 3). Deviating from 1.07 equiv. of **4** was found to be inefficient (entries 1, 2, and 4). In this study, except the quantity of **4**, all other parameters were remained the same as in Scheme 6.3 (route 3A).

Reaction of 4-hydroxy benzaldehyde **4** with 2-(5-ethyl-2-pyridyl) ethyl methane sulfonate **11** using K_2CO_3 in toluene led to the synthesis of **5**. During feasibility studies, 1.00 equiv. of K_2CO_3 was used, as shown in Scheme 6.3 (route 3A). After optimization, 1.75 equiv. of K_2CO_3 was found to be sufficient to obtain the best yield and purity of **5**, as shown in Table 6.7 (entry 2). Utilizing greater or lesser than 1.75 equiv. of the base did not offer better results.

The optimized reaction conditions involved in the preparation of **5** are shown in Scheme 6.3 (route 3B).

Table 6.7 Optimization of equivalents of K_2CO_3 in the preparation of **5**

Entry	Quantity organic layer containing 11 (mL)*	Quantity K_2CO_3 (g)	equiv.	Yield of 5 (g)	(%)	% HPLC purity
1	175	17.1	0.75	28.4	67.2	76.6
2	175	22.8	1.00	30.2	71.5	79.3
3	175	34.3	1.50	32.3	76.5	82.2
4	**175**	**40.0**	**1.75**	**33.8**	**80.0**	**85.4**
5	175	57.1	2.50	31.8	75.3	86.0

*175 ml of organic layer is equal to 33 g of **11**.

Table 6.8 Optimization of MeOH quantity in the preparation of **7**

			Yield		
Entry	Quantity 5 (g)	MeOH (vol)	(g)	(%)	% HPLC purity
1	**25**	**7.0**	**23.3**	**67.1**	**96.2**
2	25	10.0	23.5	67.7	95.9
3	25	12.0	23.3	67.1	98.2

Table 6.9 Optimization of equivalents of 2,4-thiozolidinedione in the preparation of **7**

			Yield		
Entry	Quantity 5 (g)	6 (equiv.)	(g)	(%)	% HPLC purity
1	25	0.68	22.6	65.1	95.7
2	25	0.85	23.2	66.8	96.1
3	**25**	**0.75**	**23.2**	**66.8**	**97.6**
4	25	0.94	22.9	65.9	96.8
5	25	0.61	21.7	62.5	95.2

Preparation of **7** involved the condensation of 2,4-thiozolidinedione **6** and **5** using a piperidine base in methanol (MeOH). As shown in Scheme 6.3 (route 3A), 13.0 volumes of MeOH were used. After optimization, 7.0 volumes were found to be sufficient for effective stirring and solubility of the substrates (Table 6.8, entry 1). In this study, except for volume of the solvent, all other parameters remained the same as in Scheme 6.3 (route 3A).

In feasibility studies, 0.85 equiv. of 2,4-thiozolidinedione **6** was used to prepare **7**, as shown in Table 6.9 (entry 2). After optimization, 0.75 equiv. of **6** was found to be sufficient to afford **7** in better yield and purity (Table 6.9, entry 3). In this study, except equivalents of **6**, all other parameters stayed the same as in Scheme 6.3 (route 3A).

As displayed in Scheme 6.2, preparation of **7** involved the condensation of **6** and **5** using a piperidine base in MeOH. In feasibility studies, 0.72 equiv. of piperidine was used for the preparation of **7** (Scheme 6.3, route 3A). After optimization, 0.83 equiv. of the piperidine base was found to be sufficient to effect the transformation (Table 6.10, entry 3). In this study, except

Table 6.10 Optimization of piperidine quantity in the preparation of **7**

Entry	Quantity 5 (g)	Piperidine (equiv.)	Yield (g)	Yield (%)	% HPLC purity
1	25	0.64	22.8	65.6	97.4
2	25	0.72	22.8	65.6	98.3
3	**25**	**0.83**	**25.5**	**73.4**	**94.5**
4	25	0.95	23.6	68.0	96.7

equivalents of the piperidine base, all other parameters were kept the same as in Scheme 6.3 (route 3A).

The optimized reaction conditions involved in the preparation of **7** are shown in Scheme 6.3 (route 3B).

Preparation of **1** involved reduction of **7** using sodium borohydride ($NaBH_4$) in aq. MeOH and DMF. A dimethyl glyoxime and cobalt chloride hexahydrate system was used as catalyst. In feasibility studies, 1.2 equiv. of $NaBH_4$ was used for the preparation of **1**, as shown in Scheme 6.3 (route 3A). After optimization, 1.7 equiv. of $NaBH_4$ was found to be sufficient to effect the transformation with high yield and purity (Table 6.11, entry 6).

The improved process presented in Scheme 6.3 (route 3B) and described in depth in this section has multiple advantages over the existing synthesis. These improvements include a) a high-yielding transformation from **2** to **11** with little E2-elimination impurity during the reaction (only 5–8% vs. the route described in the patent [Scheme 6.1], yielding 30–40% E2-elimination impurity); b) replacement of expensive Pd metal with Co (II) salt catalysts

Table 6.11 Optimization of quantity of $NaBH_4$ in the preparation of **1**

Entry	7 (g)	$NaBH_4$ (g)	$NaBH_4$ equiv.	Yield (g)	Yield (%)	% HPLC purity
1	20	2.1	1.0	18.2	90.4	88.8
2	20	2.6	1.2	18.5	91.9	90.7
3	20	3.0	1.4	18.5	91.9	94.2
4	20	3.2	1.5	19.0	94.4	97.2
5	**20**	**3.6**	**1.7**	**19.1**	**94.9**	**98.1**
6	20	4.0	1.9	18.2	90.4	98.7
7	20	4.9	2.3	18.0	89.4	98.1

in the reduction of **7** to **1**; c) replacement of a plant-unfriendly and pyrophoric base NaH for an alkylation analogous to that used in Scheme 6.1; d) a new cascade of reactions, avoiding expensive metal (Pd), toxic acid (HBr), and a potentially dangerous diazonium chloride intermediate, to yield intermediate **10** in analogy to Scheme 6.2; and e) replacement of expensive and partially recoverable solvents, for example, dioxane, THF, and CH_2Cl_2, with toluene, MeOH, and isopropanol.

6.6 Conclusions

There has been insufficient commitment thus far toward developing novel and greener synthetic chemistry. The generic industry has instead been driven by short-term objectives and immediate profit. At Dr. Reddy's we have realized, at an early stage, the importance of green principles in our business and, more generally, society and global sustainability. Arguably, the business environment for generic development is changing rapidly due to ever-increasing environmental constraints on multiple fronts. Relevant regulations on a global basis are leading to a more level playing field and are encouraging adoption of green principles in the generic business as well. A holistic consideration is a must in the selection of the greenest route—shifting the nongreen part to suppliers is not an option any more. It has become clear that the greenest process will always be the most cost-effective one in the long run.

Applying these ideas to our own projects has been very successful. We have learned that true implementation of green principles in the generic business is only realized by seamless amalgamation of business, science, and engineering, leading to a more sustainable world that our generation will leave behind.[14] Following these ideals has led us to an improved green process for pioglitazone **1**. The entire synthesis to the target was reevaluated to minimize waste and select greener reagents and solvents, which were screened and optimized for the route. As a result of this comprehensive reexamination, a new process to **1** was developed, which is more compatible with the industrial scale and has notable advantages over existing syntheses.[15] Incorporating green chemistry now requires a paradigm shift in thinking, but the results can be reaped for many years into the future.

Acknowledgments

We thank Dr. Reddy's Laboratories for supporting this work.

References

1. Sheldon, R. A. (1994). *CHEMTECH*, **24**, 38–47.
2. Sheldon, R. A. (2000). *C. R. Acad. Sci., Ser. IIc: Chim.*, **3**, 541–551.
3. Sheldon, R. A. (2005). *Green Chem.*, **7**, 267 (and references therein).
4. Anastas, P.; Warner, J. (1998). *Green Chemistry: Theory and Practice*. Oxford: Oxford University Press.
5. Trost, B. M. (1991). *Science*, **254**, 1471.
6. Le, A.; Pucko, W.; Szelejewski, W. (2004). *Org. Process Res. Dev.*, **8**, 157–162.
7. Colca, J. R.; McDonald, W. G.; Waldon, D. J.; Leone J. W.; Lull, J. M.; Bannow, C. A.; Lund, E. T.; Mathews, W. R. (2004). *Am. J. Physiol. Endocrinol. Metab.*, **286**, E252–E260.
8. Paddock, M. L.; Wiley, S. E.; Axelrod, H. L.; Cohen, A. E.; Roy, M.; Abresch, E. C.; Capraro, D.; Murphy, A. N.; Nechushtai, R.; Dixon, J. E.; Jennings, P. A. (2007). *Proc. Natl. Acad. Sci. USA*, **104**, 14342–14347.
9. Meguro, K.; Fujita, T.; Hatanaka, C.; Ooi, S. US Patent US54812570.
10. Huber, J. E. US Patent US5585495.
11. Momose, Y.; Meguro, K.; Ikeda, H.; Hatanaka, C.; Oi, S.; Sohda, T. (1991). *Chem. Pharm. Bull.*, **39**, 1440–1447.
12. Sohda, T.; Momose, Y.; Meguro, K.; Kawamatsu, Y.; Sugiyama, Y.; Ikeda, H. (1990). *Arzneim.-Forsch.*, **40**, 37–42.
13. Yoshioka, T.; Nishi T.; Kanai, T.; Aizawa, Y.; Wada, K.; Fujita, T.; Horikoshi, H. European Patent EP549365.
14. Dunn, P.; Wells, A.; William, M. T. (2010). *Green Chemistry in the Pharmaceutical Industry*. Wiley.
15. Madivada, L. R.; Anumala, R. R.; Gilla, G.; Alla, S.; Charagondla, K.; Kagga, M.; Bhattacharya, A.; Bandichhor, R. (2009). *Org. Process Res. Dev.*, **13**, 1190–1194.

Chapter 7

The Development of a Convergent Green Synthesis of Linezolid, an Oxazolidinone Antibacterial Agent

William R. Perrault, James B. Keeler, William C. Snyder, Christian L. Clark, Michael R. Reeder, Richard J. Imbordino, Rebecca M. Anderson, Nabil Ghazal, Stephen L. Secreast, and Bruce A. Pearlman

Global API Process Technology, Pfizer Global Manufacturing, Pfizer, Inc., PORT-91-201, 7000 Portage Rd. Kalamazoo, MI 49001, USA

wrperrault@kalexsyn.com

Pfizer has developed a novel, convergent, green synthesis of linezolid (the active ingredient in Zyvox™). The second-generation process has replaced the launch process and has numerous green chemistry benefits. Chlorohydrin imine reagent **1** contains both the chiral center and the key 5-*S*-aminomethyl moiety of linezolid. In the launch process, *S*-1-chloro-2,3-propanediol was utilized to install the oxazolidinone functionality. However, this yielded a 5-*S*-hydroxymethyl group requiring activation as 3-nitrobenzenesulfonate and displacement with excess ammonia to generate the corresponding aminomethyl group of linezolid. The second-generation process affords oxazolidinone imine **3** in a convergent step. The penultimate 5-*S*-aminomethyl oxazolidinone **4**

Scalable Green Chemistry: Case Studies from the Pharmaceutical Industry
Edited by Stefan G. Koenig

ISBN 978-981-4316-49-1 (Hardcover), 978-981-4364-81-2 (eBook)
www.panstanford.com

is then easily formed via hydrolysis with stoichiometric hydrochloric acid. Acylation of this amine with acetic anhydride, utilizing an improved Schotten–Baumann reaction, affords high-purity linezolid.

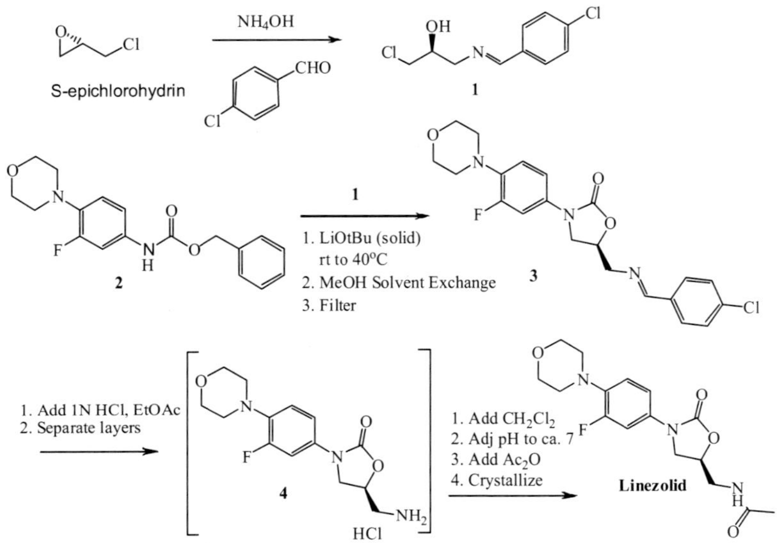

7.1 Background

Linezolid, approved by the Food and Drug Administration (FDA) in 2000, is the only member of the oxazolidinone class of antibacterials on the market, the first new class of antibiotics in over 30 years. Linezolid is approved for the treatment of antibiotic-resistant gram-positive bacterial infections, including vancomycin-resistant *Enterococcus faecium* (VREF), methicillin-resistant *Staphylococcus aureus* (MRSA), and multidrug-resistant *Streptococcus pneumonia* (MDRSP). These antibiotic-resistant bacterial infections have become an ever-increasing threat to public health. Rapid growth in global demand for this valuable life-saving drug has led us to develop a greener and more efficient convergent synthetic process in order to meet future needs, while at the same time reducing the cost and environmental impact of production.

The oxazolidinone class of antibacterials is a group of compounds that were discovered to have such activity by Dupont in 1973.[1] The initial Dupont synthesis gave 5-halomethyl phenyloxazolidinones **8** via a reaction of phenyl isocyanates **5** with the corresponding epoxides in the presence of lithium halides (Scheme 7.1). In this reaction, the lithium halide opened the halohydrin epoxide to give intermediate chlorohydrin **6**, which was then added to isocyanate **5**, followed by rapid intramolecular *N*-alkylation.

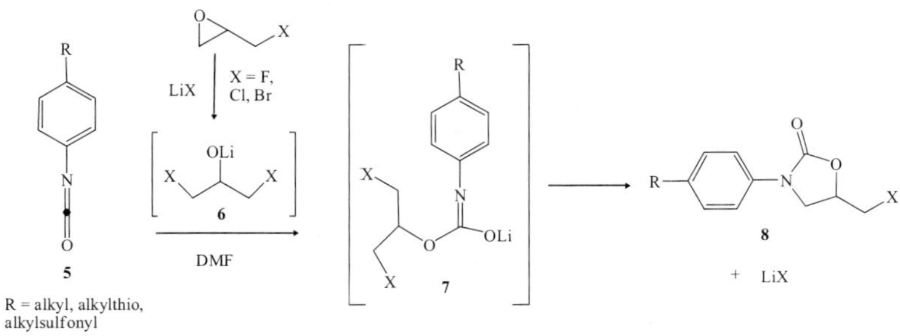

Scheme 7.1 Initial Dupont oxazolidinone synthesis.

Subsequently, Dupont discovered that the oxazolidinone 5-methylacetamide moiety in the *S*-configuration gave significantly increased antibacterial activity.[2] The desired enantiomer could be obtained directly when enantiopure *R*-glycidyl butyrate (R′ = butyryloxy, prepared via enzymatic kinetic resolution) was reacted with an isocyanate. Acetamide **14** was then prepared via butyrate transesterification, alcohol activation as mesylate **11** or equivalent tosylate, and displacement with azide to **12**, followed by reduction and acylation as in Scheme 7.2.

A significant synthetic difficulty with the isocyanate route to the oxazolidinones for both analog preparation and subsequent scale-up was the challenge of preparing isocyanates from the corresponding anilines. This reaction required the use of phosgene, which has significant toxicity concerns. To use only 1 equiv. of aniline, the hydrogen chloride by-product had to be distilled at high temperature to complete conversion to the isocyanate. Furthermore, adventitious water readily converted the isocyanate to the corresponding symmetrical urea. Peter Manninen at the Upjohn Company

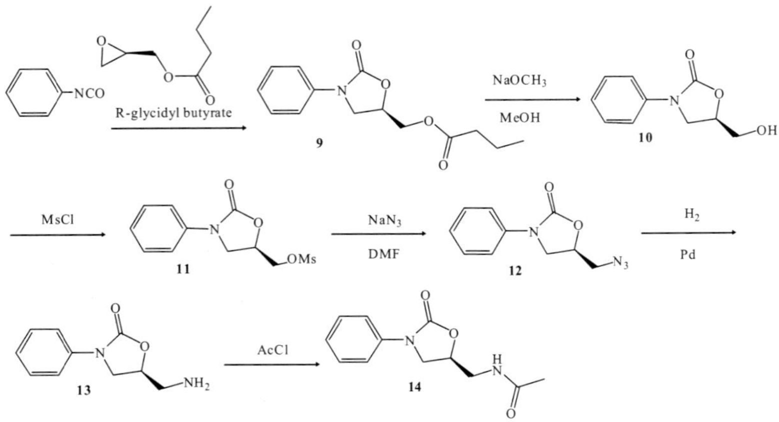

Scheme 7.2 Dupont 5-*S*-oxazolidinonyl methylacetamide synthesis.

significantly broadened the scope of oxazolidinone synthesis when he discovered that if an aryl carbamate was deprotonated with a strong base and the resultant anion reacted with enantiopure *R*-glycidyl butyrate, the desired oxazolidinone alcohol was obtained directly under very mild conditions.[3] This improved synthesis facilitated the preparation of numerous oxazolidinone analogs, including linezolid, eventually leading to FDA approval as the first and only oxazolidinone antibacterial treatment.

In the Manninen synthesis (Scheme 7.3), deprotonation of carbamate **2** was accomplished using butyl lithium and the resultant anion **15** was then alkylated by enantiopure *R*-glycidyl butyrate (commercially available at the time via enzymatic kinetic resolution in 96% e.e.). Alkoxide intermediate **16** then quickly cyclized to give oxazolidinone **17** with loss of lithium benzyloxide. Transesterification of the butyrate group gave the desired 5-methyloxazolidinyl alcohol **18** directly as the product along with benzyl butyrate. The procedure gave synthetically useful yields only when lithium was used as the cation, a strong indication that coordination of this Lewis acidic cation to the epoxide bond was essential to achieving alkylation. Amazingly, despite the number of transformations occurring in one pot, the procedure proved high yielding and applicable to numerous carbamate substrates.

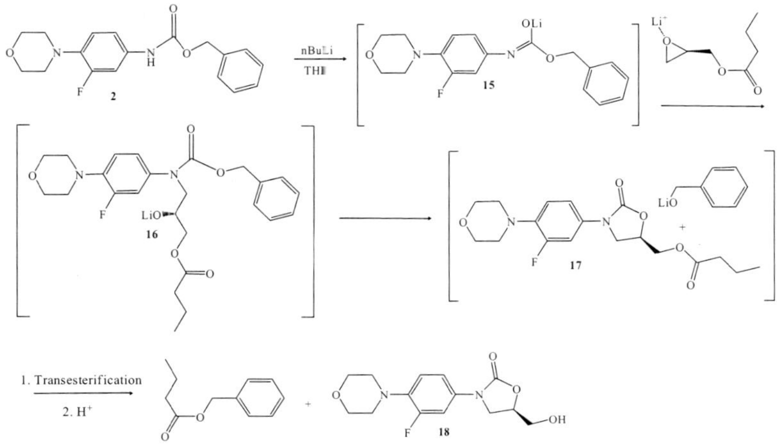

Scheme 7.3 Manninen synthesis of enantiopure oxazolidinone alcohols

Large-scale preparation of linezolid, as well as numerous other oxazolidinones targeted for potential development, led to a need to further improve the synthesis for clinical supplies as well as potential commercial production. As the Manninen reaction afforded oxazolidinone alcohol, synthesis of linezolid required activation and displacement of this group to afford the desired oxazolidinone acetamide. For expediency purposes, it was convenient to prepare the amine via activation of the alcohol as the mesylate and displacement with sodium azide, followed by reduction as in the original Dupont process. However, the potential explosive decomposition of azides made them very unattractive for larger-scale synthesis, even on the laboratory scale. A significantly improved procedure for scale-up was discovered by Dana Toops, also at Upjohn, who discovered that under pressure, mesylate **19** reacted cleanly with aqueous ammonia to give aminomethyl oxazolidinone **20** directly (Scheme 7.4). The

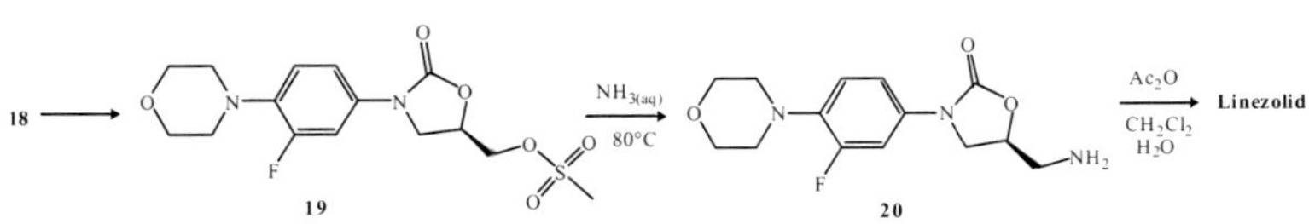

Scheme 7.4 Toops' ammonolysis reaction.

amine was then conveniently converted to the desired acetamide via Schotten–Baumann reaction with acetic anhydride.[3]

For subsequent scale-up, there were some disadvantages with the Toops' mesylate ammonolysis procedure to overcome. These included the use of at least 80 equiv. of ammonia in order to suppress the corresponding 2:1 alkylation by-product and the high pressure required to achieve reaction completion at 80°C. Numerous alternative activating groups were explored and their ammonolysis rates compared to the mesylate. It was discovered that the corresponding *m*-nitrobenzenesulfonate **23** increased the rate of the ammonolysis reaction over 40-fold relative to the mesylate. Several other sulfonates also proved useful in accelerating the reaction; however, the *m*-nitrobenzenesulfonate was the preferred substrate due to the ready preparation of the corresponding sulfonyl chloride from *m*-nitrobenzenesulfonic acid sodium salt, a large volume article of commerce used to prepare hair dyes. This allowed the displacement reaction to be run at or near atmospheric pressure, thus allowing the process to be run in general purpose equipment.

Further improvements were made to the Manninen reaction as well. For instance, it was discovered that the cheaper, more readily available, and more optically pure reagent *S*-chloropropanediol, which could be cyclized with a strong base to afford *S*-glycidol, served an equivalent role in the Manninen reaction. Also, much weaker bases, such as lithium *t*-butoxide, could be substituted for butyl lithium in the reaction, thus expanding the scope of the chemistry to more sensitive molecules.[4] For the linezolid launch process, the preparation of carbamate **2** was streamlined and the oxazolidinone formation reaction was further improved. Jeff Havens found that potassium *t*-butoxide could be used to cyclize the chloropropanediol to give glycidol and that lithium diisopropylamide (LDA), more readily available at the time than lithium *t*-butoxide, could be used to give the lithium anion of the carbamate. The resultant launch process is shown in Scheme 7.5.

7.2 Green Linezolid Process

While developing the launch process to linezolid, we explored numerous convergent approaches to the oxazolidinone acetamide

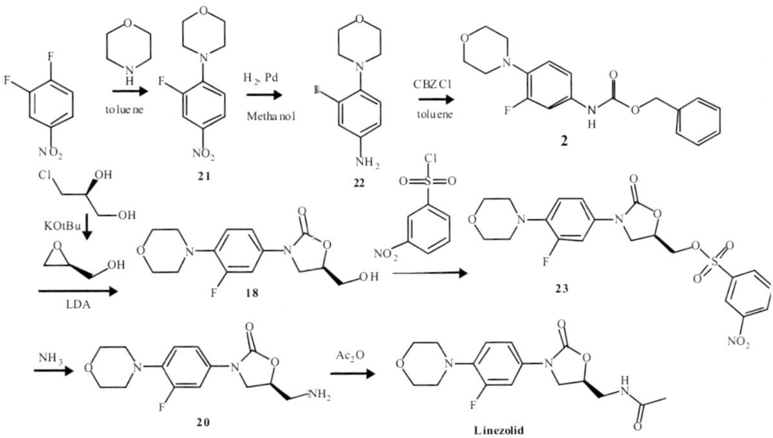

Scheme 7.5 The linezolid launch process.

functionality in which a custom three-carbon reagent was prepared incorporating an amine instead of an alcohol functionality.[5] This sort of approach was highly desirable since the ammonolysis step was the least efficient portion of the launch synthesis from a green chemistry perspective. A large excess (80 equiv.) of ammonia and dilute conditions were required in order to minimize the 2:1 adduct. Eventually, successful convergent approaches were demonstrated utilizing reagents **26** and **27** (Scheme 7.6). However, these syntheses were not shown to be practical until after the launch process had been implemented. They were, however, applied to several other oxazolidinone clinical candidates with great success on the multikilogram scale.

We then explored the direct reaction of imine **24** with carbamate **2** in order to directly prepare enantiopure oxazolidinone imines. Imine **24** is equivalent to **27** in utility but could be prepared in two fewer steps. We demonstrated that at slightly elevated temperatures, the alkylation of carbamate **2** with **24**, indeed, proceeded in high yield with lithium *t*-butoxide in polar aprotic solvents. Subsequently, we discovered that 4-chloro chlorohydrin imine analog **1** was highly crystalline and performed identically in the oxazolidinone formation, thus completing the discovery of the second-generation process (Scheme 7.7).[6]

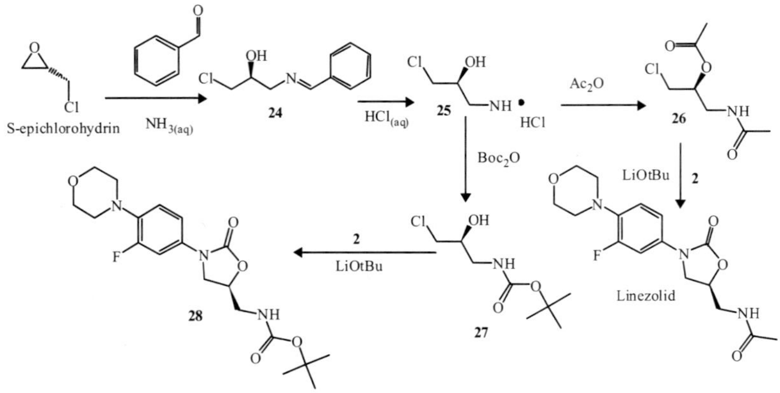

Scheme 7.6 Initial convergent oxazolidinone syntheses.

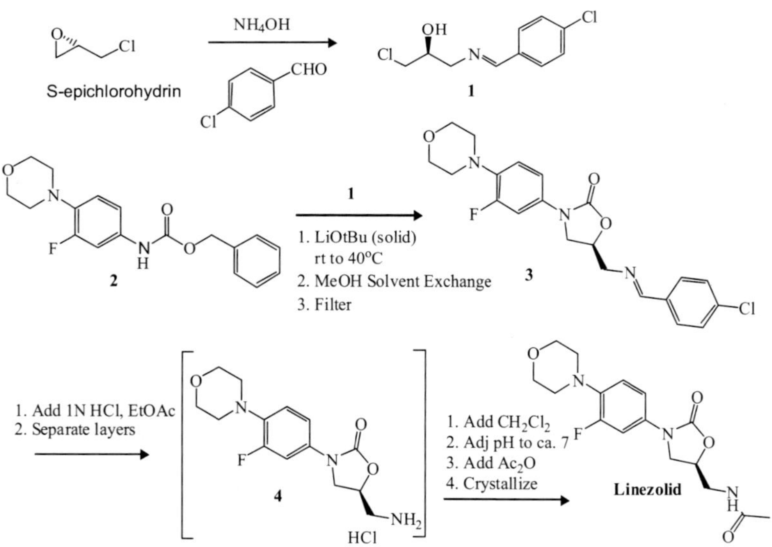

Scheme 7.7 Second-generation route to linezolid.

The second-generation process has replaced the launch process and has numerous green chemistry benefits: overall yield is increased by 8%, total waste is reduced by 56%, nonrecycled waste is reduced by 77%, methylene chloride waste is reduced by 78%, and a pressurized ammonia step is eliminated. At current volumes,

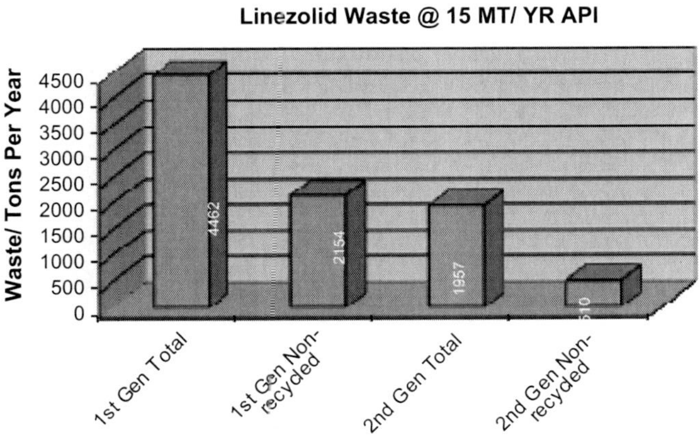

Figure 7.1 Comparison of launch (first-generation) and second-generation waste streams.

total waste will be reduced by 1.9 million kg/year, while 1.7 million kg/year of nonrecyclable waste will be eliminated (Fig. 7.1). The improved process utilizes a highly efficient, low-dilution, convergent synthesis to replace the more dilute linear synthesis utilized in the launch process.

Reagents **26** and **27** have become essential tools in the development of new oxazolidinone antibacterial agents. Since their publication in 1996 and 2002, respectively, there have been 38 references to these two compounds indexed in *Chemical Abstracts*, including 27 patents. As chlorohydrin imine reagent **1** improves on the capabilities of these two reagents, it can be expected to have an even greater impact in the field. This should result in the faster discovery of even more powerful oxazolidinone antibacterial agents and yield new weapons in the fight against antibiotic resistance.

7.3 Conclusion

The increased yield, practicality, cost-effectiveness, and other green chemistry advantages of the second-generation linezolid process have been critically evaluated and validated by numerous personnel in the course of moving the process toward full-scale implementation in production. The greatly reduced waste from the new process will result in significant reduction in the transport of hazardous waste and the consequent potential for accidental exposure of humans, animals, and plants to these toxic materials. Eliminating the pressurized ammonia reaction will lead to a reduced chance for the release of ammonia gas into the environment. The greater efficiency of the new process, both in terms of yield and total material utilization per kilogram, will greatly reduce the use of natural resources. Reducing the cost of linezolid production will make this lifesaving drug more readily available to a larger portion of humanity and significantly reduce the loss of life inherent in the emergence of new antibiotic-resistant pathogenic organisms.

References

1. (a) Fugitt, R. B.; Martinelli, L. C. (1973). *J. Pharm. Sci.*, **62**(6), 1013–1016; (b) Fugitt, R. B.; Luckenbaugh, R. W. (Dec. 5, 1978). US Patent 4,128,654.
2. (a) Gregory, W. A.; Brittelli, D. R.; Wang, C. L.; Wuonola, M. A.; McRipley, R. J.; Eustice, D. C.; Eberly, V. S.; Bartholomew, P. T.; Slee, A. M.; Forbes, M. (1989). *J. Med. Chem.*, **32**, 1673–1681; (b) Gregory, W. A. (July 24, 1984). US Patent 4,461,773; (c) Daly, J. S.; Eliopoulos, G. M.; Willey, S.; Moellering, R. C., Jr. (1988). *Antimicrob. Agents Chemother.*, **32**, 1341–1346; (d) Gregory, W. A. (Nov. 10, 1987). US Patent 4,705,799.
3. Brickner, S. J.; Hutchinson, D. K.; Barbachyn, M. R.; Manninen, P. R.; Ulanowicz, D. A.; Garmon, S. A.; Grega, K. C.; Hendges, S. K.; Toops, D. S.; Ford, C. W.; Zurenko, G. E. (1996). *J. Med. Chem.*, **39**, 673–679.
4. Perrault, W. R.; Pearlman, B. A.; Barbachyn, M. R.; Manninen, P. R.; Toops, D. S.; Houser, D. J.; Fleck, T. J. (Nov. 17, 1998). US Patent 5837870.
5. (a) Perrault, W. R.; Pearlman, B. A.; Godrej, D. B.; Jeganathan, A.; Yamagata, K.; Chen, J. J.; Lu, C. V.; Herrinton, P. M.; Gadwood, R. C.; Chan, L.; Lyster, M. A.; Maloney, M. T.; Moeslein, J. A.; Greene, M. L.; Barbachyn, M. R. (2003). *Org. Process Res. Dev.*, **7**, 533–546; (b) Perrault, W. R.; Pearlman, B. A.; Godrej, D. (Oct. 31, 2002). WO 0285849; (c) Pearlman, B. A. (Nov. 7, 1997) WO 9924393 (d) Perrault, W. R.; Gadwood, R. C. (Apr. 25, 2002). WO 0232857; (e) Lu, C. V.; Chen, J. J.; Perrault, W. R.; Conway, B. G.; Maloney, M. T.; Wang, Y. (2006). *Org. Process Res. Dev.*, **10**(2), 272–277.
6. Imbordino, R. J.; Perrault, W. R.; Reeder, M. R. (Oct. 18, 2007). WO 2007/116284.

Chapter 8

Development of a Nonaqueous Process for the Synthesis of 3-Amino-pentan-1,5-diol

Thimma Rawalpally and Thomas Cleary
Pharmaceuticals Technical Development Actives (PTDA-FL), Roche Carolina Inc., 6173 East Old Marion Highway, Florence, South Carolina 29506-9330, USA
thimma_r.rawalpally@roche.com and cleary.thomas@gene.com

The development of an efficient process for the large-scale production of water-soluble 3-amino-pentan-1,5-diol, an intermediate in the synthesis of p38 inhibitors of the pyridinylimidazole class, is presented. Various synthetic routes to 3-amino-pentan-1,5-diol are discussed in this chapter. In the second-generation telescoped process, all four chemical steps were optimized using 2-Me-tetrahydrofuran (THF) and MeOH solvents with an overall yield of 89% and a gas chromatography (GC) purity of 97–98%. The key to this process is the telescoped deprotection, purification, and nonaqueous isolation of 3-amino-pentan-1,5-diol.

Scalable Green Chemistry: Case Studies from the Pharmaceutical Industry
Edited by Stefan G. Koenig

ISBN 978-981-4316-49-1 (Hardcover), 978-981-4364-81-2 (eBook)
www.panstanford.com

8.1 Introduction

One of the primary ways that chemists have traditionally measured the efficiency of a reaction is by the percentage yield of the product. However, nowadays what one does not produce from a chemical reaction is almost as important as what one does produce. Chemists have become much more conscious about unwanted by-products and hazardous waste. A pharmaceutical green chemist must strive for the correct choice of starting material, ideal number, and order of chemical steps; the appropriate selection of solvents and reagents; and efficient strategies for isolation and purification. Pharmaceutical green chemistry is the ideal that one strives for, and the pursuit of this ideal will lead to ever-improving processes. Chemists should aim to incorporate the 12 principles of green chemistry[1] into a given process during development. The example presented in this chapter highlights some of these 12 principles.

Biological agents that selectively neutralize pro-inflammatory cytokines (tumor necrosis factors [TNFs], such as TNFα, and interleukin IL-1β) have been shown to reduce the number of swollen and tender joints and to retard the destruction of joint tissue in patients with rheumatoid arthritis.[2] The compound 3-amino-pentan-1,5-diol is useful for the preparation of active pharmaceutical ingredient (API) **3**, a mitogen-activated protein (MAP)-kinase inhibitor useful in the treatment of rheumatoid arthritis. API **3** can be prepared by coupling 3-amino-pentan-1,5-diol (**1**, referred to throughout this chapter as "aminodiol") and sulfone intermediate **2**, as shown in Scheme 8.1.

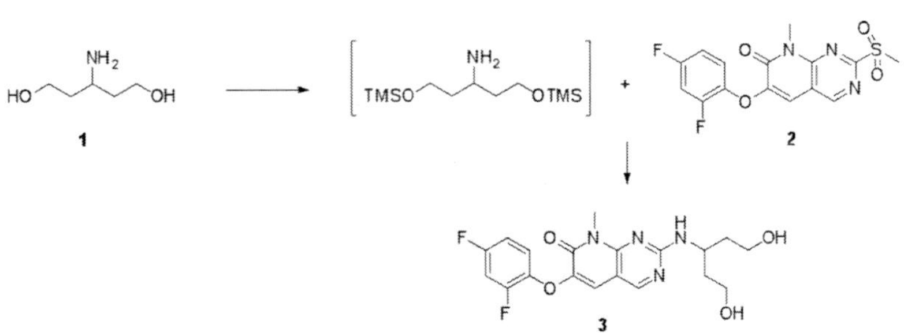

Scheme 8.1 The synthetic scheme of an API.

Aminodiol is a simple, functionalized five-carbon molecule, but its physicochemical properties make its synthesis on a large scale quite challenging. The compound is low melting (64°C) and highly hygroscopic. It also lacks an ultraviolet (UV) chromophore, making it difficult to develop a UV absorption–based analytical method. In addition, aminodiol exhibits high water solubility and is insoluble in a majority of conventional organic solvents (except methanol). Since aminodiol is coupled to sulfone intermediate **2** via its trimethylsilyl (TMS) ether to form an API (Scheme 8.1), it is essential to remove any trace of protic solvents or water. Unfortunately, removal of residual methanol or water is quite difficult and requires multiple energy-intensive solvent exchanges.

8.2 IND Synthesis

The original investigational new drug (IND) synthesis of aminodiol is shown in Scheme 8.2. The synthesis was developed by our colleagues at Roche Bioscience, and this process enabled us to make APIs for initial clinical supplies.

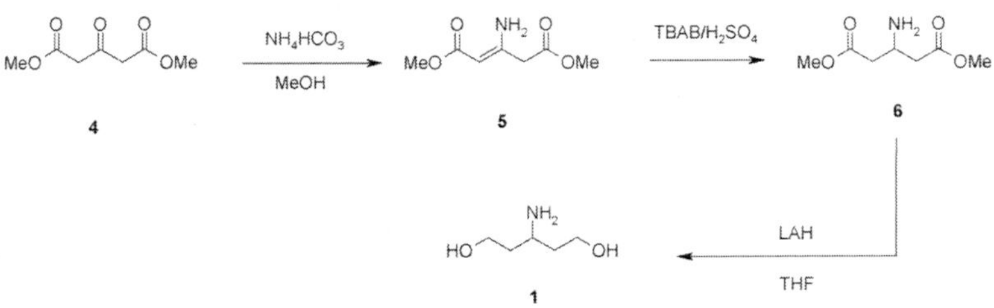

Scheme 8.2 The original IND synthesis of aminodiol.

Dimethyl 3-aminoglutaconate **5** was prepared starting from dimethyl acetone-1,3-dicarboxylate **4** via amine condensation. The enamine reduction of **5** under acidic conditions using borane-*tert*-butylamine complex (TBAB) provided **6**. Dimethyl 3-aminoglutarate **6** was reduced using lithium aluminum hydride (LAH) on the basis of a literature method.[3] However, isolation of **6** from the aluminum salts was operationally demanding, and the high water solubility of aminodiol resulted in poor recovery. Though the aluminum salts

Table 8.1 Solvent distribution in IND aminodiol synthesis

Solvent category*	Solvent	Usage (kg solvent/kg product)
Preferred	Methanol	7
	2-Propanol	25
Usable	THF	33.6
Undesirable	Dichloromethane	22.5
	Propylamine	8
	Ethylamine	8

*Solvent categorization is based on Roche internal classification.

were saturated with diethylamine and propylamine to improve the isolation of aminodiol, the initial purity was still not sufficient to carry forward. Hence, further purification by thin film evaporation was necessary prior to coupling with sulfone intermediate **2**. Following all of these maneuvers, the overall yield of the four-step process was 57%.

Evaluating the IND process in reference to green chemistry principles identified several opportunities for improvement, as follows:

- Three isolation steps and the use of thin-film evaporation to purify the product requiring high energy input
- Waste disposal of six different solvents, including undesirable dichloromethane (see Table 8.1 for solvent distribution)
- E-factor of 106
- High material input requirement due to low overall yield (57%)
- Expensive removal and disposal of waste aluminum salts along with diethyl- and propylamine, with a high environmental burden
- Use of LAH on a commercial scale inherently unsafe and requiring additional engineering controls

8.3 First-Generation Aminodiol Synthesis

The first-generation aminodiol synthesis, as shown in Scheme 8.3, was developed by our colleagues at Roche Boulder. In this route,

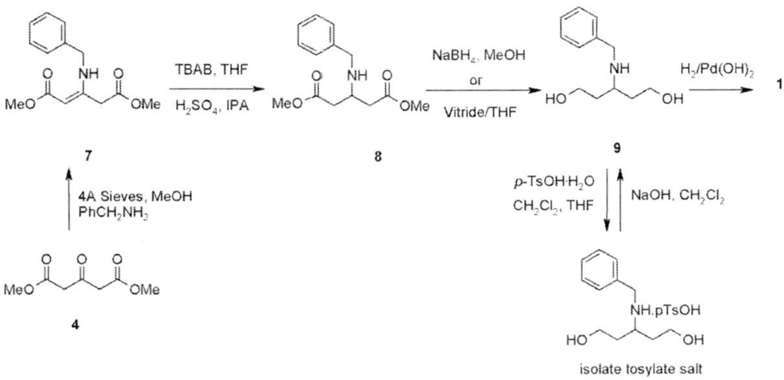

Scheme 8.3 First-generation synthesis of aminodiol (benzyl protection).

the IND synthesis was modified to include benzyl protection on the amine to decrease its water solublity.[4,5]

Dimethyl 3-*N*-benzylamino-glutaconate intermediate **7** was purchased from Chemie Uetikon. In the literature, the preparation of **7** was described as starting from dimethyl acetone-1,3-dicarboxylate **4**.[4] Direct one-pot reduction of dimethyl 3-*N*-benzylamino-glutaconate **7** to 3-*N*-benzylamino-pentan-1,5-diol **9** with sodium borohydride/AcOH (not shown) gave variable yields and resulted in lactone impurity **10** (Fig. 8.1). Instead, a stepwise process was initiated with reduction of **7** to amine **8**, easily accomplished in good yields using TBAB in THF.[5] The crucial second reduction of dimethyl 3-*N*-benzylamino-glutarate **8** to 3-*N*-benzylamino-pentan-1,5-diol **9** was achieved using either of two reducing agents, sodium borohydride or sodium bis(2-methoxyethoxy) aluminum hydride (Vitride®). Sodium borohy-

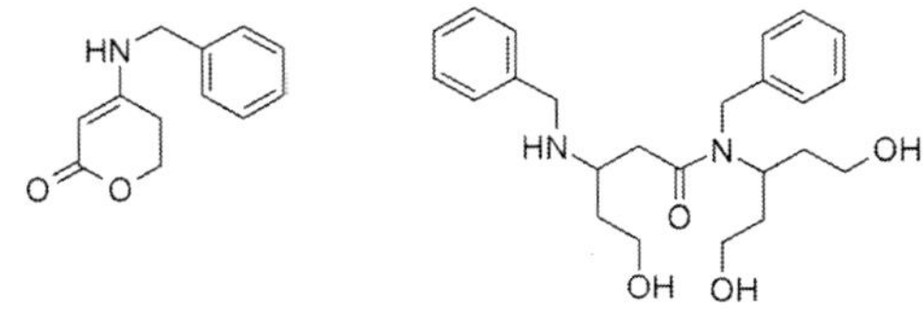

10

11

Figure 8.1 Alternate route impurities.

dride, though inexpensive, required a long reaction time (72 hr) and resulted in impurity **11** (Fig. 8.1), which was difficult to remove from the downstream process. Alternatively, the reduction employing a toluene solution of Vitride® was found to be faster (3 hr) and the process was easily scalable.

Compound **9** was purified via its crystalline tosylate salt. This purification involved salt formation, isolation, and drying of the salt prior to regeneration of the free base. Deprotection of the benzyl group to obtain aminodiol was possible under nonaqueous conditions by hydrogenolysis using $Pd(OH)_2$ in methanol. It was essential to remove the last traces of MeOH/water because the aminodiol was to be silylated in the next step before coupling to sulfone intermediate **2** (Scheme 8.1). Such a stringent requirement for removal of methanol and water required multiple azeotropic THF distillations. Yet the low solubility of aminodiol in THF caused further complications by forming a sheet of glass-like precipitate on the wall of the reaction vessel, thus trapping small amounts of unwanted water and methanol. This modified five-step process, starting from **7**, resulted in 64% overall yield of aminodiol (purity >99% by GC) and allowed an API to be produced for midstage clinical supplies.

In spite of the above process improvements, there were several outstanding issues with this first-generation synthesis. Achieving good recovery of compound **9** required six dichloromethane extractions (187 kg dichloromethane/kg product). The process required a total of five different solvents (see Table 8.2 for solvent distribution), and the use of the Vitride® reagent produced an aluminum salt-laden aqueous waste, which gelled on standing or if

Table 8.2 Solvent distribution in the first-generation aminodiol synthesis

Solvent category	Solvent	Usage (kg solvent/kg product)
Preferred	Methanol	7
	2-Propanol	25
Usable	THF	33.6
	Toluene	7.1
Undesirable	Dichloromethane	141.4

neutralized (40 kg gel/kg-product). The waste was very expensive to dispose of and had a high environmental impact (landfill). The E-factor for the first-generation synthesis was 246.

For the second-generation aminodiol synthesis, the primary criterion was to avoid isolation of aminodiol from an aqueous system, to prevent waste, and to reduce the raw material input by improving the overall yield.

8.4 Second-Generation Synthesis

Upon reviewing the literature,[6] it appeared that *N*-*Boc*-aminodiester **12** should be amenable to reduction with sodium borohydride, thus suggesting a second-generation aminodiol synthesis (Scheme 8.4). To evaluate the feasibility of this reduction of **12**,

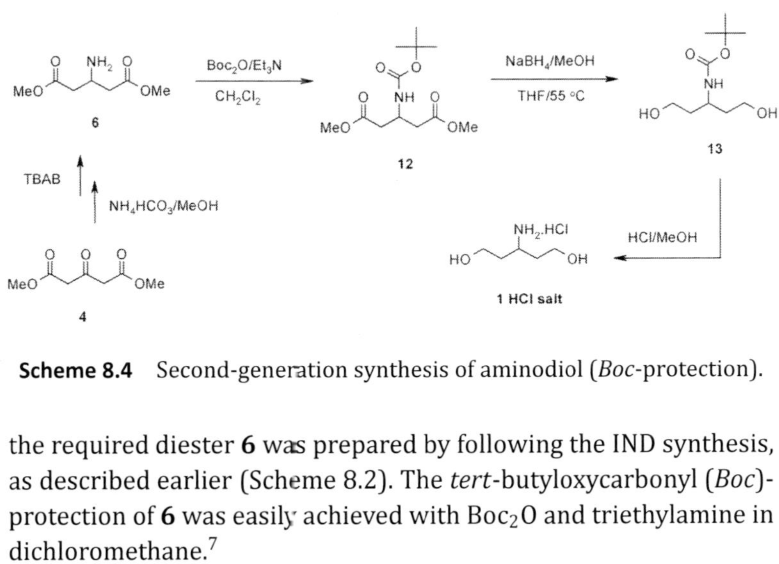

Scheme 8.4 Second-generation synthesis of aminodiol (*Boc*-protection).

the required diester **6** was prepared by following the IND synthesis, as described earlier (Scheme 8.2). The *tert*-butyloxycarbonyl (*Boc*)-protection of **6** was easily achieved with Boc_2O and triethylamine in dichloromethane.[7]

The reduction of **12** in THF with 4 equiv. of sodium borohydride afforded **13** in 83% isolated yield and 96% purity (nuclear magnetic resonance [NMR] and GC). In a typical procedure, methanol was added slowly to a slurry of **12** and sodium borohydride at 55°C in THF. After reaction completion (<0.5% of **12**), the excess reducing agent was quenched with water. The methanol and THF were distilled, and the aqueous layer was extracted with 1-butanol. After solvent exchange, *Boc*-aminodiol **13** was obtained as a solution in MeOH. Maintaining the reaction temperature between 55–65°C afforded good conversions (>99%), while lower temperatures led to incomplete reactions. In spite of high reaction conversion, the isolated yield of **13** was lower than desired because it was difficult to extract the *Boc*-protected diol **13** from the aqueous layer under the initial conditions.

Deprotection of *Boc*-protected amines is usually achieved by using an acid, such as HCl or trifluoroacetic acid (TFA). In this case, the *Boc* group was easily removed under acidic conditions (HCl/MeOH or *p*-toluenesulfonic acid [PTSA]), but the isolation of free base **1** was difficult. Aminodiol partitions preferably to the aqueous phase; hence, extraction of aminodiol as a free base with an organic solvent was not practical. In addition, residual triethylammonium salt interfered with the silylation of aminodiol and the coupling to sulfone intermediate **2** in the downstream process when attempts were made to carry-forward the free base. Neutralization of triethylammonium salt with $NaHCO_3$ or NaOMe under anhydrous conditions was unsuccessful since the by-product NaCl salt and any excess reagent also interfered with silylation (Scheme 8.1).

A basic ion-exchange resin, Amberjet 4200 (Cl), could be used to generate the free aminodiol from its corresponding HCl salt. In principle, this resin should afford a free base (aminodiol) without any by-product. In fact, the basic ion-exchange resin worked well for the generation of a free amine, and the aminodiol was separated from the resin by filtration. The recovery of the product was quite good on a small scale (Scheme 8.5). To replicate this process on

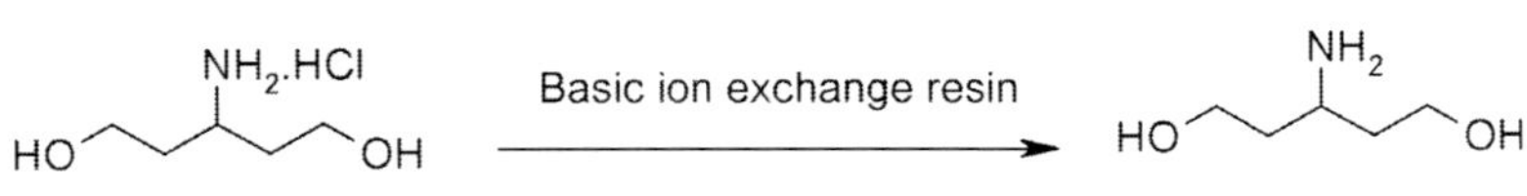

1 HCl salt

1

Scheme 8.5 Free base formation using basic resin.

scale-up, it was necessary to preactivate the resin by washing it with aqueous 1N NaOH and rinsing with methanol, which was quite labor intensive. Because the aminodiol was extremely hygroscopic and elimination of all moisture from the preactivation process was difficult, we did not pursue this method further.

Strongly acidic ion-exchange resins can be used to deprotect the *Boc* group,[8] opening up the possibility for nonaqueous isolation of aminodiol. Indeed, treatment of **13** with strongly acidic Amberlyst-15 ion-exchange resin at 40°C in MeOH afforded 36% deprotection in 24 hours (Scheme 8.6). In the second step, the resin-bound aminodiol was released by exchange with ammonia. The process was further optimized with respect to the type of resin, resin loading, reaction temperature, 7N methanolic ammonia extraction volume, and reaction time. The optimized conditions led to *Boc*-deprotection in 18 hours and utilized 150% resin loading (based on **13** w/w) in methanol at 60°C. Comparable results were obtained using pharmaceutical-grade FPC22H or FPC23H, both Amberlite resins with the sulfuric acid functional group on a styrene-divinylbenzene copolymer. Aminodiol was then released by stirring the resin in 7N methanolic ammonia solution for four hours at 4°C.

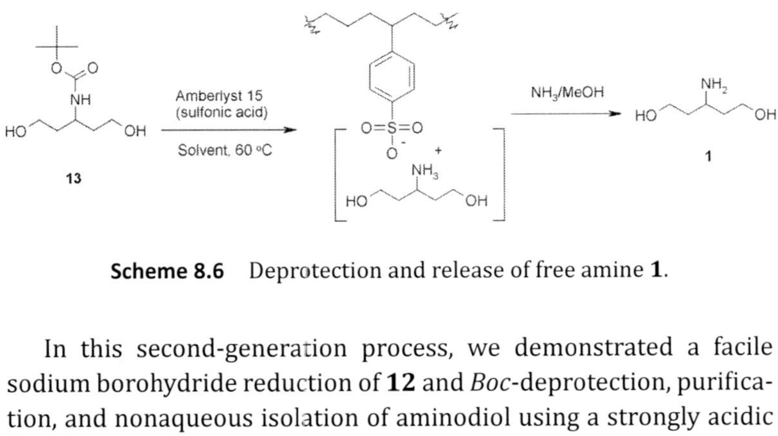

Scheme 8.6 Deprotection and release of free amine **1**.

In this second-generation process, we demonstrated a facile sodium borohydride reduction of **12** and *Boc*-deprotection, purification, and nonaqueous isolation of aminodiol using a strongly acidic ion-exchange resin, followed by exchange of the free amine from the resin with an ammonia base. An 83% overall yield of aminodiol was obtained starting from compound **4** (Schemes 8.4 and 8.6).

8.4.1 *Telescoped Second-Generation Synthesis*

Even though the second-generation process met the primary criteria for improvements, such as higher overall yield, use of a safe reducing reagent, and a nonaqueous isolation, the process still required multiple isolations, use of multiple solvents, multiple solvent exchanges, and an expensive TBAB reagent[9] for the reduction of **5**. Further development work was undertaken to address the above-mentioned issues. The resulting second-generation telescoped process, which included four chemical steps with a single isolation, is depicted in Scheme 8.7.

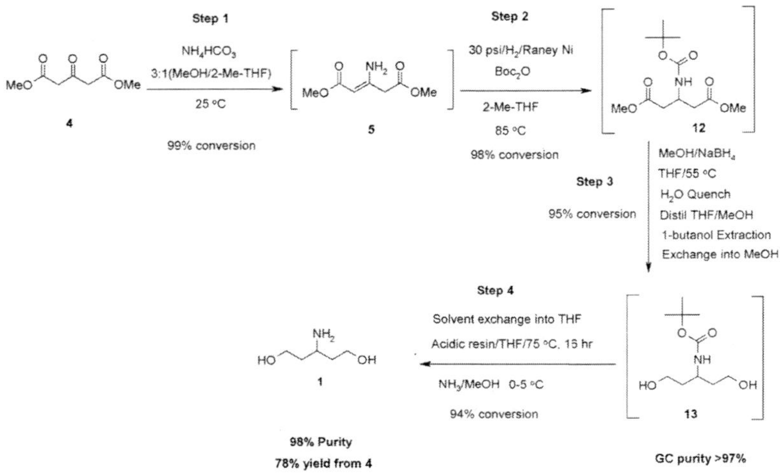

Scheme 8.7 Second-generation aminodiol process.

Dimethyl 3-aminoglutaconate **5** was prepared following the IND synthesis. The reaction was optimized with an ammonium hydrogen carbonate source of ammonia in methanol. During the course of development work, it was observed that protic residues, such as ammonia and methanol from step 1, would react with di-*t*-butyldicarbonate (Boc_2O) in the subsequent hydrogenation/*Boc*-protection step to form *tert*-butyl carbamate and methyl *tert*-butyl carbonate impurities, respectively. For step 1, 2-methyltetrahydrofuran (2-Me-THF) became the solvent of choice because it does not react with Boc_2O in step 2. Unfortunately, liberated ammonia from ammonium hydrogen carbonate had poor solubility in 2-Me-THF. To achieve complete conversion, a 3:1 mixture of methanol/2-Me-THF was required. At the end of the reaction, constant volume distillation was performed to exchange the methanol for 2-Me-THF and to remove the ammonia gas. The resulting 2-Me-THF solution of **5** was carried directly into step 2.

To improve the cost efficiency and scalability of the process and to minimize waste, we turned our attention toward the catalytic reduction of enamine **5**.[10] It was difficult to achieve complete conversion with Pd/C despite high H_2 pressure (350 psi) and high catalyst loading (10–20%). On the basis of earlier work,[11] the Raney Ni–catalyzed hydrogenation of **5** was achieved with 10% loading at 200 psi hydrogen pressure under slightly acidic conditions to afford **12** after 18 hours (<0.5% of **5**). Acetic acid was required as a co-catalyst to drive the Raney Ni hydrogenation reaction to completion via protonation of the free amine. In the next step, the free amine was to be protected with a *Boc*-protecting group anyway, so we sought to use the *Boc*-protection to drive the equilibrium forward. On the basis of this concept, we attempted to combine these steps into a one-pot process. The one-pot hydrogenation/*Boc*-protection step 2 sequence was optimized in 2-Me-THF by varying the equivalents of Boc_2O, catalyst loading, hydrogen pressure, and temperature (Table 8.3 and Scheme 8.7). Using 1.1 equivalents of Boc_2O at 85°C with 10% Raney Ni catalyst loading and 200 psi hydrogen pressure afforded complete conversion in 3.5 hours (Table 8.3, entry 2).

To allow hydrogenation to be performed in standard glass-lined vessels, reduced pressures (<30 psi) were evaluated. Accordingly, 30 psi of hydrogen gas with 1.3 equiv. of Boc_2O and a slightly higher catalyst loading (15 wt%) gave complete conversion after 11

Table 8.3 Optimization of reductive amination/*Boc*-protection

Entry	Equiv. Boc_2O	H_2 pressure (catalyst % loading)	Temp (0°C)	Time (hr)	IPC[1]
1	1.0	200 psi (10 wt%)	65–85	7	3%
2	1.1	200 psi (10 wt%)	85	3.5	0.5%
3	1.1	30 psi (10 wt%)	85	21	3%
4	1.3	30 psi (15 wt%)	85	11	0.3%
5	1.3	30 psi (15 wt%)	95	3.5	9%
6[2]	1.3	30 psi (15 wt%)	85	4	0.3%

[1]Area % of **5** remaining by HPLC
[2]System was purged with hydrogen after 2 h.

hours (Table 8.3, entry 4). An attempt to improve the reaction rate by increasing the reaction temperature to 95°C was unsuccessful, presumably from decomposition of the Boc_2O reagent (entry 5). Purging the reaction vessel with hydrogen after two hours improved the reaction rate (entry 6), presumably by sweeping away the CO_2 by-product of the reaction. To avoid handling the dry, pyrophoric catalyst on a large scale, the effect of water on step 2 was evaluated. Wetting the Raney Ni catalyst with up to 50% water was well tolerated. After filtering to remove the catalyst, the 2-Me-THF solution of intermediate **12** was carried forward into step 3 without isolation.

8.4.2 *Opportunities for Process Improvements after the Campaign*

After the pilot plant scale-up of step 3 using sodium borohydride for reduction, several opportunities for improvement were identified. The methanol addition to $NaBH_4$ generated hydrogen gas, and the reaction became viscous due to the continuous hydrogen evolution. The viscosity of the reaction mixture also led to agitation, sampling, and foaming problems. In addition, quenching residual borohydride with water was slow, and hydrogen evolution persisted for more than two days. Removal of the THF solvent via distillation to facilitate good layer separation and product recovery was difficult due to the continuing hydrogen effervescence. Additionally, due to the high boiling point and viscosity of 1-butanol, its solvent exchange with the lower boiling methanol was challenging.

8.4.3 *Postpilot Plant Campaign Process Optimization*

To address the above step 3 scale-up issues and to telescope steps 1 and 2 with step 3, we decided to explore the use of 2-Me-THF as the solvent in the step 3 reduction. On a small scale, sodium borohydride reduction of **12** was complete within 30 minutes after methanol addition. Good layer separation was achieved at room temperature after the aqueous quench, and intermediate **13** was isolated in 93% yield. On scale-up, initial recovery was only 76%, but the yield was improved to 91% by extracting the aqueous layer at 60°C.

Separation of layers at higher temperature improved the solubility of by-product salts, thereby reducing the amount of water required to dissolve these salts. In addition, 2-Me-THF has lower solubility in water at higher temperature than at ambient temperature (14.4 g/100 g in water at 19.3°C vs. 6.6 g/100 g at 60°C), facilitating the extraction.[12]

Methanol reacts with borohydride to generate reactive borane and hydrogen gas. Controlled addition of a small quantity of methanol on a large scale is difficult, so to have a better control of hydrogen evolution, methanol was diluted with 2-Me-THF for well-controlled addition. The reduction was optimized with 3 equiv. of sodium borohydride instead of the previous 4 equiv. Excess borohydride was quenched with acetone prior to quenching with water; acetone reacts with borane instantaneously to form 2-propanol and therefore generates less hydrogen than treatment with water alone. By performing a reverse addition at 60°C, that is, adding the reaction mixture to an acidic aqueous solution (pH 4) of 2-Me-THF solution, rapid but controllable hydrogen evolution was achieved.

Incorporating these modifications made step 3 more easily scalable and addressed major safety concerns. Hydrogen evolution was more controllable, both during methanol addition and upon quench. Heat and mass transfer were much more efficient, and the long-lasting effervescence after the water quench was effectively eliminated, giving better layer separation and good product recovery. To further streamline the process, we explored the use of 2-Me-THF as a solvent in the step 4 *Boc*-deprotection reaction as well. In fact, complete deprotection was achieved in 4 hours at 75°C (as compared to 18 hr at 60°C in methanol) to provide aminodiol in 89% overall yield (based on **4**) and 96–98% GC purity. This telescoped second-generation process (Scheme 8.8) was successfully demonstrated on the pilot plant scale.[13,14]

Typically, the aminodiol produced in the lab had greater purity compared to pilot plant batches (Table 8.5). Handling the viscous and hygroscopic aminodiol was difficult where a distilled aminodiol sample of greater than 99% purity solidified upon standing at room temperature. We thus focused our attention on developing a crystallization process to resolve the handling issue. After

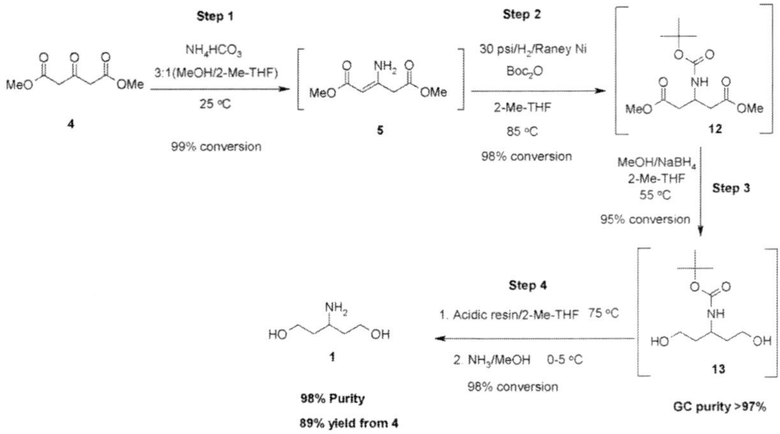

Scheme 8.8 Telescoped second-generation aminodiol process.

some experimentation, a preliminary crystallization process was developed using a methanol/THF solution, giving 80% recovery. Alternatively, a short-path distillation process was also developed on a small scale. Accordingly, Table 8.5 outlines the purity profile of aminodiol derived from these various processes. Unfortunately, neither the crystallization nor the distillation process was scaled up due to project termination.

8.5 Environmental Assessment of the Synthetic Routes

This one-pot, telescoped second-generation synthesis of aminodiol involves a single isolation with four chemical reactions starting from the readily available and inexpensive dimethyl acetone-1,3-dicarboxylate **4** (Scheme 8.8). The overall yield for this four-step process was improved significantly (~30%; Table 8.8) compared to IND (Scheme 8.2) and first-generation (Scheme 8.3) routes, and reduction in the material input is clearly reflected in the E-factor of these routes (Table 8.7). The raw material cost was reduced by 70% compared to the first-generation process (Scheme 8.3).

Despite still requiring four steps and a single isolation, this elegant one-pot deprotection/purification process using an acidic resin rendered high-quality aminodiol (>98%). Washing the resin

Table 8.4 Aminodiol purity profile

Identity	Lab batch	Pilot plant batch	Distilled[a]	Cryst.[b]
Aminodiol1, %	98.8	97.3	99.1	99.4
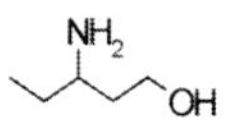				
Amino impurity, %	0.4	0.6	0.3	0.03
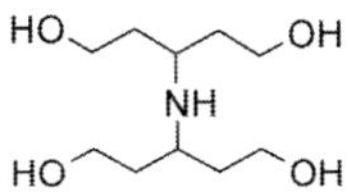				
Dimer 1, %	0.2	0.8	0.01	0.1
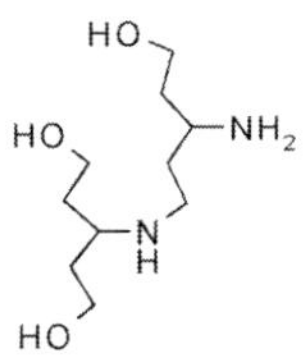	0.2	0.9	ND	0.2
Dimer 2, %				

[a] Distilledaminodiol subsequently crystallized.
[b] Crystallized from MeOH/THF.

with methanol before extracting the free amine with methanolic ammonia removed neutral impurities. Only amine-based by-products were carried forward into the aminodiol, as shown in Table 8.5.

All four chemical steps were optimized using the greener 2-Me-THF (available from renewable resources) and MeOH solvents. Optimizing all four steps with 2-Me-THF was quite challenging but enabled us to telescope the four steps. Though not demonstrated due to project termination, in principle the 2-Me-THF can be easily recycled (Table 8.6). Additionally, the one-pot catalytic reduction and *Boc*-protection of enamine **5** with the Raney Ni catalyst

Table 8.5 Comparison of solvent usage

Aminodiol	Solvent	IND synthesis	First generation	Second generation
Preferred solvents	Methanol	7.0	2.3	36.2
	2-Propanol	25.0	2.5	-
	2-Me-THF	-	-	12.6
Usable solvents	Toluene	-	7.1	-
	THF	33.6	28.6	-
Undesirable solvent	Dichloromethane	22.5	141.4	-

Table 8.6 Comparison of E-factors

Aminodiol process	Solvents (kg)	Raw materials (kg)	Total input (kg)	Product (kg)	E-factor
IND	788	139	927	8.7	106
First generation	6600	300	6900	28	246
Second generation	64	13	77	1	77

eliminated the need for stoichiometric quantities of the expensive TBAB reagent.[8]

Sodium borohydride is an inherently safer reducing agent, easily scalable, and inexpensive where the residual boric acid salts can be used elsewhere[15] without impacting the environment (landfill).

Using a single solvent, 2-Me-THF, for four chemical steps eliminated the need for energy-intensive multiple solvent exchanges, while energy-intensive multiple centrifugations/drying were avoided by the single isolation (Table 8.7).

Acknowledgments

We would like to thank our colleagues at the Roche Bioscience process group (IND synthesis) and the Roche Boulder development group (first-generation synthesis) for their initial development efforts.

Thanks to Dr. Yong Jiang, Ms. Joye Collins, and Mr. Wade Blackmon for providing analytical support. Thanks to Drs. Thomas Williamson and Stephen Chan for providing NMR and GC/mass spectrometry (MS) support, respectively. The authors would like to thank

Table 8.7 Comparison of operational process parameters

Chemistry	IND process	First generation	Second generation telescoped
Number of steps	4	6	4
Isolated intermediates	3	6	1
Overall yield	57%	63%	89%
Solvents used	MeOH, THF, CH_2Cl_2, 2-PrOH, diethylamine & propylamine	MeOH, THF, CH_2Cl_2, Toluene & 2-PrOH	MeOH& 2-Me-THF
Catalyst used	None	None	Raney Ni (50% wet)
Reducing agent	LAH	Vitride®	$NaBH_4$

Dr. Robert Huber and Mr. Brandon Shealy for developing the initial crystallization procedure and evaluation of aminodiol in the API synthesis. Thanks to Dr. Joe Veits and the Basel development group lead by Drs. Rolf Fisher and Thomas Osswald for developing a distillation procedure for aminodiol.

References

1. Anastas, P. T.; Warner, J. C. (1989). *Green Chemistry: Theory and Practice.* Oxford University Press, 30.
2. Arzeno, H. B.; Chen, J. J.; Dunn, J. P.; Goldstein, D. M.; Lim, J. A. WO 2002018379.
3. Grob, G. A.; Krasnobajew, V. (1964). *Helv. Chim. Acta.*, **8**, 2145.
4. Prugh, J. D.; Deana, A. A. (1988). *Tetrahedron Lett.*, **29**, 37.
5. Cain, R. O.; Moorlag, H.; Tucker, C. E.; Wong, J.-W. (2008). US Patent 2008154063 A1 20080626.
6. Tomori, H.; Hibutani, K.; Ogura, K. (1997). *Heterocycles*, **44**, 213.
7. Liu, Y.; Zhao, C. D.; Bergbreiter, E.; Romo, D. (1998). *J. Org. Chem.*, **63**, 3471.
8. Augustine, R. L.; Bellina, R. L. (1968). *J. Org. Chem.*, **33**, 1287.
9. The bulk commercial cost of TBAB is US$220/kg (US$20/mol).
10. Saliou, C.; Fleurant, A.; Celerier, J. P.; Lhommet, G. (1991). *Tetrahedron Lett.*, **32**, 3365.

11. For a similar approach to this problem, see Zhu, W.; Ma, D. (2003). *Org. Lett.*, **26**, 5063.
12. PENN Specialty Chemicals, Inc.www.pennakem.com/pdfs/methf5.pdf
13. Cleary, T.; Ji, Y.; Lee, L.; Rawalpally, T.; Sarma, K. (2008). *PCT Int. Appl.* WO 20080151992 A2 20081218.
14. Rawalpally, T.; Ji, Y.; Cleary, T.; Edwards, B. (2009). *Org. Process Res. Dev.*, **13**, 478.
15. It is a weak acid often used as an antiseptic, insecticide, flame retardant, etc.

Chapter 9

Development of a Robust, Environmentally Responsible Process for the Manufacture of Tofacitinib Citrate

Rajappa Vaidyanathan*
Chemical Research and Development, Pfizer, Inc., Eastern Point Road, Groton, CT, USA
rajappa.vaidyanathan@bms.com

Process development efforts that resulted in a robust, environmentally responsible, commercializable process for tofacitinib citrate will be presented in this chapter. The retrosynthetic strategy to this potent immunosuppressant/immunomodulator, synthetic approaches to key building blocks, and development of the end game will be discussed.

9.1 Introduction

There is a critical need for safer and more convenient treatments for organ transplant rejection and autoimmune disorders, such as rheumatoid arthritis. Consequently, new targets that offer selective immunosuppression/immunomodulation are of great interest. Due

*Current address: Process Research and Development, Bristol-Myers Squibb, Biocon Park, Jigani Link Road, Bommasandra IV, Bangalore 560099, India

Scalable Green Chemistry: Case Studies from the Pharmaceutical Industry
Edited by Stefan G. Koenig

ISBN 978-981-4316-49-1 (Hardcover), 978-981-4364-81-2 (eBook)
www.panstanford.com

to their role in cytokine signaling, Janus tyrosine kinases (JAKs) influence both innate and acquired immunity, amongst other functions. JAK3 and JAK1 are expressed in lymphoid cells and are involved in the signaling of multiple cytokines important for various T cell functions. Blockade of the JAK1-3/STAT pathway with a small molecule was therefore anticipated to provide therapeutic immunosuppression/immunomodulation. As a result, the Pfizer compound library was screened against JAK3, resulting in the identification of a series of inhibitors containing a pyrrolo[2,3-*d*]pyrimidine as the key pharmocophoric feature. Modifications within this chemical series led to the identification of tofacitinib citrate (**1**), which was recently approved by the Food and Drug Administration (FDA) for the treatment of rheumatoid arthritis. Tofacitinib citrate is also under development for the treatment of a variety of autoimmune diseases, such as psoriasis, and for the prevention of organ transplant rejection.[1,2]

In this chapter, chemistry efforts aimed at developing a commercializable process for this exciting compound will be discussed.

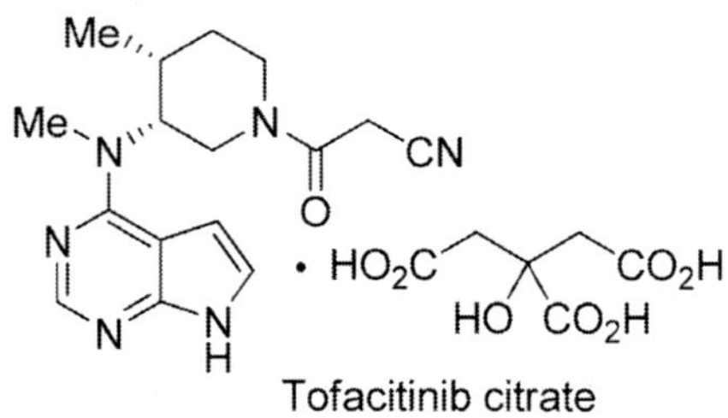

Tofacitinib citrate
1

9.2 Retrosynthetic Strategy

As can be seen in Scheme 9.1, CP-690550 (**2**), the free base form of **1**, contains a 3-methylaminopiperidine moiety attached to a pyrrolopyrimidine nucleus. Furthermore, the structure contains a methyl substituent at the 4-position of the piperidine ring and a cyanoacetamide group on the piperidine ring nitrogen. Due to the potential sensitivity of the cyanoacetamide moiety to a variety of reaction conditions, it seemed prudent to install it in the last bond-

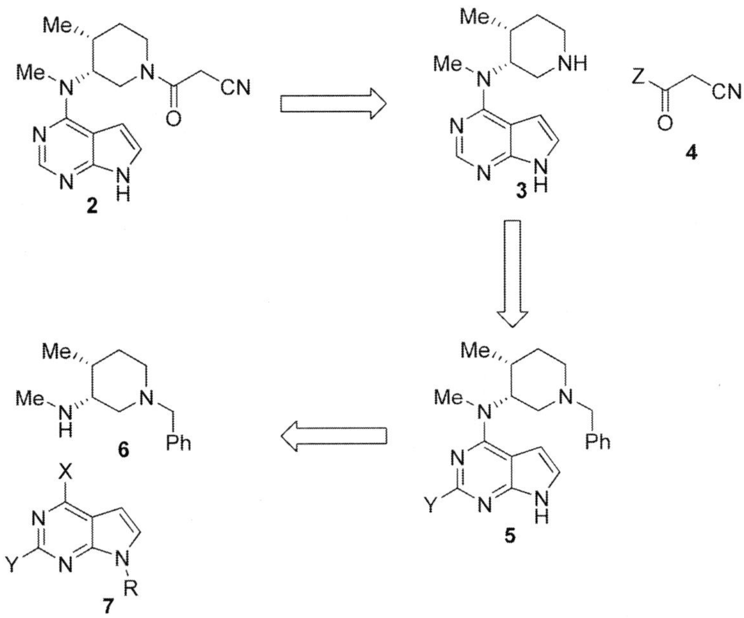

Scheme 9.1 Retrosynthesis.

forming step by treatment of amine **3** with a suitably activated form of cyanoacetic acid (**4**). Amine **3** could be formed from **5**, which, in turn, could be synthesized from piperidine **6** and an appropriately substituted/activated pyrrolopyrimidine coupling partner **7**. The rationale behind the choice of the benzyl-protecting group in **6** will become apparent later in the chapter (Table 9.1).

While several other bond disconnection approaches were considered, the strategy depicted in Scheme 9.1 was deemed the most viable and was utilized to synthesize tofacitinib citrate from discovery through the development phases. While the bond formation sequence remained virtually unchanged over the duration of the program, the reactions and manufacturing processes evolved over time. The advances that culminated in a commercializable process for tofacitinib citrate will be presented in this chapter as follows:

1. Approaches to piperidine **6**
2. Choice and synthesis of the coupling partner

3. Development of the end game (i.e., the coupling, deprotection, and amidation steps)

9.3 Approaches to Piperidine 6

Several routes were initially explored for the synthesis of headpiece **6**, and the most promising (Scheme 9.2) was scaled up to manufacture material for preclinical and early clinical studies. This route involved the benzylation of 4-methylpyridine (**8**), followed by partial reduction of the aromatic ring to produce intermediate **9**. This tetrahydropyridine was subjected to a hydroboration/oxidation sequence, followed by Parikh–Doering oxidation and reductive

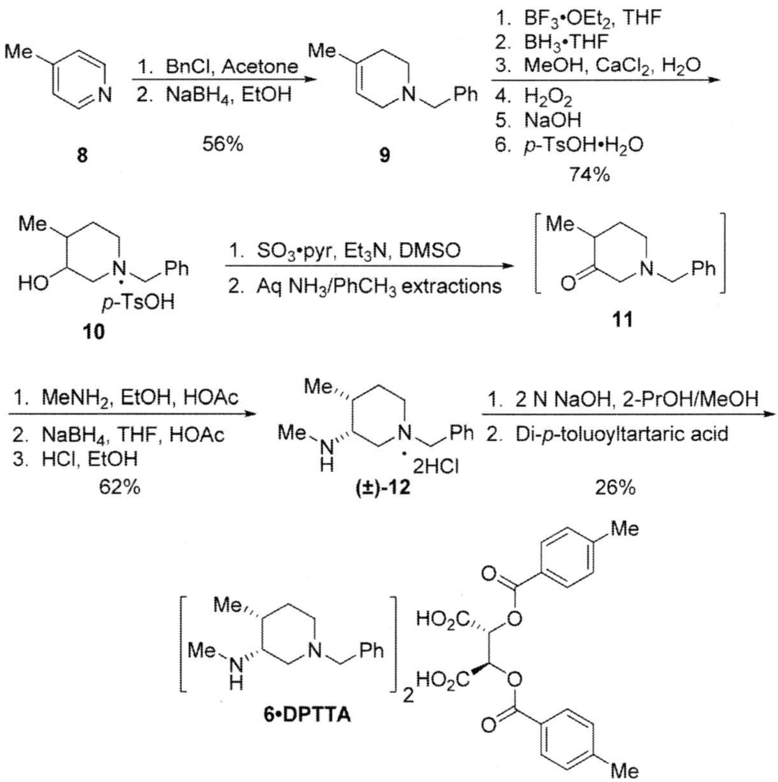

Scheme 9.2 Early scale-up route.

amination to form **12** as the racemic dihydrochloride salt. A classical resolution delivered the headpiece as its di-*p*-toluoyl-(*L*)-tartaric acid salt (**6·DPTTA**) in 6.6% overall yield from **8**.[3]

While this route furnished **6** from inexpensive, readily available starting materials in the early stages of development, the following drawbacks prompted the quest for a better synthesis: 1) the multiple manipulations to the oxidation state at the 3-position of the piperidine ring were tedious; 2) the hydroboration/oxidation sequence required the use of large volumes of solvent during the workup and isolation; 3) the Parikh–Doering oxidation resulted in a product that was contaminated with dimethyl sulfide; and 4) the overall yield from this sequence was less than ideal.

In the next-generation approach to **6**, it was envisioned that the methylamino moiety at the 3-position could be installed via reduction of a suitably functionalized pyridine such as **14**, which could be accessed from commercially available 4-methyl-3-aminopyridine **13**. This approach would circumvent some of the pitfalls of the first-generation synthesis (vide supra). Hydrogenation of **14** was expected to predominantly produce the desired *cis*-diastereomer **15**, thereby setting the relative stereochemistry at the 3- and 4-positions.

Treatment of **13** with dimethyl carbonate led to carbamate **14**, which was reduced by hydrogenation over 5% Rh/C to afford the corresponding piperidine **15** as a 17:1 mixture of *cis:trans* diastereomers. A reductive amination using benzaldehyde provided **16** as the hydrochloride salt. In the first iteration of this synthesis, the pyridine reduction was carried out using 20 wt% catalyst loading using acetic acid as the solvent. As the compound progressed through development, it became apparent that the Rh/C catalyst was one of the biggest contributors to the cost of **6**. Therefore, subsequent process development efforts focused on lowering the catalyst loading. After a systematic investigation, it was determined that the reduction could be efficiently accomplished with 3 wt% of Rh/C using HCl as an additive and methanol as the solvent. Furthermore, the use of methanol simplified the workup at the end of the reduction and led to a 15% yield enhancement in the two-step ring reduction/reductive amination sequence relative to the acetic acid process (Scheme 9.3).[4,5]

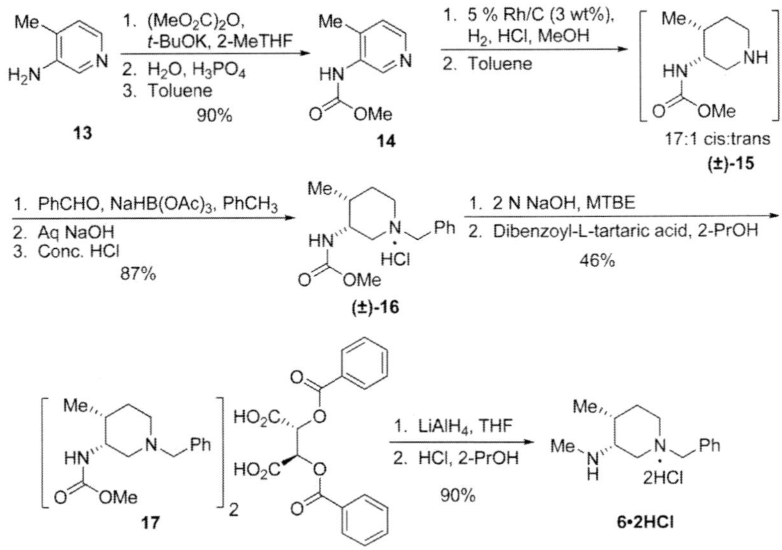

Scheme 9.3 Synthesis of chiral piperidine.

Piperidine **16** was resolved using dibenzoyl-(L)-tartaric acid, and the resulting optically enriched enantiomer **17** was reduced with $LiAlH_4$ to give **6** as its di-hydrochloride salt.

In the early stages of development, **6·DPTTA** was used as the salt form of **6**; however, further process research, as described before, led to the use of **6·2HCl** as the preferred salt form. Therefore, in the following sections, both of these salts will be used in different scenarios. The counterion simply reflects the stage of development and was found to have little bearing on the outcome of the coupling reaction with the pyrrolopyrimidine unit.

9.4 Choice and Synthesis of the Coupling Partner

In the first-generation approach to **1**, it was envisaged that an S_NAr reaction could be used to construct the key carbon-nitrogen bond between the pyrrolopyrimidine and piperidine **6**. Metal-mediated cross couplings were also considered, but the inherent advantages of S_NAr reactions (i.e., avoiding the use, and subsequent removal, of metals and ligands) appeared attractive. In its simplest manifestation, such a transformation would involve **6** and 4-chloropyrrolopyrimidine **24** as the reactants. The 4-chloropyrrolopyrimidine was synthesized from ethyl cyanoacetate (**18**) and acetal **19** using known procedures.[6] Alkylation of **18** with **19**, followed by treatment with thiourea, led to **21**. A three-step sequence involving desulfurization, cyclization, and chlorination provided **24** in acceptable yield (Scheme 9.4).

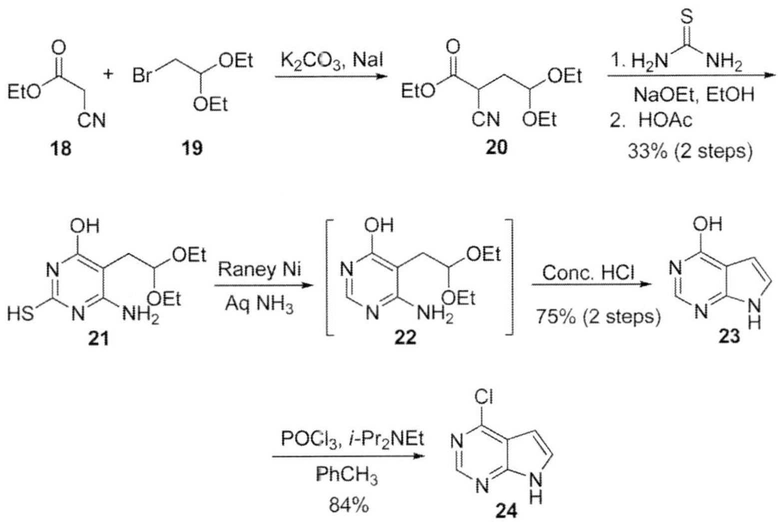

Scheme 9.4 Synthesis of coupling partner 24.

Attempts to couple **24** with **6** were met with modest success. The desired product **25** was isolated in 47% yield after prolonged heating in the presence of excess **24** (Scheme 9.5). The low yield and the need for excess **24** prompted further investigation. It was not clear whether our approach to form the key bond utilizing an S_NAr reaction was inherently flawed or whether that goal could still be achieved by a more judicious choice of coupling partners. Therefore, a few model systems were examined in an effort to understand the factors that contributed to the sluggish coupling reaction.

While the desired reaction of **6** with **24** produced the product in 47% yield after four days, the corresponding reactions of **26**

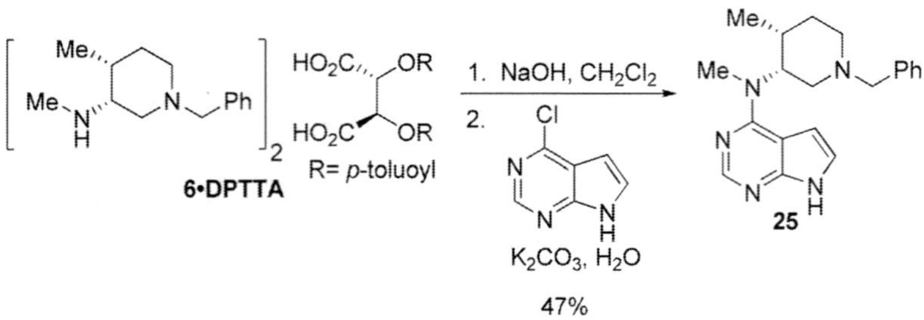

Scheme 9.5 Initial amination reaction.

and **27** with **24** proceeded to completion within six to eight hours under identical conditions. This was gratifying for two reasons: 1) it suggested that our S_NAr approach was indeed reasonable, and 2) neither the methyl substituent on the exocyclic nitrogen nor the methyl at the 4-position hampered the S_NAr reaction. Interestingly, the presence of electron-withdrawing substituents on the piperidine ring nitrogen (**28**) virtually shut down the reaction (Table 9.1).

On the basis of these results, it was hypothesized that the nucleophilicity of the exocyclic nitrogen was diminished in **6** (compared to **26** and **27**) due to the inductively electron-withdrawing nature of the ring nitrogen. If this were the case, the coupling reaction might be facilitated by enhancing either the nucleophilicity of **6** or the electrophilicity of **24**. The former appeared challenging, so we decided to explore the latter approach first.

We envisioned that the electrophile could be activated by incorporation of an electron-withdrawing substituent on the pyrrolopyrimidine. Accordingly, **24** was treated with *p*-toluenesulfonyl chloride to provide the corresponding sulfonamide **29**. When **29** was treated with **6**, the reaction proceeded to completion within eight hours and provided the product **30** in 77% yield. The sulfonamide was then removed by treatment with hot aqueous sodium hydroxide (Scheme 9.6).

The success of this strategy confirmed our hypothesis that activation of the electrophile by the incorporation of an electron-

Table 9.1 S_NAr reaction of amines with 24

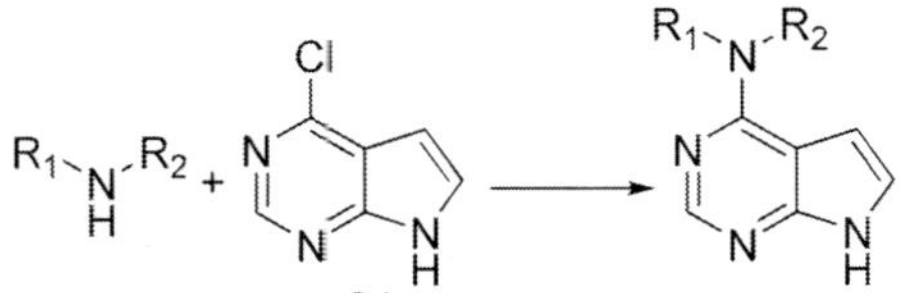

Amine substrate	Reaction time
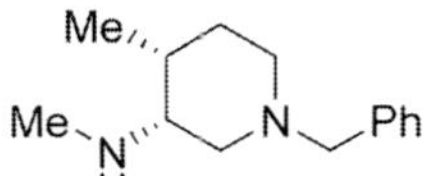 **6**	4 + days
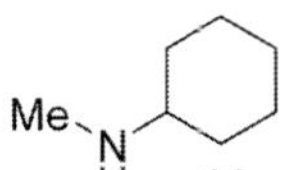 **26**	6 h
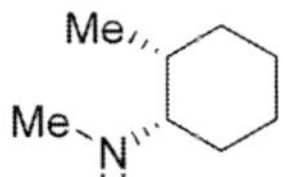 **27**	6-8 h
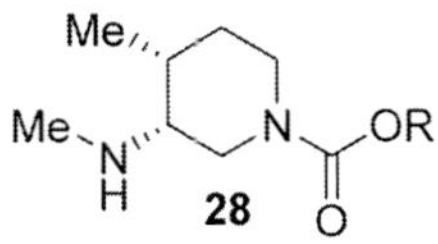 **28** R = *t*-Bu, Bn	Very Slow

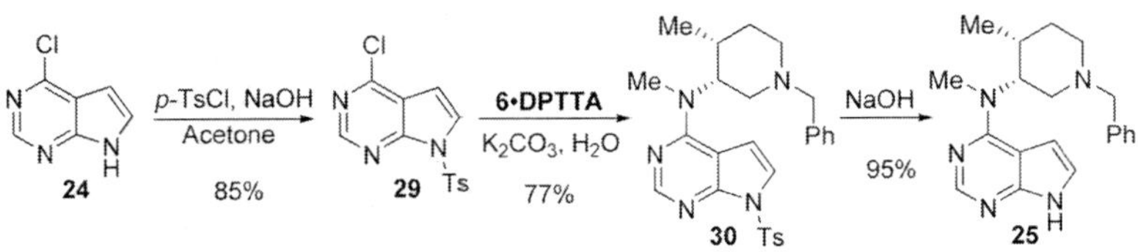

Scheme 9.6 Original coupling reaction.

withdrawing functional group would facilitate the S_NAr reaction. Furthermore, this approach allowed us to utilize our stock of **24** to rapidly provide material to support clinical demands. Nonetheless, two major areas of improvement were identified in the coupling reaction: 1) the synthesis of 4-chloropyrrolopyrimidine was cumbersome (Scheme 9.4), and 2) the sulfonamide formation/removal sequence was deemed wasteful. Hence, further efforts were aimed at identifying a better-activated pyrrolopyrimidine.

Our work thus far had proven that the coupling partner would have to be more reactive than 4-chloropyrrolopyrimidine. The product of the coupling reaction would contain a benzyl group (on the piperidine ring nitrogen) that would have to be reductively removed at some point downstream. Therefore, it seemed logical to choose a reducible activating group (for the pyrrolopyrimidine fragment) that could be removed during the subsequent debenzylation reaction. One such plausible coupling partner containing a reducible activating group was 2,4-dichloropyrrolopyrimidine (**34**). This compound was synthesized from commercially available 6-aminouracil (**31**) via condensation with chloroacetaldehyde (**32**), followed by chlorination of the intermediate 2,4-dihydroxypyrrolopyrimidine (**33**) using $POCl_3$ (Scheme 9.7).[7]

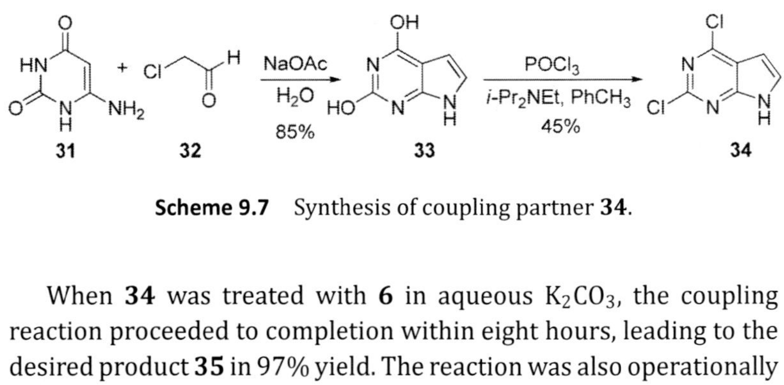

Scheme 9.7 Synthesis of coupling partner **34**.

When **34** was treated with **6** in aqueous K_2CO_3, the coupling reaction proceeded to completion within eight hours, leading to the desired product **35** in 97% yield. The reaction was also operationally simple: all of the ingredients (**6**, **34**, K_2CO_3, and water) were combined and heated to reflux. The resulting solid was isolated by filtration to provide the product.

As mentioned earlier, **6·DPTTA** was used in the early stages of development; subsequently, it was replaced with **6·2HCl**. In either case, the reaction proceeded equally well. In summary, our systematic investigation culminated in a robust process for the coupling reaction that provided the desired product (**35**) in virtually quantitative yield (Scheme 9.8).

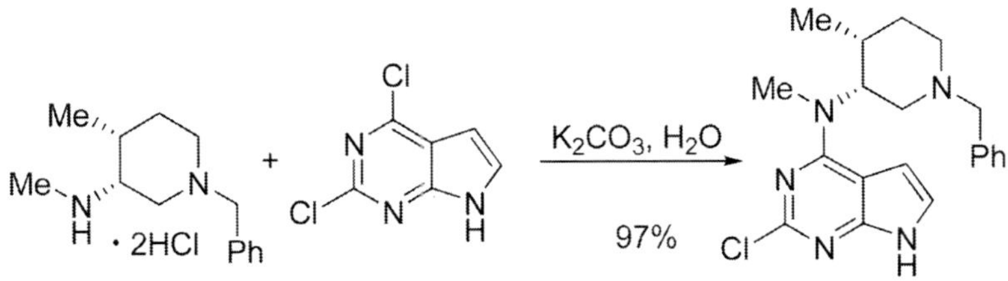

6•2HCl 34 35

Scheme 9.8 Robust coupling reaction.

9.5 The Debenzylation Step

In the initial approaches to tofacitinib citrate, 4-chloropyrrolopyrimidine (**24**) and its tosylated version (**29**) were utilized as the coupling partners, and consequently, both approaches shared the same substrate for the debenzylation step (**25**). With this substrate, the debenzylation was effected by hydrogenation over 20 wt% of 20% $Pd(OH)_2/C$ in a mixture of 2-propanol and acetic acid, providing **3** in 84% isolated yield (Scheme 9.9).

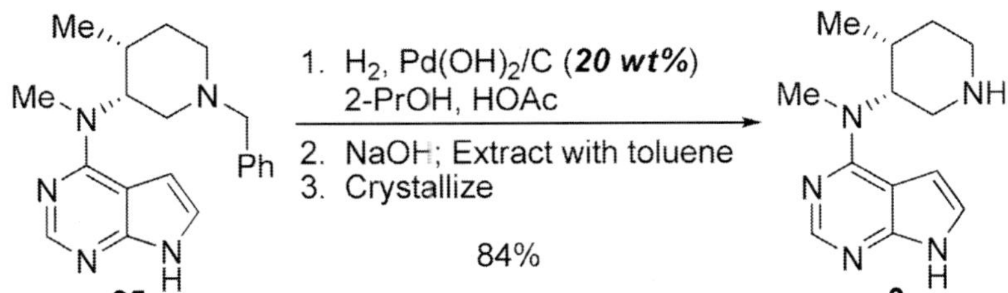

25 3

Scheme 9.9 Early piperidine debenzylation.

When the coupling partner was changed to **34**, it was found that the debenzylation could be carried out in water using 10 wt% of 20% $Pd(OH)_2/C$. Since 1 equiv. of HCl was liberated in the reaction (due to reduction of the chloro group in **35**), the product of the hydrogenation was actually the hydrochloride salt of **3**, which was soluble in water (Scheme 9.10). This was then carried on to the amidation reaction (vide infra).

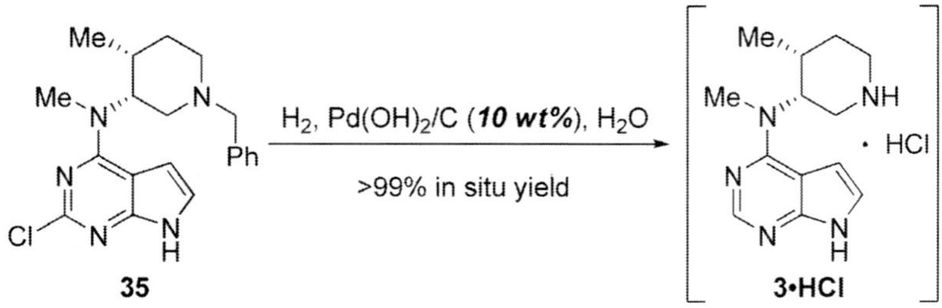

Scheme 9.10 Modified piperidine debenzylation.

The debenzylation itself worked very smoothly under these conditions. However, the reaction mixture was initially a thick slurry, with the solids going into solution as the reaction progressed, presumably due to formation of hydrochloride salts. It was suspected that the thick slurry could cause complications during scale-up (mixing issues), and hence options to mitigate such risks were explored. We postulated that the addition of HCl at the beginning of the reaction would increase the solubility of the starting material and consequently circumvent the thick slurry phase. Interestingly, the addition of 1 equiv. of HCl at the beginning of the reaction not only improved the physical properties of the reaction mixture but also provided a dramatic rate enhancement, to the point that the catalyst loading could be lowered to about 1 wt%. With these subtle, yet significant process modifications, the product was isolated as an aqueous solution of its dihydrochloride salt (**3·2HCl**) in >99% yield (Scheme 9.11).

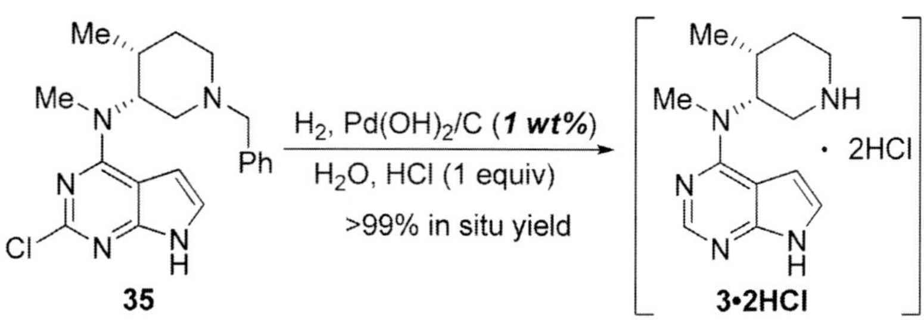

Scheme 9.11 Final piperidine debenzylation.

9.6 Amidation, Salt Formation, and Isolation

This step can be divided into three distinct parts: the workup of the reaction mixture posthydrogenation, the amidation reaction, and isolation of the active pharmaceutical ingredient (API) as the citrate salt. The early iterations of the final step will be briefly discussed, followed by a description of development efforts that addressed the program's long-term needs and resulted in a commercializable process. Past experience had shown that isolation of free base **2** was not a viable option on a large scale. Further, since the first two steps were carried out under aqueous conditions, it was imperative that the final transformation and isolation be designed to produce an API that would meet its acceptance criteria. The amidation, therefore, was the most crucial step in the process and, as it turned out, was mechanistically the most interesting one as well. Consequently, this step will be discussed in greater detail compared to the rest of the synthesis.

In the early stages, the substrate for the amidation reaction was isolated **3**. This compound was treated with **38** (obtained by reaction of cyanoacetic acid **37** with *N*-hydroxysuccinimide **36** and dicyclohexylcarbodiimide [DCC]), followed by salt formation to afford crude **1**. The product was then recrystallized from aqueous ethanol to provide tofacitinib citrate (**1**, Scheme 9.12).

Several other amidation reactions were explored, but most of the conditions led to either incomplete reactions or complex mixtures of products. While this was a bit frustrating, it was not necessarily unexpected, given the fact that the acylating agent was an activated form of cyanoacetic acid. Deprotonation at the activated methylene of the acylating agent would lead to decomposition via the corresponding ketene. In essence, most groups that activated the carboxyl group toward nucleophilic attack by the amine would also increase the acidity of the activated methylene and promote decomposition via a ketene intermediate. It was important to strike a balance between these two competing pathways.

One of the few conditions that worked satisfactorily was amidation using mixed anhydride **40** derived from cyanoacetic acid **37** and pivaloyl chloride **39** (Scheme 9.13).[8] The reaction

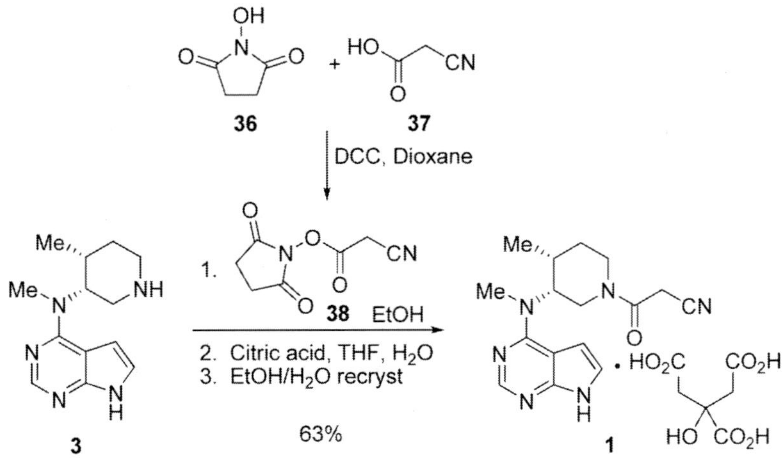

Scheme 9.12 Early piperidine amidation.

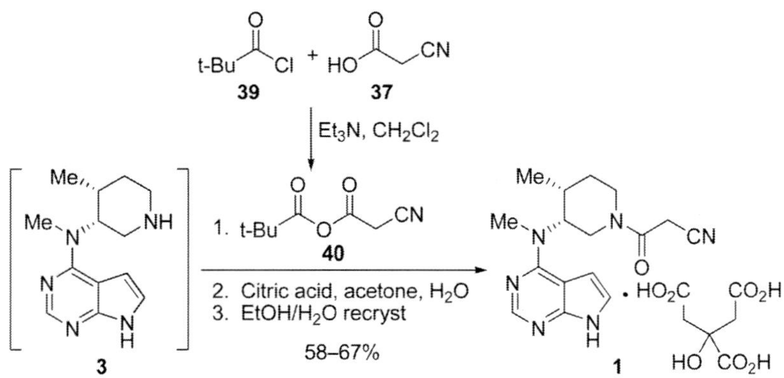

Scheme 9.13 Final piperidine amidation.

mixture after the hydrogenation (an aqueous solution of **3·HCl**) was basified, extracted into methylene chloride, and treated with mixed anhydride **40**. Upon reaction completion, the mixture was treated with citric acid to provide crude **1**. Recrystallization from aqueous ethanol furnished pure **1**.

The main limitations of the two amidation approaches described in Schemes 9.12 and 9.13 were as follows: 1) in the first case, the isolated crude product was contaminated with residues from the coupling reagents that necessitated recrystallization; 2) in the case of the transformation depicted in Scheme 9.13, the reaction of **3** with residual pivaloyl chloride led to the formation of the *t*-butyl amide of **3**, which could be removed only by recrystallization; 3) the fact that these amidations were stepwise activation/amidation protocols made it imperative that reliable in-process control tests be established to confirm the complete formation of the activated species prior to introduction of the amine into the reactor; 4) the solvents utilized for the activation/amidation reactions (methylene chloride and dioxane) were less than desirable for a commercial process, and there was a strong need to replace them with more acceptable alternatives;[9] and 5) the yield in the final step needed improvement.

As we moved toward a long-term synthetic solution for the manufacture of **1**, several factors were considered: 1) a single-step process would be preferred over a two-step activation/amidation sequence, both for process efficiency and to avoid decomposition of the activated cyanoacetic acid, and 2) it would be preferable to use a single solvent for the extractive workup (posthydrogenation), amidation, and salt formation. Such a solvent would have to be environmentally benign and polar enough to extract **3** from an aqueous phase.

An extraction screen was initiated to determine the best solvent for the extractive workup of the reaction mixture posthydrogenation. From this screen, 1-butanol emerged as the solvent of choice for this operation. 1-Butanol is an excellent, but rarely used, solvent for extractive workups involving polar organic compounds. Water has ~18% solubility in 1-butanol at room temperature (i.e., the organic phase is ~18% water in 1-butanol), and the phase separations are generally clean (and can be improved by performing the extractions at elevated temperatures). Furthermore, the water in the organic layer can be effectively removed by distillation (the azeotrope contains 43% water and boils at ~93°C).[10] Amine **3** had excellent solubility in 1-butanol and could be extracted almost quantitatively into the organic layer via phase separation at high pH. The organic layer was washed with water to remove any salts and then distilled to provide a solution of **3** in 1-butanol.

While the workup was being streamlined, we were simultaneously examining other options for carrying out the amidation. One of the most efficient and atom-economical options for amide formation is the reaction of an amine with a simple ester. We envisioned that amine **3** could be converted to cyanoacetamide **2** by direct treatment with an alkyl cyanoacetate. Accordingly, a solution of **3** in 1-butanol was treated with ethyl cyanoacetate (**18**) and ethylene glycol at 117°C to provide free base **2**.[11] The reaction reached ~80% conversion within the first six hours but required prolonged heating to go to completion. Not surprisingly, this led to substantial degradation of the product and resulted in an isolated overall yield of ~60% after citrate salt formation and recrystallization. The overall yield was similar to the activation/amidation protocols, but the fact that the amidation proceeded using ethyl cyanoacetate was gratifying and prompted further exploration.

At this point, we faced a conundrum: the reaction had to be performed at elevated temperature to push it to completion; at the same time, lower temperatures were required in order to minimize thermal degradation of the product. As we examined the likely causes of the stalled reaction, we surmised that complex equilibria and side reactions were operative. For example, the activated methylene groups of ethyl cyanoacetate and the product could be deprotonated (presumably by amine **3**) and subsequently undergo Thorpe–Ziegler-type reactions. These side reactions would consume the ethyl cyanoacetate and render it unavailable for the desired reaction. Since the reaction was carried out in 1-butanol, transesterification of ethyl cyanoacetate was also likely. The relative rates of these competing reaction pathways, and the equilibria involved, would impact product formation (and degradation).

It was hypothesized that the desired reaction could be promoted by either base or acid catalysis. Several "acid" catalysts were screened, yet none appeared to enhance the reaction rate. The reaction of **3** with ethyl cyanoacetate (**18**) was examined at 55°C with various bases in 1-butanol.[12] In this screen, 1,8-diazabicyclo[5.4.0]undec-7-ene (DBU, **41**) afforded 26% conversion after 30 minutes, while other bases provided only 1–6% conversion. After 18 hours at 55°C, all reactions showed some decomposition. A marginal rate enhancement was observed in the presence of

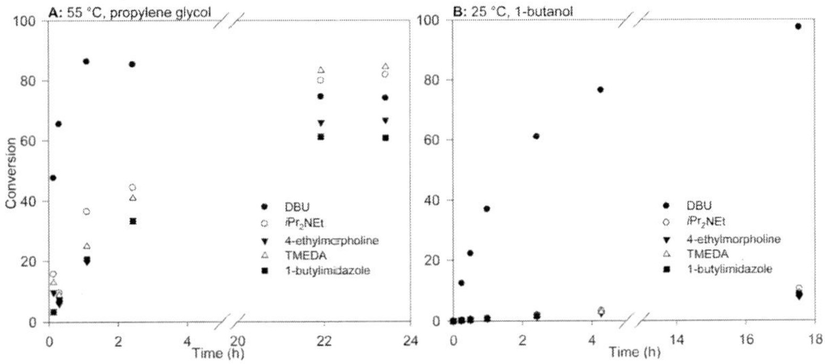

Figure 9.1 Reaction of 3 with 18 in the presence of various bases in (A) propylene glycol at 55°C and (B) 1-butanol at 25°C.

Hunig's base (*N,N*-di-*iso*-propylethylamine) in propylene glycol as the solvent. This led to a systematic screen of amine bases in propylene glycol at 55°C (this temperature was chosen to strike a reasonable balance between the enhanced reaction rate and decreased thermal decomposition). A cursory glance at the data after 24 hours revealed that all of the reactions had progressed to approximately the same extent (60–80%).

A closer look at the data indicated that the reaction with DBU had proceeded to about 90% conversion within the first 30 minutes, and the product degraded slowly over time from then onward (Fig. 9.1A). This lead was instructive and exciting—instructive because it reinforced the importance of kinetic profiling of reactions (i.e., monitoring conversion/progress over time rather than at a single time point) and exciting because it suggested that the reaction could perhaps be carried out at a lower temperature to provide complete conversion while minimizing decomposition.

When the base screen was repeated at ambient temperature in 1-butanol (the solvent of choice based on the extractive workup development) to minimize decomposition, the DBU-promoted reaction was still much faster than those using other bases and led to high conversion (Fig. 9.1B). After further refinement, the process was streamlined to effect clean conversion of **3** to **2** within 12 hours, using 1 equiv. DBU at 25°C. The scope of this DBU-catalyzed amidation was expanded to include other amine nucleophiles,[13] and acyl imidazole electrophiles.[14] On the basis of our work and other literature evidence,[15] it has been postulated that the reaction proceeds via the pathway outlined in Scheme 9.14.

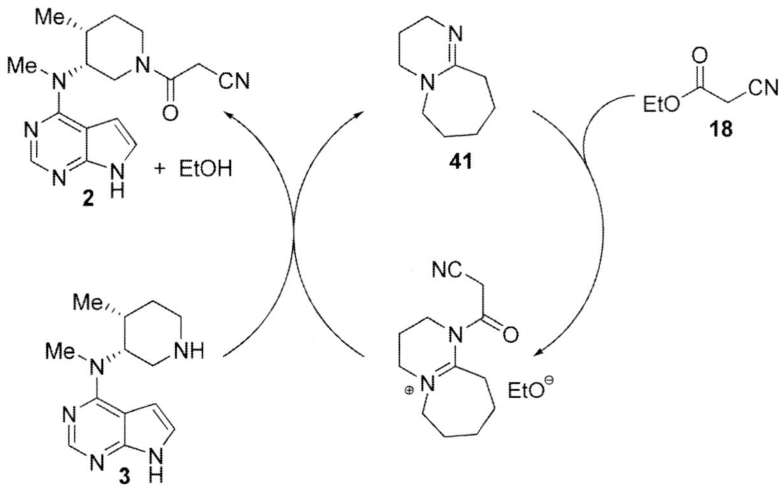

Scheme 9.14 Postulated reaction pathway.

After the amidation reaction was complete, citric acid and aqueous acetonitrile were added to the reaction mixture to form tofacitinib citrate. The solvent system for the final isolation was chosen after careful experimentation and provided material that met the acceptance criteria for tofacitinib citrate. The overall yield from the hydrogenation step onward (i.e., hydrogenation, workup, amidation, salt formation) was ~85% on a multikilogram scale.

In summary, the final process for the manufacture of tofacitinib citrate is depicted in Scheme 9.15. The two starting materials, **6·2HCl** and **34**, were treated with K_2CO_3 in water to produce the coupled product **35**, which was isolated. Compound **35** was hydrogenated over Pearlman's catalyst in water to give piperidine **3** as its dihydrochloride salt. An extractive workup using 1-butanol, followed by azeotropic distillation, provided a solution of **3** in 1-butanol. This was subjected to the amidation reaction using ethyl cyanoacetate and DBU. Addition of citric acid and aqueous acetonitrile furnished tofacitinib citrate.

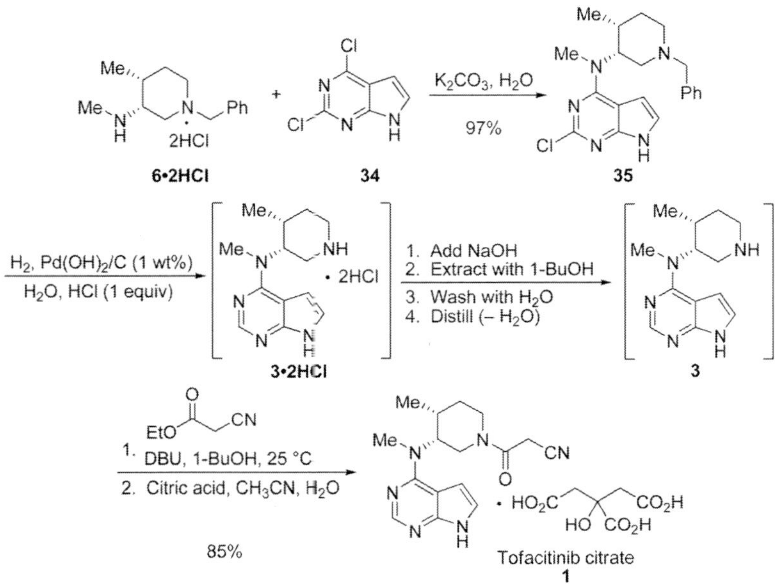

Scheme 9.15 Final manufacturing process.

9.7 Conclusions

A robust, commercializable three-step process was developed for the manufacture of tofacitinib citrate. The synthetic routes to the two starting materials and the API were iteratively improved to increase efficiency, while minimizing cost and environmental impact.

In the ultimate process, the first two steps to manufacture tofacitinib citrate are carried out in water, while the third step involves a novel DBU-catalyzed ester to amide transformation at ambient temperature. The process includes two isolations (the intermediate at the end of the first step and the API), with both being "direct drop" isolations. Steps 2 and 3 are telescoped, increasing overall efficiency, and the undesired solvents, methylene chloride and dioxane, have been successfully replaced.

We believe that this process exemplifies Pfizer's continued commitment to developing safe and efficient chemical processes that have a minimal impact on the environment, adhering to the established principles of green chemistry.

Acknowledgments

The success of projects of this magnitude hinges on unstinted engagement and dedication from all team members. While the list of colleagues who contributed to this project is too long to be included here, it is my pleasure to acknowledge the individuals who were most intimately involved with the work described. This exciting drug candidate was discovered by Mark Flanagan *et al.* at Pfizer's Groton Research Laboratories. Sally Gut Ruggeri spearheaded the early process development efforts. David Ripin, Frank Busch, Nathan Ide, Frank Urban, Marcus Ewing, Stephen Hubbs, Teresa Makowski, Dennis Bourassa, Jian Jin, and Tim Norris made significant contributions to the synthesis of the two starting materials. Brett Lillie, Robert McLaughlin, Timothy White, Kevin Hettenbach, Brian Chekal, and Peter Rose were the key researchers involved in the end-game development work. Jason Mustakis and Kevin Doyle led the engineering and analytical disciplines, respectively, while Joel Hawkins and Kristin Price provided reaction screening support. The Sandwich pilot plant team headed by James Long deserves special mention for executing multiple campaigns in its facility. Stéphane Caron, Juan Colberg, and Robert Dugger provided strategic input at different stages of this project. A special word of thanks to Nathan Ide for helpful suggestions during the preparation of this manuscript.

References

1. Flanagan, M. E.; Blumenkopf, T. A.; Brissette, W. H.; Brown, M. F.; Casavant, J. M.; Chang, S.-P.; Doty, J. L.; Elliott, E. A.; Fisher, M. B.; Hines, M.; Kent, C.; Kudlacz, E. M.; Lillie, B. M.; Magnuson, K. S.; McCurdy, S. P.; Munchhof, M. J.; Perry, B. D.; Sawyer, P. S.; Strelevitz, T. J.; Subramanyam, C.; Sun, J.; Whipple, D. A.; Changelian, P. S. (2010). *J. Med. Chem.*, **53**, 8468.
2. Changelian, P. S.; Flanagan, M. E.; Ball, D. J.; Kent, C. R.; Magnuson, K. S.; Martin, W. H.; Rizzuti, B. J.; Sawyer, P. S.; Perry, B. D.; Brissette, W. H.; McCurdy, S. P.; Kudlacz, E. M.; Conklyn, M. J.; Elliott, E. A.; Koslov, E. R.; Fisher, M. B.; Strelevitz, T. J.; Yoon, K.; Whipple, D. A.; Sun, J.; Munchhof, M. J.; Doty, J. L.; Casavant, J. M.; Blumenkopf, T. A.; Hines, M.; Brown, M.

F.; Lillie, B. M.; Subramanyam, C.; Chang, S.-P.; Milici, A. J.; Beckius, G. E.; Moyer, J. D.; Su, C.; Woodworth, T. G.; Gaweco, A. S.; Beals, C. R.; Littman, B. H.; Fisher, D. A.; Smith, J. F.; Zagouras, P.; Magna, H. A.; Saltarelli, M. J.; Johnson, K. S.; Nelms, L. F.; Des Etages, S. G.; Hayes, L. S.; Kawabata, T. T.; Finco-Kent, D.; Baker, D. L.; Larson, M.; Si, M.-S.; Paniagua, R.; Higgins, J.; Holm, B.; Reitz, B.; Zhou, Y.-J.; Morris, R. E.; O'Shea, J. J.; Borie, D. C. (2003). *Science*, **302**, 875.

3. Ripin, D. H. B.; Abele, S.; Cai, W.; Blumenkopf, T.; Casavant, J. M.; Doty, J. L.; Flanagan, M.; Koecher, C.; Laue, K. W.; McCarthy, K.; Meltz, C.; Munchhoff, M.; Pouwer, K.; Shah, B.; Sun, J.; Teixeira, J.; Vries, T.; Whipple, D. A.; Wilcox, G. (2003). *Org. Process Res. Dev.*, **7**, 115.
4. Cai, W.; Colony, J. L.; Frost, H.; Hudspeth, J. P.; Kendall, P. M.; Krishnan, A. M.; Makowski, T.; Mazur, D. J.; Phillips, J.; Ripin, D. H. B.; Ruggeri, S. G.; Stearns, J. F.; White, T. D. (2005). *Org. Process Res. Dev.*, **9**, 51.
5. Busch, F. R.; Mustakis, J.; Makowski, T. M.; Lucas, J.; Hawkins, J. M.; Bourassa, D.; Jin, J. Unpublished results.
6. Davoll, J. (1960). *J. Chem.* Soc., 131.
7. (a) West, R. A.; Ledich, K.; Hitchings, G. (1959). British Patent 812366; (b) West, R. A.; Ledich, K.; Hitchings, G. (1960). *Chem. Abstr.*, **54,** 5921.
8. Ruggeri, S. G.; Hawkins, J. M.; Makowski, T. M.; Rutherford, J. L.; Urban, F. J. (2007). WO 2007/012953 A2.
9. Alfonsi, K.; Colberg, J.; Dunn, P. J.; Fevig, T.; Jennings, S.; Johnson, T. A.; Kleine, H. P.; Knight, C.; Nagy, M. A.; Perry, D. A.; Stefaniak, M. (2008). *Green Chem.*, **10**, 31.
10. McConville, F. X. (2002). *The Pilot Plant Real Book*. Worcester, MA: FXM Engineering and Design.
11. Ashford, S. W.; Henegar, K. E.; Anderson, A. M.; Wuts, P. G. M. (2002). *J. Org. Chem.*, **67**, 7147.
12. This screen was run in the Conjure Flow System, Accendo Corporation, Tucson, AZ.
13. Price, K. E.; Larrivée-Aboussafy, C.; Lillie, B. M.; McLaughlin, R. W.; Mustakis, J.; Hettenbach, K. W.; Hawkins, J. M.; Vaidyanathan, R. (2009). *Org. Lett.*, **11**, 2003.
14. Larrivée-Aboussafy, C.; Jones, B. P.; Price, K. E.; Hardink, M. A.; McLaughlin, R. W.; Lillie, B. M.; Hawkins, J. M.; Vaidyanathan, R. (2010). *Org. Lett.*, **12**, 324.
15. Kiesewetter, M. K.; Scholten, M. D.; Kirn, N.; Weber, R. L.; Hedrick, J. L.; Waymouth, R. M. (2009). *J. Org. Chem.*, **74**, 9490.

Chapter 10

Selective Nitration under cGMP Conditions

Peter Poechlauer, Sascha Braune, and Roland Obermüller
DSM Fine Chemicals Austria, St.-Peter-Str. 25, A-4021 Linz, Austria
peter.poechlauer@dsm.com, sascha.braune@dsm.com,
and roland.obermueller@dsm.com

10.1 Introduction: Pharmaceuticals—Successful Products, Inefficient Manufacture?

Among producers of chemicals, the pharmaceutical industry holds a position that is special in several ways: arguably, the pharmaceutical industry is among the most innovative branches when it comes to inventing and developing new chemical lead structures, as well as embodying new therapeutic approaches and concepts. For the past 50 years this industry has developed cures for the large majority of widespread conditions, with few exceptions still waiting to find adequate forms of treatment.

Consequently, the innovative power of this industry has been judged by its ability to develop ever-better, more selective medicines with fewer side effects and more convenient forms of application.

Scalable Green Chemistry: Case Studies from the Pharmaceutical Industry
Edited by Stefan G. Koenig

ISBN 978-981-4316-49-1 (Hardcover), 978-981-4364-81-2 (eBook)
www.panstanford.com

The patient's response to a treatment, key figures such as drug tolerance, and "survival rates" or "progression-free survival rates" have been the near-exclusive drivers of improvement. With respect to these key figures, the pharmaceutical industry has made impressive progress, saving and improving the lives of innumerous people. Only lately, with improvements in many fields dropping to small increments—albeit still on a high level—further parameters related to the development, production, and application of medicines have come into the focus of public discussion.

First and foremost, the discussion around health care costs has centered on the price of pharmaceuticals.[1] The industry has tried to show that high prices are justified by expensive clinical trials for new drugs. Unfortunately, highly publicized cases of perceived marginal improvements (e.g., in the *Nexium*® *v. Prilosec*® case[2]) or insufficient clinical testing, with consequences such as withdrawal of drugs like Vioxx®,[3] belie the industry's overall efforts. The oversimplified rationale that "the smaller the benefit of a new drug over existing drugs, the higher the effort to show this benefit in the clinic" (e.g. in the *Plavix*® *v. Aspirin*® case[4]) has heated up the discussion around pharmaceutical pricing.

The focus of this discussion has, nota bene, always been on the price of originator pharmaceuticals. Generics are seen as a way to keep health care costs at bay. If one accepts the large price difference between an original drug and its generic equivalent as justified by costs of research and clinical trials, one should also think that at the time a drug goes generic its research costs should have been easily paid. Nevertheless the originator of a drug hardly ever lowers the price of the prescription to the level of a typical generic equivalent.[5] The reason for this might be the production process or the bet that the established brand holds great cachet with customers.

Second, a discussion around production safety and production efficiency has now been initiated. Spectacular incidents linked to pharma production, while they do occur,[6] are scarcely referenced. In the public discussion they are usually dwarfed by incidents in chemical sectors handling larger volumes of materials, such as refineries. However, when it comes to pollution of soil, groundwater, etc., pharmaceutical companies have been named as polluters, for example, several pharmaceutical companies operating in Puerto

Rico were recently named on the US Environmental Protection Agency's (EPA) list of the top 10 toxic chemical releasers in Puerto Rico for 1994.[7] In many cases, it is rather the toxicity or at least the unpredictable effect of the spill released to the environment than its absolute amount that causes public concern.

The multitude of effluents and their frequently hazardous properties again reflect the employed production methods of active pharmaceutical ingredients (APIs). At first sight, these processes are enormously complicated and tedious. A closer look shows that many of them are also rather wasteful. An analysis of many of these processes shows that significant improvements of process yield and waste balance are feasible.[3]

10.2 Pharmaceutical Production

10.2.1 *Identification of Leads*

Given the focus of the pharmaceutical industry on finding new therapeutic concepts, this industry has improved its capabilities and capacities in this part of the development trajectory considerably over the past 30 years. After attempts at "rational drug design" in the 1980s, the focus has been on defining lead structures, then creating a large number of structural variations of such lead structures at maximum efficiency, and testing their effect in a screen modeling the conditions in an organism.

Presently, tens of thousands of structures are produced, selected, and discarded or else taken to the next phase in a highly optimized procedure. Methods of combinatorial synthesis are employed to generate the required structural variations.[9] The chemical transformations employed have to be tolerant of structurally diverse substrates and reagents. The synthesis of a successful compound entering clinical trials has many features initially required to create diversity on a very small scale. Consequently, it is hardly ever the best way to synthesize a potentially successful compound on a larger scale. Nevertheless, even a clumsy synthesis does not mean a bottleneck or a roadblock to continued, early-stage pharmaceutical development. Almost every milligram-scale synthesis will, with acceptable effort, also deliver single kilogram amounts of a desired substance.

10.2.2 *Poor Cousin Process Development*

Little focus has been placed on finding highly efficient syntheses: in the early phases of drug development, the required amounts are seen as too low and the risk of failure of a desired compound too high to justify the development of a robust, material- and energy-efficient synthesis. As a successful compound enters clinical trials, its volume demand rises and its synthesis route is fixed. Compared to the whole development process very little time and resources are dedicated to optimize the route: the beneficial effect of a well-developed API synthesis to the overall profit or loss of a pharmaceutical development is much smaller than any delay in the launch date or a potential change in product quality. The chemicals and the route chosen for a manufacturing process are fixed at the beginning of an API's lifetime. A change in the chemistry requires re-registration with authorities, which is costly and may need additional clinical testing. Therefore, originators of APIs have hardly ever changed the chemical route of manufacturing. As the patent life of an API comes to an end, an industry with a different cost structure and competencies takes the now generic API to market—frequently via a much simpler route and at a fraction of the price.[10] Frequently, this ends the originator's business based on this API.

10.2.3 *A Need to Change toward Sustainability: Well Recognized*

A very general definition of sustainability or sustainable development is the definition by the World Commission on Environment and Development: "forms of progress that meet the needs of the present without compromising the ability of future generations to meet their needs." However, a manufacturer will typically strive only to "meet the needs of the present" and do this in a way that will keep the organization alive and in business—by staying competitive.

The above themes—the price of pharmaceuticals, production safety, waste, and noncompetitive syntheses—show that pharmaceutical manufacturing has to change to become more sustainable. Both route selection and the process technologies employed in API synthesis can contribute a lot to reach this goal.

10.3 Process Intensification

10.3.1 *Basic Principles*

A. Stankiewicz gives a definition of the term "process intensification,[11]" referring to a "dramatic reduction in the size of a plant at a given production volume," with obvious benefits, such as increased safety. He derives three categories of methods to intensify a process:[12]

- Novel processing methods (integration of reactions in multifunctional reactors)
- Use of alternative forms and sources of energy
- Novel methods of process/plant development and operation

These methods are applications of a more general, underlying principle: "Everywhere and at any time provide ideal conditions to a chemical reaction and allow it to proceed as fast as reasonable." The phrase "everywhere and at any time" refers to the reaction mixture: every volume element of the reaction mixture has to have the same stoichiometry and the same processing history. The term "ideal conditions" refers to parameters such as temperature and stoichiometry of reactants leading to optimal use of starting materials, energy, and time. Limitations caused by energy and mass transport have to be minimized. As both are linked to transport distances, these distances have to shrink.

"As fast as reasonable" refers to the reaction rate. Almost every reaction rate increases as the temperature increases, but frequently selectivity suffers, as subtle differences in activation energy allowing a reagent to select one reaction pathway over a different one vanish. So there will be a maximum temperature, above which an intolerable proportion of a by-product appears. "As fast as reasonable" may also relate to a reaction rate that allows suppressing unwanted follow-up reactions, for example, by removing the product from the reaction mixture. Ultimately, the "dramatic reduction in the size of a plant at a given production volume," quoted by Stankiewicz, is best achieved by increasing the rate and concentration of a chemical reaction.

10.3.2 *How to Implement an Intensified Process in an Existing Plant*

The above view of a chemical process implies that we start with an analysis of the reaction and define the required properties of the plant, instead of slowing down and diluting the reaction, etc., to fit it into an existing multipurpose installation. Modifying the reaction conditions will not leave the existing plant idle. Usually the first, "hot" phase of a reaction profits most from process intensification because the reaction must be slowed down significantly to permit processing in a batch reactor. Any stirred tank can handle slow phases of the reaction (until it goes to completion) very well where the reaction mixture is already perfectly mixed and needs or develops little heat. The overall productivity of the plant (its daily output) will increase, so campaign times required to produce a certain amount of product will shrink. Further increases in throughput are easily possible by adding further stirred tanks to build a continuous stirred tank cascade, with the additional benefits of reduced back-mixing.

10.3.3 *Plant Reconfiguration*

Every multipurpose plant is a functionally logical arrangement of batch reactors, vessels, tanks, distillation columns, crystallizers, filters, dryers, and similar equipment. A major task of manufacturers using multipurpose plants is to find a way to reconfigure such plants to run intensified processes in a short period of time. The investment costs for the new equipment must be minimized to allow for acceptable payback time in a business environment of frequently changing products, many of which are short lived. Therefore it is important to add changes to the existing installation in a "minimally invasive" manner. Among the simplest ways to intensify a process is to identify the operation with the largest potential for reduced production costs (saving raw materials or reducing waste or energy input) and focus on improving this operation.

10.3.4 *General Safety Considerations*

To fulfill the safety requirements for performing dangerous processes in a multipurpose plant, where two, three, or even more processes run in parallel, it is advantageous to keep the holdup of the equipment small. First of all, if a certain space in a plant is available, one can run more processes in parallel with smaller equipment. And, in the case of an incident, which theoretically should never occur following a hazard and operability (HAZOP) study or other safety analyses but occasionally does, the impact of such an incident is minimized as the concerned volume is minimal. The shift from a large scale to "intensified" equipment is only possible with a change in production methodology. One way to effect such a change is by switching to continuous processing. In a simple example, let us presume that we typically work with a stirred reactor capable of processing 10 m^3 of a reaction mixture, and we have a reaction time of three hours. Therefore, we can convert 10 m^3/3 h, or 3,33 m^3/h, of the reaction mixture, not taking into account the time required for feeding starting materials, taking the reactor to the desired temperature, dosing reagents, and transferring the reactor contents to the next operation that might be performed in a different device.

Any continuous process will obviously have to be at least as productive as the batch process. In fact, it has to be significantly better or nobody would switch from an existing and well-known process concept to a "new" one. Therefore we use the opportunity to take the reaction closer to its chemical limits, for example, by increasing the reaction rate through running the reaction at higher temperatures. Let us assume that as a rule of thumb an increase in temperature by 10 K doubles the reaction rate.[a] In our example, the limited heat transfer in the 10 m^3 reactor batch forces us to run the process at a temperature below 5°C to keep it under control. If we

[a]This assumption is derived from the equation $k(T) \sim \exp\,(-\Delta H^{\#}/(R^{*}T)$, where $\Delta H^{\#}$ is the activation enthalpy, R the gas constant, and T the temperature in Kelvin.

can operate the process in suitable equipment at 45°C (40°C hotter), we increase the reaction rate by a factor of $2^4 = 16$. The reaction time of 3 hours is reduced to about 11 minutes. We skip the slow dosing and increase other process parameters, such as concentration, to end up with a reaction time of 1 minute. A continuous reactor that can handle an equivalent throughput can be easily designed.

People tend to claim that a small, structured, and continuously operated reactor is safe equipment by definition. This is definitely not the case. Even in a microreactor, a sufficient amount of hazardous chemicals or reaction mixture might be present to cause detonation, leading to rupture of the device in which it takes place. However, an incident leading to a spill from a microreactor will have a significantly smaller impact on the environment than an incident in a large-scale reactor, if only the reactor holdup is concerned. Therefore any safety concept has to contain a safe shutdown of chemical supply in the case of rupture, as identified either by a difference in flows entering and exiting the device or by a liquid sensor placed below it.

The equipment and infrastructure (pumps, sensors) needed to run a small, structured, and continuously operated reactor pose a further risk in the case of failure during continuous operation. The keyword in this context is "backflow." If we operate a classical batch reactor it is usually quite simple to avoid any backflow or unwanted mixing of chemicals by constructive measures: one starting material is dosed or added to a solution of another chemical via the top of the reactor. Backflow of the solution is only possible through an immersion pipe in the case of pressure reversal due to pressure buildup in the reactor. If the starting material is not added via a dip pipe but is freely flowing into the eddy of the stirrer, backflow will simply not occur.

In a continuously operated plant, all parts are connected and this scheme offers the possibility for chemicals or reaction mixtures to move along a pressure gradient caused by a leakage or a pump failure: sudden movement of chemicals in the wrong direction, for example, via the bypass channel of the pump. Check valves, while imperative, are per se not hermetically tight. This leads us to one of the worst-case scenarios in a continuous plant. Let us assume an exothermic chemical reaction of two starting materials and one

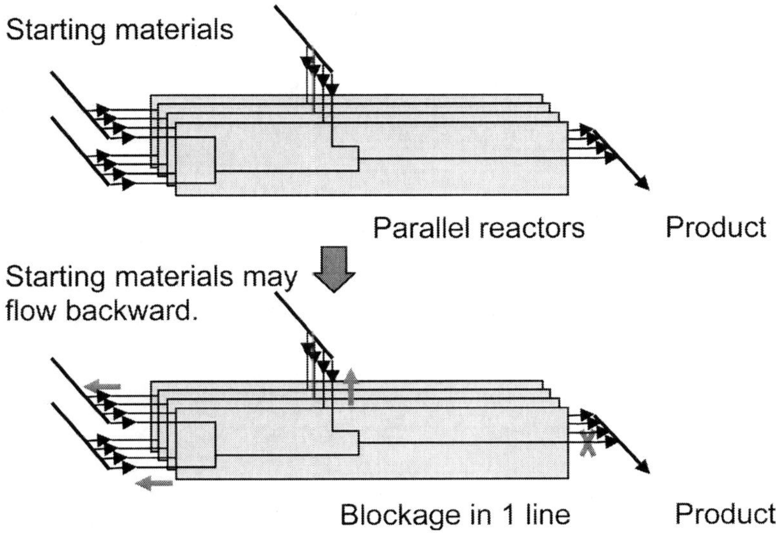

Figure 10.1 Blockage of a line affects all other lines and may lead to backflow.

solvent. In this case, the setup exists of at least two feed lines, perhaps even four, as shown in Fig. 10.1.

In a classical batch reactor, the solvent is not only necessary to dissolve the reagents, ensure sufficient mixing, and effect transport of reagents and the product. In the case of an exothermic reaction, the solvent is also a very important safety element to control the reaction: a rise in temperature will, depending on solvent heat capacity, be accommodated by removal of a certain amount of energy. Once the "storage capacity" of the solvent is exhausted, the solvent starts to boil and removes energy from the system by evaporation, which is in many cases the most effective way to remove heat from a reaction mixture. If a pump fails or a blockage occurs in a continuously operated system, one reagent might be fed directly into the feed tank of the second reagent in the absence of a solvent. The reaction enthalpy will be released immediately and without any moderation. A skeptic to a continuous processing concept would conclude that the problems the microreactor promised to solve have been shifted to its surroundings. A safety analysis shows, however, that it is easier to find technical solutions to avoid these unwanted and exceptional conditions in a continuous process than to find a solution to control a very exothermic reaction in a classical batch reactor. It is nevertheless important to be aware of these risks in a continuously operated installation.

10.4 Process Intensification in Practice

We now describe a practical example—the formation of a mononitrate ester of a diol. We had to produce this synthon as a starting material for a nitric oxide (NO)-donating API: under physiological conditions the nitrate ester moiety is cleaved and transformed to NO, which causes vascular relaxation and thus lowers blood pressure.[13] Adding an NO-donating moiety to an API consequently adds to the API a blood pressure–lowering effect. To develop a safe procedure to selectively form the mononitrate ester of a diol, one of two hydroxyl functions should react, while the second one should remain untouched. The reaction should be achieved by direct esterification of the diol with nitric acid, comparable to the formation of nitroglycerin (see Scheme 10.1).

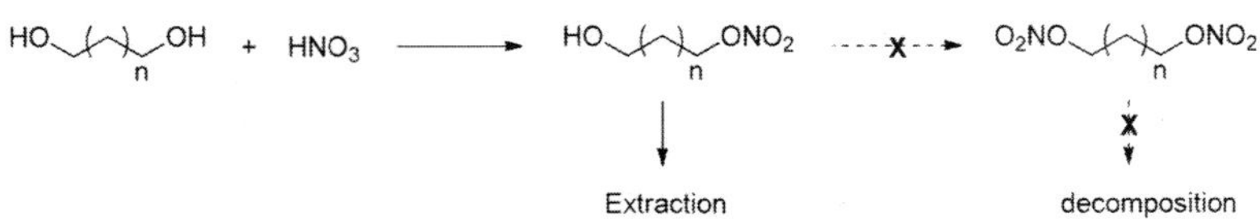

Scheme 10.1 Selective nitrate ester formation of a diol.

In direct analogy to nitroglycerin, the mononitrated product is an explosive that could not be handled *in substance* in a multipurpose plant. The only option was to handle solutions of the product with a maximum concentration of 15%. This limitation was difficult to solve in a production plant, especially as we had to employ the chemically stable, but volatile, dichloromethane as the solvent. Having a nonmiscible solvent present during the nitrate ester formation allowed extraction of the mononitrated product, once formed, into the organic layer, thus reducing further nitration.

The handling of nitric acid presents severe problems in a manufacturing environment, from the corrosive properties of nitric acid to the dangers and problems of waste disposal. Next to these obvious problems we had to solve several issues related to the chemical conversion of the diol into the mononitrate ester. Mixing the diol with nitric acid quickly released a lot of energy, which might be described as energy of protonation. If this energy is not removed the temperature will rise, leading to rapid oxidative decomposition that cannot be controlled.

After mixing of the starting materials, sufficient residence time and sufficient mixing of the emulsion of aqueous (nitric acid and diol) and organic (dichloromethane and product) phases must be ensured. If the residence time is not controlled exactly, the process will fail to reach a sufficient degree of conversion. If the mixture is then allowed to leave zones of strict temperature control, it will start to decompose. Furthermore, if the emulsion consisting of dichloromethane and the nitrating mixture of nitric acid and the diol is not kept up by sufficient shaking, the nitrated products will stay in the nitrating mixture and additional nitration or oxidative decomposition will proceed. The exact residence or reaction time is ensured by the addition of water to stop the desired reaction and to diminish the oxidative power of the nitric acid.

The addition of water to white, fuming nitric acid is a very exothermic process, and the heat may also cause oxidative decomposition of the reaction mixture. Therefore, strict temperature control during the dilution step is imperative. However, even with these countermeasures, the reaction mixture still shows explosive properties. Aqueous mixtures of nitric acid with organic ingredients may suffer exothermic decomposition due to the oxidative power of the acid.

If the concentration of nitric acid is reduced below 10%, it can be handled in classical equipment, with some precautions. The mixture leaving the reactor after the dilution step may still decompose within seconds. It is crucial to neutralize the strongly acidic reaction mixture to a pH of approx. 7. However, while excess base may solve the safety issue, in return a quality problem occurs: under strongly alkaline conditions, the desired product degrades. Continuously neutralizing the reaction mixture in a reliable way to reach a narrow pH range of the crude reaction mixture provides an element of the overall safety concept and has to be engineered to a high safety integrity level.

Another issue that needs to be solved is good solubility of the mononitrated product in the aqueous phase. It is important not to use too much water or diluted bases to stop the reaction. With excess water, large amounts of the product will be lost into the aqueous inorganic salt waste. In contrast, with only small amounts of water and highly concentrated bases, the process must cope with a tremendous energy release at the mixing point of the reaction mixture and the base.

Likewise, it is not advantageous to dilute the organic layer too strongly. For downstream processing, a sufficient product concentration is required and it must not exceed a critical concentration, above which the crude product shows explosive properties. In the worst case, an additional concentration step (careful removal of solvent by distillation) is required, which again causes safety problems due to the explosive properties of nitrated alcohols.

To solve the slew of aforementioned problems, process chemists and engineers analyzed the chemical process step by step in detail. The general procedure of this analysis is shown in Scheme 10.2. It is

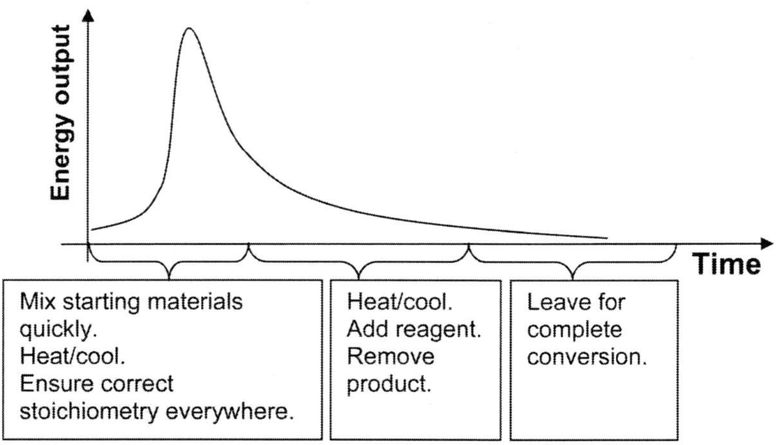

Scheme 10.2 Step-by-step analysis.

important to note that within a continuous process all operations take place at the same time but at different locations in different reactors or devices. Therefore, it is possible to use any reactors or devices that best fit the specific operation.

10.5 Safety First: A thorough Analysis

The process was first implemented in 2007 at DSM Fine Chemicals Austria as a batch process. Initially, a complete thermal process safety analysis was done for the batch process. Later, this process was changed to run continuously in a small, structured reactor (an advanced flow or microreactor). To understand the big advantages of microreactor technology regarding thermal process safety, we will first discuss the batch process.

10.5.1 *Selected HAZOP Scenarios*

The systematic approach in thermal process safety is strongly connected with single-failure scenarios derived from HAZOP. The following safety considerations are based on selected scenarios. These do not represent the complete picture, but they give a deeper insight into the major hazards of this nitration process.

- Explosive properties of the final product
- Detonable properties of nitric acid/dichloromethane (reaction) mixtures
- Cooling failure scenario
- Influence of impurities on secondary reactions
- Influence of NO_x on secondary reactions
- Overcharging of the diol—effect on thermal stability of the reaction mixture

10.5.2 *Stability of Raw Materials and Decomposition Behavior of the Final Product*

The pure raw materials used in this process are very stable. No critical safety exothermal event was found in differential scanning calorimetry (DSC) measurements up to 300°C (closed high-pressure crucibles). On the basis of the chemical structure (unstable functional groups within the molecules), the thermal decomposition behavior and explosive properties of both diol mononitrate (DMN) and diol dinitrate (DDN) were investigated.

The oxygen balances of DMN and DDN are calculated to be −101% and −53%, respectively. These values and the unstable functional groups within the chemical structure are the first indication of possible explosive properties of these compounds. The exothermal decomposition energies of DMN (gas chromatography [GC] purity 99.5%) and DDN (GC purity 99.9%) were measured to be 3041 J/g and 3573 J/g, respectively. According to the Registration, Evaluation, Authorisation and Restriction of Chemicals (REACH) legislation, explosive properties are present if the substance is impact sensitive (limiting impact energy <39 J) or friction sensitive (limiting load $\leq$ 360 N) or the limiting diameter of the Koenen test is greater than or equal to 2.0 mm. Interestingly enough, even pure DMN and DDN are not impact- or friction sensitive in terms of the testing methods. However, the limiting diameter (Koenen test) is greater than or equal to 2.0 mm for both DMN and DDN. These results are equal to explosive properties of the pure compounds. The limiting diameter (Koenen test) is less than 2.0 mm for solutions of DMN in dichloromethane containing $\leq$ 15% DMN and $\leq$ 20% DMN and containing $\leq$ 15% (DMN + DDN). A limiting diameter of <2.0 mm is equal to nonexplosive properties.

10.5.3 *Nitric Acid (Safety)*

Nitric acid with concentrations >70% has strong oxidizing properties. Violent reactions are described in the literature with numerous organic and inorganic compounds. Special care is needed during handling of this substance even at lab scale on the basis of the entry in Bretherick's:[14] "Dichloromethane dissolves endothermally in concentrated nitric acid to give detonable solutions." Detailed investigations according to test code A.1[15] were done in terms of possible detonable properties of concentrated nitric acid (65%, 85%, 90%, 95%)/dichloromethane mixtures. These experiments should exclude the ability of a substance to propagate a detonation by subjecting it to the detonation from a booster charge.

Detonable properties were found using 90% or 95% nitric acid/dichloromethane mixtures in the mass ratio of 1:1, but it was unequivocally shown that mixtures with nitric acid concentrations of 85% or 65% are not able to detonate even under cavitative conditions. Investigations on the organic phase and the biphasic reaction mixture after the end of the synthesis reaction (using 65% nitric acid) have also shown that these materials do not have detonable properties even under cavitative conditions.

Concentrated nitric acid decomposes in the presence of light or temperatures in the range of its boiling point to nitrogen dioxide (NO_2) and oxygen. The NO_2 colors the nitric acid yellow to red at higher concentration. To avoid secondary reactions during the synthesis reaction, NO_2 must be removed by reaction with urea. However, excess urea leads to the formation of explosive urea nitrate and must also be avoided.

10.5.4 *Synthesis Reaction*

The heat of reaction of the synthesis step was found to be 4.2 kJ/kg reaction mass. This is equal to an adiabatic temperature rise of 2 K. According to Stoessel,[16] this is related to low severity in the case of the runaway of the synthesis reaction. Starting from the process temperature (25°C) the maximum temperature of the synthesis reaction (MTSR) was calculated to be 27°C. Even if the synthesis reaction were to get out of control, the boiling point of the low-boiling organic phase (boiling point 40°C) would not have been reached.

In the microreactor process the situation was slightly different. Due to the different composition of the reaction mixture, the heat capacity of the reaction mixture decreased down to 1.5 kJ/kg · K. Therefore an adiabatic temperature raise of 3 K could be expected. Furthermore, the reaction was performed at higher temperatures to increase the reaction rate, allowing reaction temperatures of 40°C to 50°C to be applied. These temperatures are applicable because the small channels of the microreactor create a significant hydrodynamic pressure drop depending on the flow rates applied. The process must be run with sufficiently high flow rates to avoid boiling. In our specific example, the reaction mixture experienced pressures starting from 15 bar down to 5 bar, depending on the position in the microreactor. Consequently, taking the worst case into account—a 5 bar pressure drop and a reaction temperature of 50°C—an MTSR of 53°C could be reached. Thus, by using the process-intensified equipment for the desired esterification reaction, the boiling point of organic phase would not be attained.

10.5.5 *Secondary Reactions*

10.5.5.1 Using stabilized nitric acid

After the end of the synthesis reaction, the biphasic reaction mixture was analyzed using DSC (Fig. 10.2). Two exothermal events were found in the temperature range of 0–300°C (peak maximum/exothermal heat of decomposition): 145°C/629 J/g and 237°C/432 J/g. Using an estimated c_p of the mixture of 1.8 kJ/(kg · K), the adiabatic temperature rise of these exothermal events was calculated to be 349 K and 132 K, respectively. Even the first exothermal decomposition led to high severity in the case of a runaway.

To get a clear picture of the thermal behavior of the single layers, the organic and aqueous phases were analyzed separately. For the organic layer, two exothermal events were found in the

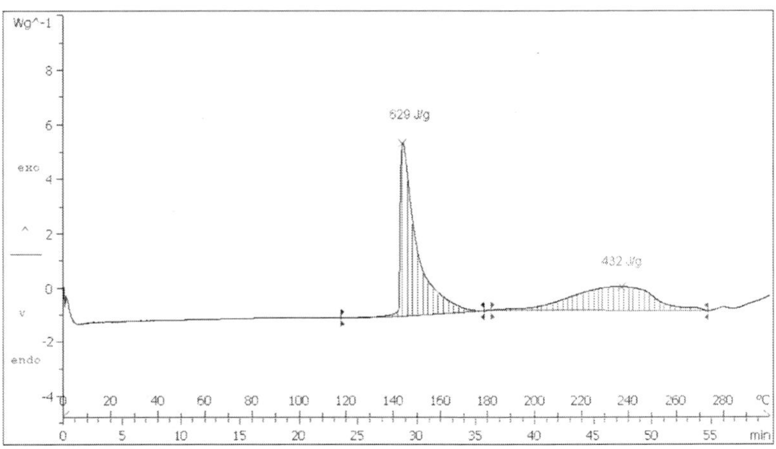

Figure 10.2 DSC results of the biphasic reaction mixture after 2.5 h reaction time. Both phases were weighed separately. Organic phase:Aqueous phase weight ratio = 1:0.59 (ratio in original reaction mixture = 1:0.5).

temperature range of 0–300°C: (peak maximum/exothermal heat of decomposition) 154°C/22 J/g and 209°C/183 J/g. Using an estimated c_p of 1.1 kJ/(kg · K), the adiabatic temperature rise of these exothermal events was calculated to be 20 K and 166 K, respectively. The severity was low for the first decomposition event and medium for the second. For the acidic aqueous layer, a single exothermal event was found in the temperature range of 0–300°C: (peak maximum/exothermal heat of decomposition) 134°C/678 J/g. Using the measured c_p of 3.0 kJ/(kg · K), the adiabatic temperature rise was calculated to be 226 K. The severity was high for this exothermal decomposition reaction. In addition, the sharp peak was strong evidence for an autocatalytic decomposition pathway. Further measurements were required to characterize autocatalytic decomposition reactions.

To estimate the probability for a runaway reaction after the end of the synthesis reaction, the time to maximum rate under adiabatic conditions (TMRad) had to be determined. Because of the safety-critical decomposition behavior of the acidic aqueous phase—the worst case—it was decided to focus on this matter. Using the thermal activity monitor (TAM) to calculate the TMRad, assuming zero-order decomposition, was a successful approach. The TMRad of the acidic aqueous phase at 25°C was calculated to be 4,887 hours (estimated activation energy: 50 kJ/mole). Even at 68°C, the TMRad was 24 hours. However, this temperature could not be reached because the system was tempered at 40°C by the boiling of dichloromethane. From this data, it can be concluded that the risk for a thermal runaway is low at the process temperature (25°C) or at the MTSR (27°C) in the case of an open system.

10.5.5.2 Autocatalytic decomposition reactions

Autocatalytic decompositions are common in the fine chemical industry.[17] A reaction is called autocatalytic if a reaction product acts as a catalyst on the reaction course.[18] They are considered hazardous because they give rise to sudden heat evolution, often with unexpected initiation and unknown external causes,[19] and are consequently perceived as unpredictable. The first indication of this behavior is given by a DSC thermogram. Narrow or sharp peaks are suspicious for this kind of decomposition. The thermoanalytical evidence is an isothermal DSC experiment (or a simulation derived by the AKTS® software) where the heat release rate passes through a maximum and then decreases again. In other words, the peak maximum is not measured at $t = 0$ minutes. As discussed before, this is the case for the acidic aqueous phase of the reaction mixture. Autocatalytic decomposition reactions could be sensitive in a safety-critical way to impurities like rust and reactor metal or thermally aged material from previously produced batches.

10.5.5.3 Using nonstabilized nitric acid: NO_x triggers the decomposition reaction

In the early development phase, we assumed that nitrogen oxides (NO_x, comprising NO or NO_2) were the reason for the instability of the reaction mass. To check this hypothesis, we performed DSC experiments before and after treatment (5 min) of the acidic aqueous phase with gaseous NO_x at ambient temperature. At the end of the treatment, self-accelerating decomposition was observed, which resulted in strong formation of NO_x within the reaction mass itself. The observed effect was also visible in the DSC thermograms. Starting from the well-known narrow peak (peak maximum: 124°C, exothermal decomposition energy: 978 J/g), the peak shape and decomposition energy changed dramatically after treatment with NO_x to a relatively broad heat release curve with significantly reduced exothermal decomposition energy (peak maximum: 93°C, exothermal decomposition energy: 183 J/g). After stirring the reaction mixture overnight, no exothermal decomposition in the temperature range of 0–300°C could be observed. This experiment was clear evidence that NO_x played a catalytic function in the decomposition of the acidic aqueous phase and would need to be removed from the nitric acid (using urea) before starting the nitration of 1,4-butanediol.

10.5.5.4 Overcharging the diol

Overcharging the diol led to a significant increase in the decomposition energy of the acidic aqueous phase. Upon doubling the amount of alcohol, the decomposition energy was determined as 1,917 J/g (peak maximum 103°C).

10.5.6 *Neutralization of the Reaction Mass*

From a chemical point of view, the strong exothermal decomposition of the acidic aqueous phase was probably connected with the oxidizing properties of nitric acid and the dissolved organic material, that is, the diol. This hypothesis was supported by the DSC data after neutralization of the acidic organic layer. No exothermal event was detected up to 300°C.

10.5.7 *Comparison of the Batch and Microreactor Process in Terms of Thermal Process Safety*

A batch process for a nitration reaction has some disadvantages. The resulting reaction mass, or, more precisely, the acidic aqueous phase, shows strong exothermal decomposition behavior with high severity in the case of a runaway. To mitigate the consequences of a thermal runaway, a catch tank has to be installed. Following neutralization of the nitric acid, the aqueous phase becomes thermally stable. However, even during the neutralization of concentrated nitric acid a lot of heat is released. In a batch process this usually leads to long dosage times of the sodium hydroxide solution to account for a limited heat exchange capacity. Using a microreactor, these thermal hazards are easier to control. Because of the small reactor volume, even a thermal runaway has no significant consequences—a quench vessel is not needed. The extraordinarily high surface to volume ratio guarantees excellent heat exchange, so the acidic aqueous phase can be neutralized within minutes—and stabilized against decomposition.

10.6 Consequences for Plant Design

The process engineers spent much time deliberating which reactor or device to choose for each operation. Especially for the first operations—consisting of mixing and heat removal—a small, structured device (or even a microreactor) was a very good option as it provided sufficient heat exchange surface per reaction volume. However, a microreactor is classically operated in a laminar flow regime. A laminar flow would not effect mixing of the two immiscible phases but rather lead to quick phase separation, which would result in the aforementioned danger of exothermic oxidative decomposition of the reaction mixture. Therefore, it was necessary to adapt the classical microreactor design.

Working jointly with a microreactor supplier, it was possible to find a solution for this problem. The manufacturer was able to provide small flow units with channels containing mixing zones and baffles. With these miniaturized baffles in miniaturized channels, it was possible to create sufficient turbulence to effect quick mass transport between the aqueous and organic phases. Simultaneously, the turbulence did not result in too large a residence time distribution. With the large surface to volume ratio of small, structured equipment, the heat evolved upon mixing of the starting materials was easily removed and the temperature controlled exactly during the reaction everywhere in the reactor.

Efficient mixing and heat removal were also needed for diluting the reaction mixture with water. In contrast to the reaction zone, high turbulence was no longer required since the desired esterification had stopped and the oxidative power of the nitric acid was reduced. For the neutralization, it was even advantageous to reduce the mixing between the phases as the heat release in the aqueous layer should not harm the product, which sits mainly in the organic layer. Therefore, the neutralization part could take place in standard microreactor design with simple straight channels.

The next difficult step was the adjustment of the pH. A pH meter is not available in a safety integrity level high enough to fulfill the safety requirements at this point of the process. The safe adjustment of the pH had to be ensured by additional controls. In a continuous plant there are further options to ensure safe operations: control of the continuous feeding of starting materials. In our example we set the feed streams of the nitric acid and the base in correlation. With this correlation it was possible to ensure a maximum nitric acid concentration as the reaction mixture left the microreactor of below 5% free nitric acid content. As mentioned before, the energy developed by the diluted and partly neutralized reaction mixture can be removed in a classical reactor. The neutralization was completed in a flow-through stirred tank reactor in which a pH meter controlled dosing of the base. In this way the safety of the process was guaranteed by the dosing line controls for the microreactor, and the quality of the product was ensured by the stirred tank.

As for a batch process, HAZOP analyses[20] were performed to avoid dangerous situations following pump breakdown and other incidents. Within these HAZOP analyses, the above-mentioned back-flow scenarios were also discussed. The solutions worked out by the groups engaged in the HAZOPs were realized in the plant. A complex system of special pumps, check valves, control units, and safety relief valves was implemented. This complex construction might discourage others to use continuous operating equipment, particularly if the installations are "overengineered," where a change from batch to continuous mode can get expensive. However, on top of the reactor expense, the costs of the surrounding equipment might kill a project through financial considerations on payback time, flexibility, cost of reconfiguration, reuse, and the like (Fig. 10.3).

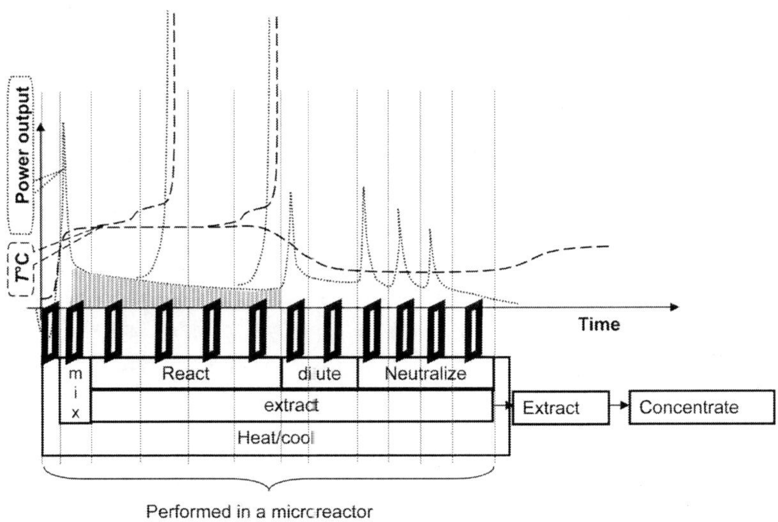

Figure 10.3 Process setup.

10.7 Quality

A further point of eminent importance in developing a pharmaceutical process is the quality aspect: the lives and health of people are in danger if mistakes occur. Several regulations are in place and summarized in the International Conference on Harmonisation (ICH) guidelines issued by the FDA. The guidelines describe the required quality standards in pharmaceutical production, and they are updated regularly. A pharmaceutical company or manufacturer of pharmaceuticals must be compliant with the so-called current good manufacturing practices, or cGMP. One important part of these guidelines is to ensure that the equipment used for production of the pharmaceutical substance shows no impact on the quality of the product. For this purpose a quality risk analysis (QRA) has to be performed to evaluate the possible impact of the equipment on the product quality.

In the same way a proven acceptable range (PAR) program has to be performed in the laboratory. The parameters, describing the process under development, are identified during the lab program. We have experienced that using continuously operated microstructured devices both in the lab and in the plant improves the reliability of PAR values considerably. How can we state this? Because in the best case, the conditions will be transferred 1:1 into the production scale by parallelization instead of up-scaling. Each volume element of the reaction mixture in the parallel channels of the production reactor experiences the same conditions as in the single channel in the lab setup. In summary, equal investigations can be performed in the case of a continuous process and in a batch process. Differences occur in the cleaning process and in the batch definition. In most continuous devices cleaning can be performed in the same way as in a batch reactor or equipment. Differences occur if the reactor or device cannot be opened by construction. In this case a swab test cannot be performed and only the last rinse methodology can be applied.

In theory, a continuous process runs nonstop to deliver a certain amount of product per hour or day. In the best case, the continuous process is initiated at the beginning of the production campaign and is shut down at the end of the campaign, delivering an in-spec product over the whole time in operation. However, this "quality by design" behavior might be interrupted by pump damage or a contaminated starting material. In such a case, it is necessary to ensure the traceability of the off-spec material. Therefore a prudent batch definition is required. In contrast to a batch process it is not possible to define a certain amount of the key starting material as a batch. In a continuous process it is not advisable to empty a supply vessel or tank completely or even stop the plant to signify the end of the previous batch and a starting point of the new batch. Especially the starts and shutdowns of a continuous plant are complicated and prone to deviations. During this phase of the process, deviations from PARs are possible and side products are formed.

To avoid start-ups and shutdowns of a continuous plant, the supply vessels or tanks must be refilled before they become empty. In this way, a certain amount of mixing of subsequent lots of starting materials will occur, making the traceability complicated. In the same way, this mixing takes place in every buffer vessel within the process and in the product-collecting devices. In a continuous process, two product-collecting devices are typically needed and used in alternating modes of operation. One device is actually collecting the product at a time, while the other one is being emptied to feed the subsequent processing step of the synthesis. Especially in the latter situation, it is not possible to empty the device completely if, for example, only a part of the material can be used in a batch of the next step. A certain amount of product might remain in the device because the switch from one device to the other is not timed exactly enough. If the collecting device is switched back to the collecting device defined as empty, mixing of different batches occurs.

One might state that this is not good cGMP practice. In terms of a batch process this is definitely right, but in contrast to a batch process, the quality of a continuous mode is established by stationary states at every position of the process as well as with continuous control of the parameters, which are quality determining. As an example, flows and temperatures are continuously determined. In this way, the *in-spec* product is produced nonstop. In the case of malfunction of an instrument, for example, breakdown of a pump, this deviation will be recognized immediately and corrective actions executed. Therefore, a classical product or intermediate release by the IPC, comparable to the release of a batch, is no longer required. The only important thing that needs to be ensured is the sufficiently high stability of the collected product. This means that stability of the product must be several times longer than the time needed for storage of the product while emptying the collecting vessel, switching to the second vessel and emptying it, switching back to the first vessel, and refilling. In the worst case, the collecting vessel shows perfect mixing like a continuously stirred reactor. As a rule of thumb, the content of the reactor is completely exchanged if the volume of the reactor is exchanged seven times. So the stability of the product should be in the range of 5 to 10 times longer than the time needed for the mentioned operations, taking into account that a storage vessel is not perfectly mixed.

As a consequence of this reasoning, the classical batch definition via a certain amount of starting material has lapsed. Instead it is necessary to switch to a time-based definition of a batch to ensure traceability. In this way it is reproducible which lots of the starting material were charged to the supply vessels of the continuous plant and which lots were used in the next chemical step. A certain blurring due to the holdup of the plant and different filling grades of buffer vessels cannot be avoided. However, this disadvantage is more than compensated for by the advantages of the continuous process on the product quality, as outlined before.

10.8 Continuous Improvement and Its Consequences on the Environmental Impact of a Plant

We have shown that a potentially dangerous nitration can be performed in a safe way in a conventional vessel but with a high price: to allow a conventional vessel to remove the reaction heat, we had to dilute and slow down the reaction. Note that this does not change the overall amount of heat managed; only its release is slowed to a rate fitting the reactor capabilities. In many cases, the heat also has to be removed at a lower temperature level. This renders the released heat worthless for further use. If the heat has to be removed at a level below ambient temperature, additional energy is required to pump it out of the system and into the environment. In our case we had to concentrate the highly diluted product to make further workup acceptably efficient. Such a process has higher energy costs and a low space-time yield translating into high installation costs, and it also produces a lot of waste. So during process development we investigated options to improve on the safety and the sustainability of this process by running it in a continuous plant. Recently, we described how to take this nitration process into a continuously operated pilot installation[21] und subsequently into a full-scale plant.[22] Next to the improved safety performance, we could also considerably reduce the E-factor of the process. The main parameters contributing to the reduction of the E-factor were:

- The amount of solvent, by running the reaction at higher concentration, and the degree of recycling, increased with a smaller number of different recycling flows.
- The stoichiometry of the nitrate ester formation altered to make more efficient use of nitric acid.

The main parameter in reducing the energy input was the option to run the reaction at higher-than-ambient temperature allowing conventional water cooling instead of refrigeration via a cooling system.

Figure 10.4 describes the change in the E-factor for the nitration step as we developed the process from batch-wise to continuous operation and further optimized the continuous process.

It is notable that the process concept, while still following the rules of GMP, has allowed for further improvements in the E-factor after the first installment. According to our plans, we will continue to improve the process and further reduce its E-factor. This example shows that the carefully planned filing of a process with authorities, such as the FDA, also determines the extent to which a later process improvement and improvement of its environmental balance is possible without re-registration.

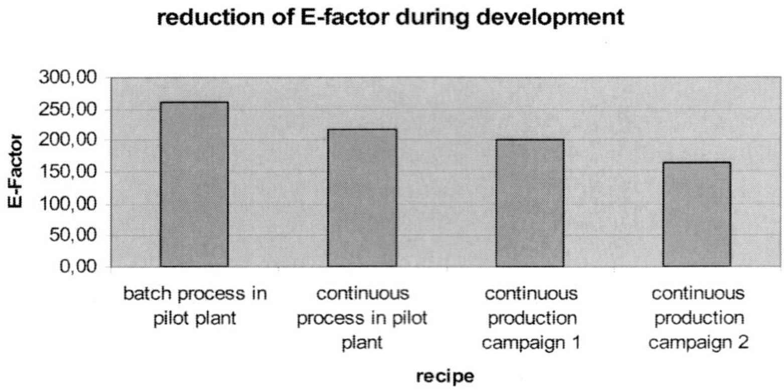

Figure 10.4 Reduction of the E-factor during development.

10.9 Conclusion

Manufacturing of pharmaceuticals has until recently only focused on the final product quality and the reliability of supply. The absolute amount of waste compared to other chemical branches and the portion of manufacturing cost of the API compared to the drug product—the medicine—have been considered too small to justify any extensive process optimization of API production. Under this paradigm, less-than-optimal processes have been employed to manufacture pharmaceuticals. As consumers become increasingly conscious of the potential environmental effects for the products they use, pharmaceutical production will come under further scrutiny. Though the absolute effects of pharmaceutical production are small, the great potential for locally occurring contamination has entered the commercially relevant sphere of public discussion.

Efficient manufacturing processes, while always desirable, become even more significant once a pharmaceutical loses its patent protection and generic production volumes increase. There are hardly any unique generic medicines, and therefore, choices to use a certain medicine are frequently made by price—which is seen as a means to keep health care costs at bay. Companies engaged in custom pharmaceutical synthesis have traditionally been selected by their customers on the categories of the price and quality of their services. The best vendors have developed skills in both chemical synthesis and process technology to develop and perform highly effective syntheses capable of withstanding competition. The implemented processes make smart use of resources with E-factors much lower than those of the original synthesis, and they have potential for further optimization. This continued development and application is supported by regulatory agencies, such as the FDA and the EPA. Process intensification has just started to show its potential to increase the efficiency and reduce the environmental burden of pharmaceutical-manufacturing processes.

References

1. Gladwell, M. (Oct. 25, 2004). High prices—how to think about prescription drugs, *The New Yorker.*
2. Marketing Nexium As Superior To Prilosec Was OK // Pharmalot
3. http://www.drugs.com/news/merck-research-chief-details-background-vioxx-3535.html
4. http://www.cbsnews.com/stories/2006/03/13/health/webmd/main 1393091.shtml
5. Aspirine® price on idealo.de: 500mg pill by originator: 0.09€; 500mg by generics manufacturer: 0,03€.
6. http://chem.info/News/2010/04/Safety-Pharmaceutical-Plant-Explosion-Poisons-1000/
7. Environmental Protection Agency. (May 20, 1997). *EPA Reports on Level of Toxic Releases to Air, Water and Land in Puerto Rico,* US EPA Region 2 News Release. http//www.epa.gov/r02earth/epd/97085.htm
8. Poechlauer, P.; Braune, S.; De Vries, A.; May, O. (2010). *Chim. Oggi/Chem. Today,* **28**(4), 14–17.
9. Salemme, F. R.; Spurlino, J.; Bone, R. (1997). *Structure,* **5**(3), 319–324, ISSN 0969-2126, DOI: 10.1016/S0969-2126(97)00189-5. http://www.sciencedirect.com/science/article/B6VSR-4CJB8KR-5W/2/5ff408665a66338b054decb7ff6ee954
10. http://www.pharmamanufacturing.com/articles/2005/294.html
11. Stankiewicz, A.; Moulijn, J. (2002). *Ind. Eng. Chem. Res.,* **41**(8), 1920–1924. DOI: 10.1021/ie011025p
12. Stankiewicz, A.; Moulijn, J. (2004). *Re-engineering the Chemical Processing Plant—Process Intensification,* CRC Press, ISBN-13: 9780824743024.

13. This discovery earned Dr. Robert F. Furchgott the Nobel Prize in Physiology or Medicine.
14. Bretherick, L. (2007). *Handbook of Reactive Chemical Hazards* (7th ed.). Butterworths.
15. United Nations. (2003). *United Nations: Recommendations on the Transport of Dangerous Goods, Tests and Criteria* (4th ed.). New York.
16. Stoessel, F. (2008). *Thermal Safety of Chemical Processes: Risk Assessment and Process Design* (1st ed.). Wiley-VCH.
17. Grewer, T. (1994). *Thermal Hazards of Chemical Reactions. Industrial Safety Series* (Vol. 4). Amsterdam: Elsevier.
18. Oswald, W. (1970). *Physikalische Organische Chemie*. Hammet: Louis Plack.
19. Brogli, F.; Frimm, P.; Meyer, M.; Zubler, H. (1980). Hazards of self-accelerating reactions, in *3rd International Symposium Loss Prevention and Safety Promotion in the Process Industries*. Basel: Swiss Society of Chemical Industry, 665–683.
20. Kletz, T. A. (1986). *Hazop & HAZAN Notes on the Identification and Assessment of Hazards.* Rugby, UK: Institution of Chemical Engineers.
21. Braune, S.; Pochlauer, P.; Reintjens, R.; Steinhofer, S.; Winter, M.; Lobet, O.; Guidat, R.; Woehl, P.; Guermeur, C. (2009). *Chim. Oggi/Chem. Today*, **27**(1), 26–29.
22. Braune, S. From Lab to Production—An Example from Industry for the Use of a Microreactor in a Multi-Ton Scale. ISPE Milan Congress on Science and Risk Assessment for Business Success, March 22–25, 2010, Milan, Italy.

Chapter 11

Going Green Using Combined Real-Time Analytics and Process Automation

Dominique Hebrault and Terry Redman

Mettler-Toledo Ingold Inc., 36 Middlesex Turnpike, Bedford, MA 01730, USA

dominique.hebrault@mt.com and terry.redman@mt.com

11.1 Introduction

In today's world, the pharmaceutical and chemical industries face major challenges, including globalization, environmental regulation, and shrinking product life cycles. Meeting these challenges requires the development of innovative technologies and alternative approaches geared toward reducing costs and improving the environmental and economic profiles of chemical processes. Breakthroughs in chemical process operations and modeling are necessary for achieving energy and material efficiency gains.

Because of the impact on public health, the pharmaceutical industry has been the subject of more scrutinized government oversight than any other industry. Since 2002, the Food and Drug Administration's (FDA) Guidance on Process Analytical Technology (PAT) has had a significant impact on the way potential medicines

Scalable Green Chemistry: Case Studies from the Pharmaceutical Industry
Edited by Stefan G. Koenig

ISBN 978-981-4316-49-1 (Hardcover), 978-981-4364-81-2 (eBook)
www.panstanford.com

are discovered and developed.[1] The latest FDA Guidance for Industry, released in November 2009, defines Quality by Design (QbD) as *"a systematic approach to development that begins with predefined objectives and emphasizes product and process understanding and process control, based on sound science and quality risk management."*

Proper integration of PAT and process automation, together with the use of multivariate tools for design, data acquisition, and analysis, is vital for the design of information-rich experiments. Real-time in situ process monitoring is increasingly considered fundamental by the pharmaceutical industry because it offers an opportunity to obtain information about a reaction system much faster than traditional methods, leading the way to high-throughput process optimization.[2,3] PAT has been bringing value to three areas of drug development:

- Chemical research and development: Spectroscopy and small-scale heat flow offer alternatives to withdrawing samples for off-line analysis.[4] The numerous benefits range from increased productivity, decreased solvent consumption, and the capability to monitor reactions under extreme conditions (temperature, pressure, unstable compounds), to the basic ability of accurately monitoring reaction start-point, endpoint, stall-point, and component concentrations during the transformation.
- Reaction engineering: Advanced spectroscopy and calorimetry techniques, in combination with data treatment software, provide the necessary tools for process characterization and optimization. Full equipment automation allows researchers to carry out unattended experiments during downtime, for example, during meetings, nights, and weekends. This increased productivity can be a key element in reducing overall development timelines and reducing time to market.[5]
- Pilot plant manufacturing: Real-time detection of reaction completion is used to directly minimize cycle time, improve product quality, and simplify purification steps. In addition, the use of in situ monitoring eliminates the need for

sampling of toxic reaction mixtures for off-line analysis, that is, reducing worker exposure to hazardous materials. In some cases, real-time monitoring of critical process variables can be used for direct control of the process to ensure safe and efficient operations. For example, real-time monitoring of the heat of a reaction tied to the controlled addition rate of a key reagent can be used to avoid levels of thermal accumulation that could lead to a runaway reaction.

In situ particle size monitoring (focused beam reflectance measurement [FBRM®]), in-process video microscopy (particle video microscope [PVM®]), attenuated total reflection (ATR)–Fourier transform infrared (FTIR)-based reaction analysis (ReactIR™), and reaction calorimetry (RC1*e*™) are innovative technologies that respond to the need for real-time process monitoring, resulting in increased process throughput, consistency, and reliability.[6] Case studies will be presented in this chapter, showing how these technologies have been used in pharmaceutical development to minimize waste, improve reaction output, increase energy efficiency, decrease the formation of by-products, and minimize the potential for accidents. Over the past few years, these key achievements, leading to improved manufacturing sustainability, have often been labeled as green chemistry and green engineering principles.[7] In that respect, the use of PAT is a key enabling technology in the adoption of green chemistry and green engineering practices.[8] Often, the primary objectives of faster time to market, assured drug quality, and manufacturing sustainability contradict each other, but PAT and QbD are powerful tools that can help find the best compromise between these overlapping, though sometimes conflicting, goals.

11.1.1 *In situ ATR-Based Fourier Transform Infrared Spectroscopy with ReactIR™*

ATR-based, in situ FTIR spectroscopy for reaction analysis has, over the past 15 years, become a powerful tool for scientists in the pharmaceutical and chemical industries to extract critical information about key reaction species and their rates of transformation.[9,6]

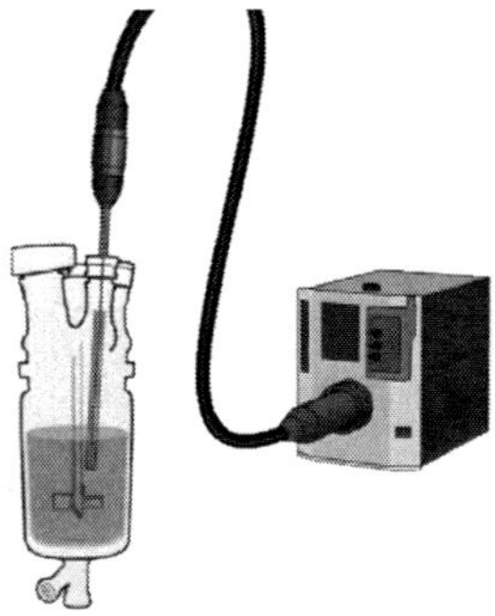

Figure 11.1 ReactIR™ probe technology with the sampling probe immersed in a reaction vessel.

Hence, the acquisition of chemical and biochemical knowledge, and the improvement of the quality and robustness of related processes, has been facilitated. In the pharmaceutical industry, the range of application extends from discovery chemistry to process research and development and beyond into manufacturing scale and quality control.

Probe-based, real-time, in situ measurements allow reactions to be monitored constantly, eliminating the need for vessel sampling for off-line analysis. The potential for transfer contamination and sample degradation between the reaction vessel and the analytical instrument is therefore reduced (Fig. 11.1).

ATR technology has an advantage over transmission technology in that even darkly colored and heavy suspensions or slurries can readily be analyzed (Fig. 11.2).

The typical spectral range goes from 4,000 to 650 cm^{-1} but depends on the sampling technology. A spectrum representative of reaction mixture mid-IR absorbance would be collected as per the user's choice from every second to every few minutes, depending on reaction kinetics. Simultaneous scanning through wavelength range and reaction time allows displaying continual and cumulative changes in the spectra against time.[10] Rates of formation and disappearance of starting material, intermediates, and final product are obtained as a result (Fig. 11.3).

Familiar two-dimensional plots representing concentration change as a function of time will be obtained from the waterfall

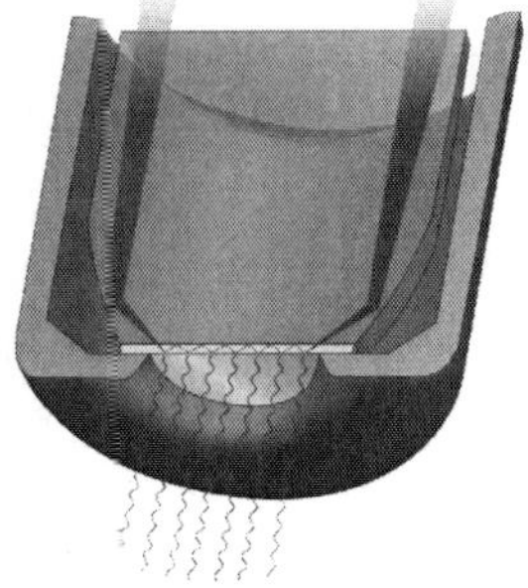

Figure 11.2 Cutaway of ReactIR™ probe tip on the basis of ATR.

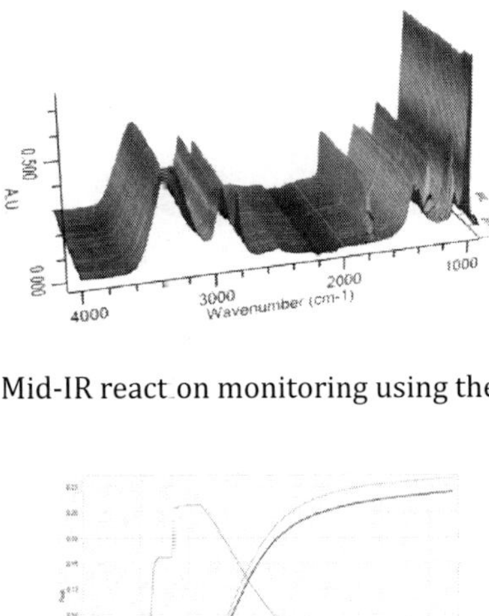

Figure 11.3 Mid-IR reaction monitoring using the waterfall plot.

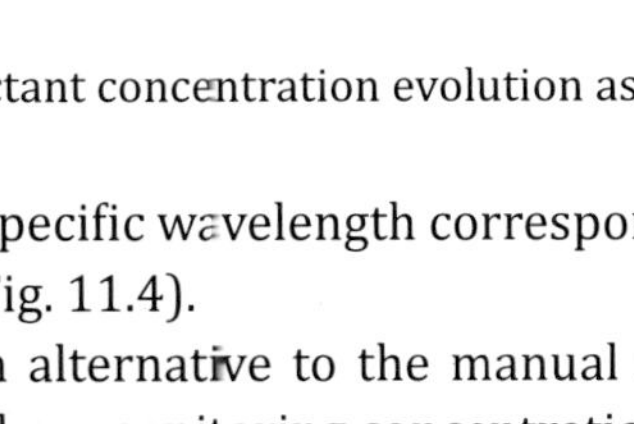

Figure 11.4 Reactant concentration evolution as a function of time.

plot by trending a specific wavelength corresponding to a functional group of interest (Fig. 11.4).

A more modern alternative to the manual selection of specific bands of interest when monitoring concentration changes over time is the use of multivariate analysis, available within the ReactIR™ software under the name of ConcIRT™. This algorithm automatically generates quantitative and qualitative information from dynamically changing spectroscopic data. As a result, it speeds up and simplifies the understanding of complex chemical reactions and automatically identifies the number of chemical components and their relative concentration profiles. In principle, ConcIRT™ searches the reaction data for changes as a function of time, even in the presence of severely overlapping spectral absorptions. Frequencies that change at the same rate are assumed to belong to the same component and are mathematically resolved as a pure component spectrum. The resolved spectrum is then used to calculate the relative concentration profile.

Another example of the increasing power of software is the iC Kinetics™ module designed after the well-known *Reaction Progress Kinetic Analysis* (RPKA) developed by Pr. Donna Blackmond.[11] This kinetic analysis approach simplifies kinetic studies of organic reactions. It leverages the extensive data available from accurate in situ monitoring of chemical reaction progress, like spectroscopy and calorimetry, and uses simple graphical manipulation to extract vital kinetic information in far fewer experiments than using a traditional kinetic approach. RPKA and iC Kinetics™ help process development chemists optimize reactions faster, investigate catalyst performance, ensure process robustness, and conduct a driving force analysis of chemical processes (Fig. 11.5).

ReactIR™ has established itself as a leading technology for mid-IR reaction monitoring. Over the past few years, the trend has been to head toward compact instrumentation, plug-and-play probe technology, and intuitive software interface (Fig. 11.6).

The latest and most acclaimed development is undoubtedly the micro flow cell technology (Fig. 11.7), which combines the high

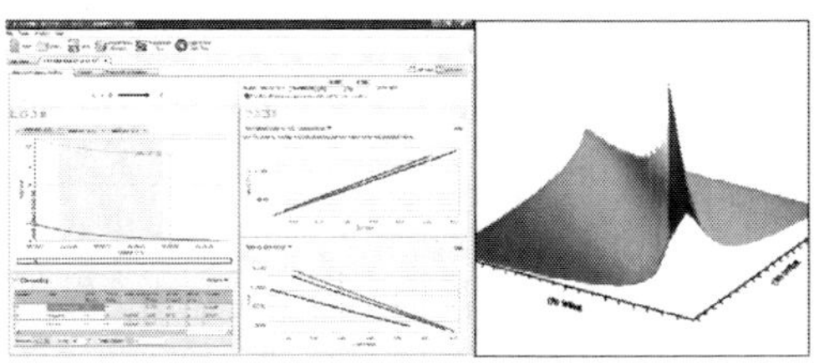

Figure 11.5 The iC Kinetics™ software module is based on Pr. Blackmond's RPKA.

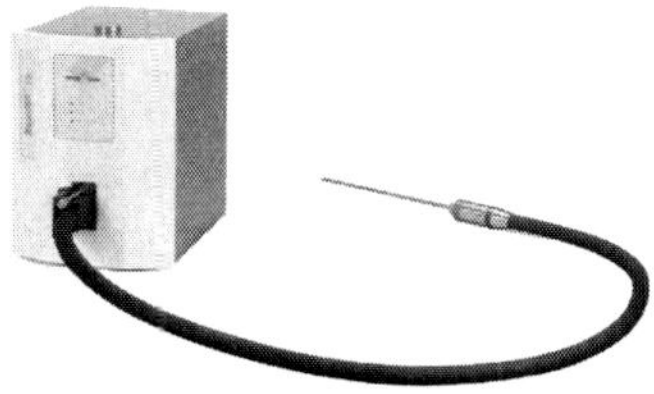

Figure 11.6 The ReactIR™ 10 instrument.

Figure 11.7 DS micro flow cell with an integrated ATR sensor (left), connected to a ReactIR™ 10 instrument.

molecular specificity of FTIR to continuous flow chemistry and enables chemists and engineers to develop chemical reactions under mild conditions that would be unthinkable using a traditional batch method. The simplicity, flexibility, and robustness of the micro flow cell technology for ReactIR™ has contributed to a fast adoption rate and is now routinely used in academia and industry to monitor chemical components in real time down to the submillimolar concentration.

Whether the goal is to prevent pollution or more generally enhance material and energy balance efficiency, ReactIR™ clearly finds its place among the arsenal of analytical methodologies for real-time, in-process monitoring, one of the 12 green chemistry principles.

11.1.2 *Reaction Calorimetry with RC1e™*

The use of reaction calorimetry to develop and bring a process into production quickly and safely can be traced back 35 years to when the first liter-scale reaction calorimeter, the WFK75 from Ciba Basle, was manufactured and used with notable success in the company's own process development. Ever since, hardware and software have

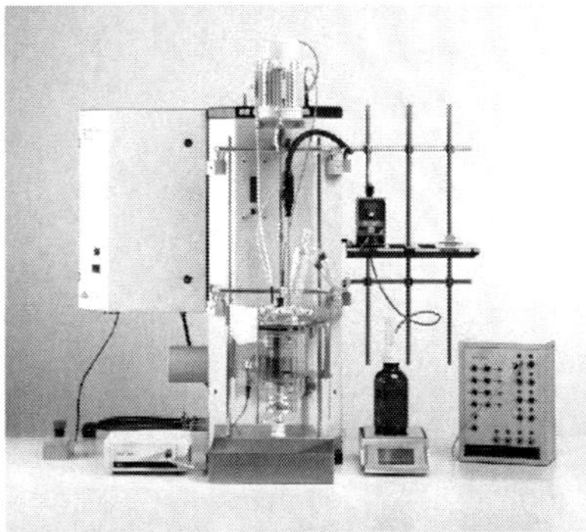

Figure 11.8 Modern version of the RC1e^{TM} outfitted with automated pumps and balances.

been continually improved to serve the interests of scientists around the world and help them improve process safety and product quality, as well as the profitability and environmental profiles of chemical and biochemical processes.[12,6]

The modern version of the first calorimeter, the RC1e^{TM}, now recognized as an industry standard, routinely provides scientists with basic, though valuable, data such as reaction enthalpy, heat flow profiles, maximum heat flow, heat transfer data, and specific heat of the reaction mass. In addition to ensuring the quick and safe transfer of economically sound processes from bench scale to production, reaction calorimetry provides of a wealth of knowledge about process kinetics and reaction mechanisms (Fig. 11.8).

For many years, heat flow has remained the reference calorimetry technology, thanks to its robustness and accuracy. Heat flow measurement is based on the simultaneous and frequent measurement (every two seconds) of reaction temperature (T_r) and jacket temperature (T_j) (Fig. 11.9). Under isothermal conditions, the difference between reaction temperature and jacket temperature (T_r–T_j) can easily be correlated to the reaction heat (Q_r) upon measurement of the heat transfer coefficient (UA). The heat transfer coefficient depends on a variety of factors, among them vessel, mixing, reaction mixture constitution, viscosity, and volume. As a result, UA needs to be measured at least twice, once before the reaction occurs and once after. Ultimately, the reaction heat can

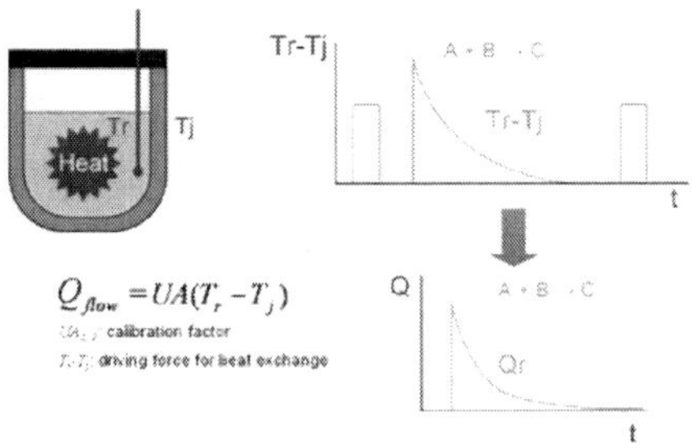

Figure 11.9 Reaction heat (Q_r) measurement principle using heat flow technology.

be determined once the last calibration value has been obtained. Under nonisothermal conditions, the reaction heat equation needs to take accumulated heat into account as well. Reagent dosing can also influence heat balance and thus needs to be factored in.

In 2007, a superior heat measurement method (RTCal™) was introduced to market to provide easy access to heat information data online in real time, without calibrations (Fig. 11.10).

RTCal™ rests on the use of sensor bands on the vessel itself that measures heat in real time as it flows through the reactor wall, without recourse to time-consuming calibration steps. As a result, and unlike the heat flow technology, it allows one to:

- obtain heat flow instantly
- make adjustments to the actual chemistry in real-time
- optimize reaction and process parameters online on the basis of heat flow data
- enable orthogonal data validation with heat flow

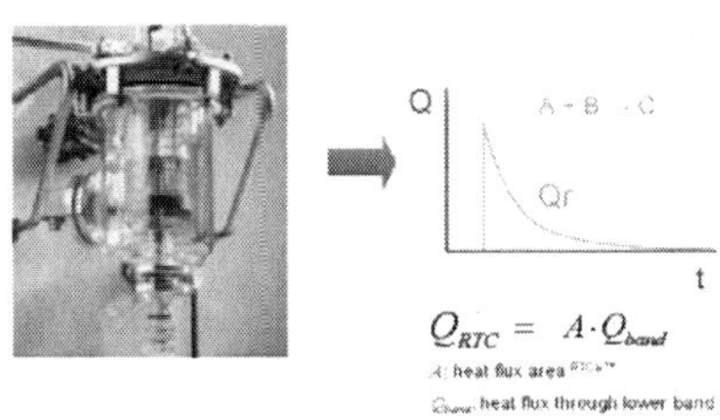

Figure 11.10 RC1*e* vessel equipped with the RTCal™ technology that provides process heat in real time.

Today, real-time calorimetry (RTCal™) is widely used in industrial laboratories for high-precision research and development activities as diverse as thermal data determination, process understanding and control, and kinetic analysis. Although not limited to it, process safety remains the core application area for liter-scale RC1*e* reaction calorimetry.[13] Evaluating crude thermal data from RC1*e* runs was often limited by the knowledge and experience of the evaluator. The advancement of software development has changed this. Now, advanced software modules like iC Safety™ offer automated conversion of reaction calorimetry data into safety knowledge, providing the basis for correct thermal risk assessment. Without in-depth knowledge of reaction calorimetry, a user can combine real-time calorimetry with iC Safety™ to access fundamental process parameters such as:

- heat of reaction and molar enthalpy
- heat removal rate for scale-up, reaction rate, and dosing rate
- thermal conversion and thermal accumulation
- adiabatic temperature rise
- maximum temperature of synthesis reaction (MTSR)

Modern computing technologies allow users to manipulate intuitive graphical windows and acquire a better understanding of other abstract reaction safety concepts, including thermal accumulation and thermal conversion (Fig. 11.11).

Reaction calorimetry clearly helps address at least two green chemistry principles, principle 12 (inherently safer chemistry

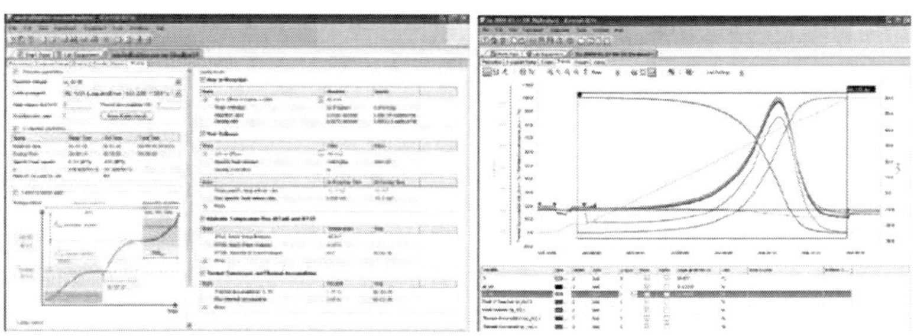

Figure 11.11 iC Safety™ graphical interface for automated conversion of thermal data into process safety information.

for accident prevention) and principle 11 (real-time, in-process monitoring to control chemical processes and prevent the formation of hazardous substances).

11.1.3 *Real-Time Particle System Characterization for Crystallization*

11.1.3.1 Focused beam reflectance measurement

Focused beam reflectance measurement (FBRM®) is a highly precise and sensitive probe-based measurement technology that tracks changes to particle dimension, particle shape, and particle count. FBRM® probes are installed directly into crystallization vessels for direct measurement of the crystal population. Measurements are acquired and recorded in real time, tracking dynamic changes in the crystal size distribution during nucleation, growth, breakage, and agglomeration—providing feedback information that can be directly used for process optimization and control. No sampling or sample preparation is required—even in highly concentrated (70% and higher) and opaque suspensions.

How does FBRM® work?

The FBRM® probe (Fig. 11.12) is immersed into a dilute or concentrated flowing slurry, droplet emulsion, or fluidized particle system. The instrument is installed directly into a crystallizer so that the crystal slurry will flow directly past the sapphire window of the probe. A laser light source is focused to a fine spot at the sapphire window interface, and a set of rotating optics scans the focused laser beam in a fixed circular path at the interface where the probe window is in contact with the crystal slurry. Where the focused laser light intersects the surface of a particle or droplet, light will be scattered from the surface. Light that is back-scattered at an angle of 180° is detected and recorded by the probe to provide measurements of individual particles.

A simulated view from the probe window (Fig. 11.13) shows individual particle structures back-scattering laser light back to the probe. These pulses of back-scattered light are detected by the probe

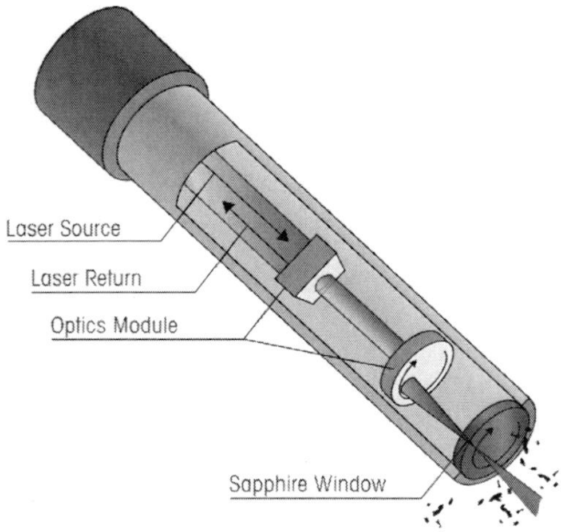

Figure 11.12 Simplified schematic of the FBRM® probe.

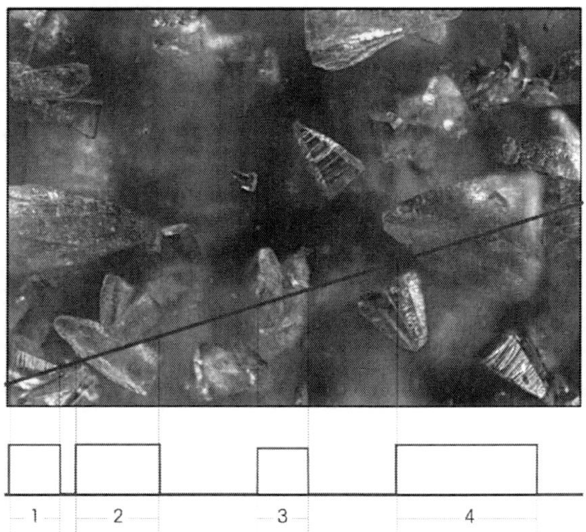

Figure 11.13 Simulated scan of a focused laser beam through a crystal slurry and the measured chord lengths detected using FBRM®.

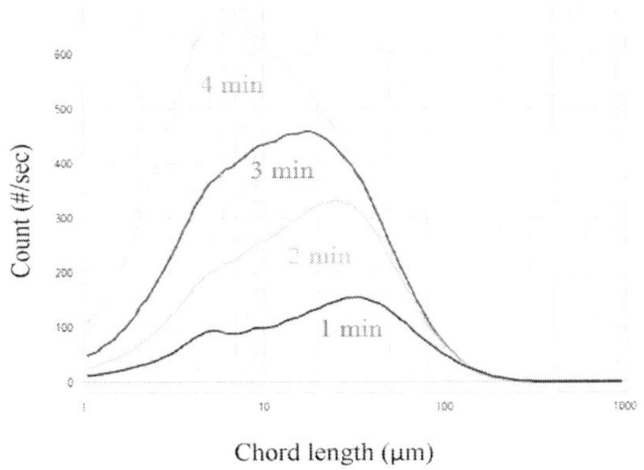

Figure 11.14 The CLD reports the number of particles detected as a function of the measured chord length. (Each curve represents a unique CLD at a given point in time.)

and translated into chord lengths based on the simple calculation of the scan speed (velocity) multiplied by the pulse width (time).

A chord length (a fundamental measurement of particle dimension) is defined as the straight line distance from one edge of a particle or particle structure to another edge. Thousands of individual chord lengths are typically measured each second to produce the chord length distribution (CLD) (Fig. 11.14). The CLD is a "fingerprint" of the particle system and provides calculated statistics to detect and monitor changes in particle dimension and particle count in real time (Fig. 11.15). Unlike most other particle analysis techniques, FBRM® measurements assume no particle shape. This allows the fundamental measurement to directly track changes in the particle dimension, shape, and count.

11.1.3.2 Particle video microscope

A PVM® is a real-time, probe-based vision tool that provides instant critical insight into crystal, particle, and droplet systems. PVM® enables chemists and engineers to detect and understand process changes that could take months to discover with traditional off-line microscopy techniques. PVM® uses a high-resolution charge-coupled device (CCD) camera (Fig. 11.16) and internal illumination source to obtain high-quality images, even in dark and concentrated

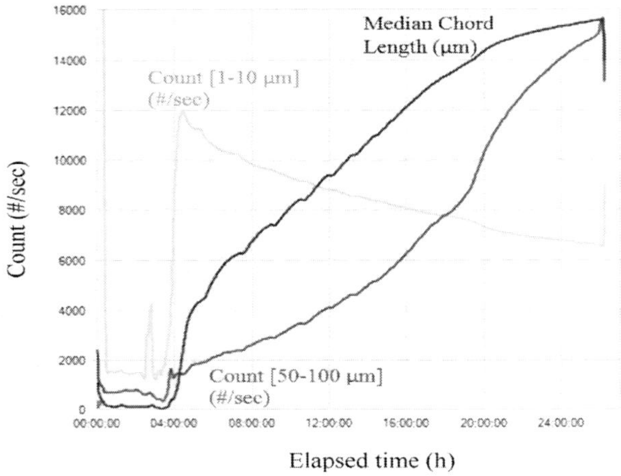

Figure 11.15 Real-time monitoring of changes in the CLD provide trend lines that track changes in particle count and dimension.

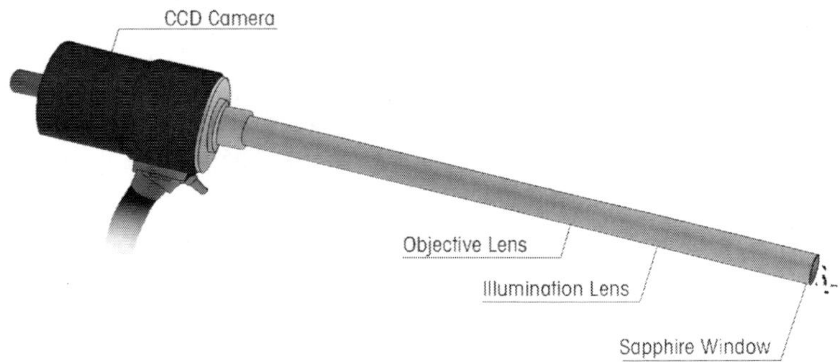

Figure 11.16 Schematic diagram of PVM probe for in situ microscopy.

suspensions or emulsions. With no calibration needed and easy data interpretation, PVM® quickly provides critical knowledge of crystal, particle, and droplet behavior.

The benefit of in situ monitoring is illustrated in Fig. 11.17. In this example, the crystal product filters very poorly and off-line microscopy reveals a wide distribution of fine crystals, potentially caused by many different issues and therefore not easily solved. In situ imaging quickly reveals that the crystals are actually being produced in a dendritic crystal habit—which maintain their form in

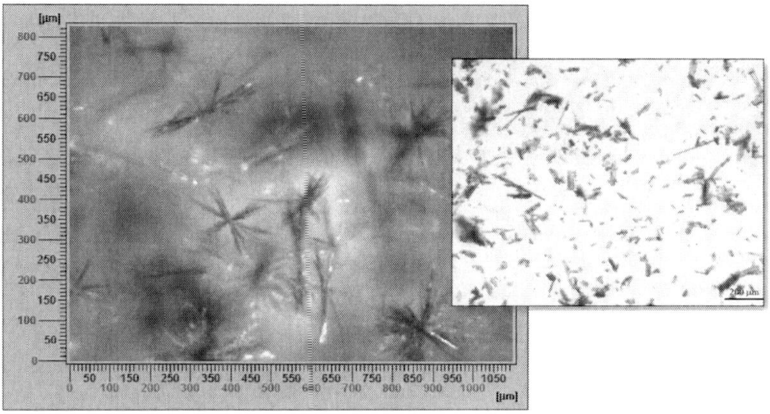

Figure 11.17 The in situ PVM image (left) provides evidence that crystals are forming with a dendritic crystal habit. Off-line microscopy (right) only reveals crystal fragments after sample withdrawal and slide preparation.

the vessel but are crushed to fine dust during sampling, filtration, and slide preparation. The dendritic crystal form, which in this case was caused by poor seed preparation, was more easily remedied with the additional information provided by in situ microscopy.

11.2 Case Studies

11.2.1 *Process Analytical Technologies for Continuous Processing and Microreaction Technology*

Continuous production has been applied for a very long time in most petrochemical plants in the oil and gas industries. Some processes in the chemical industry have also been manufactured in a continuous manner when it provides the benefit of better control and cost reduction. It is only relatively recently that the fine chemicals and pharmaceutical industries have invested time and effort in cost reduction and efficient manufacturing practices. As a result, scientists have started looking carefully at the strengths and limitations of continuous and microreaction technology.[14] Still, very few published examples of continuous drug manufacturing are available from the pharmaceutical industry. Considering the many green engineering attributes of continuous flow production along with the increasing focus of industry on green practices, it is likely that many more examples will soon be available in the literature.

One of the hindrances preventing a faster and broader development of flow chemistry has been the low availability of convenient, specific, in-line monitoring techniques to take full advantage of flow technology. One of the preferred in-line techniques now becoming widely implemented is ATR-based FTIR spectroscopy. The main perceived value for the development of flow chemistry is its structural specificity and convenient software control.[15] Literature examples are now becoming available, some of them covered later, that show the use of in-line IR for faster reach of steady state, more time-efficient screening of process conditions, and the reduction of waste due to real-time measurements of product quality and concentration. Today, efforts to produce drug and drug intermediates in continuous mode can be found:

- in chemical development at production scale, to avoid scale-up issues, enhance the safety profile, and improve yield. The kinetic and thermodynamic properties of processes to be developed for continuous manufacturing are typically studied in a batch mode.
- in drug discovery. Microflow and small-scale flow reactors are safe and space-efficient alternatives to round-bottom flasks. They can be used to prepare grams to kilograms of material involving the use of highly energetic transformations (nitration, diazotization, hydrogenation) or chemistries that require high temperatures ($<200^\circ$C).

The following examples were chosen to cover both aspects.

11.2.1.1 Development of a safe, scalable, and continuous process for the preparation of 6-hydroxybuspirone

This example from Bristol-Myers Squibb, describing the three-step preparation of 6-hydroxybuspirone, began with a first publication[16] explaining how FTIR was successfully used as a PAT to gain a good understanding of key process variables and their impact on process output (quality, yield). A subsequent article[17] described the development of a continuous process to 6-hydroxybuspirone.

6-Hydroxybuspirone is an active metabolite of buspirone, itself manufactured and marketed as Buspar, employed for the treatment of anxiety disorders and depression. The work described in the articles originated when several kilograms of 6-hydroxybuspirone were needed for toxicological studies, formulation development, and further clinical trials. The effort was justified because the initial drug discovery route lacked the required robustness for further scale-up, resulting in unreliable product quality.

6-Hydroxybuspirone **2** was prepared from buspirone **1** via deprotonation with potassium hexamethyldisilazide (KHMDS) followed by oxidation of the resulting anion with oxygen, and final reduction with triethylphosphite (Scheme 11.1).

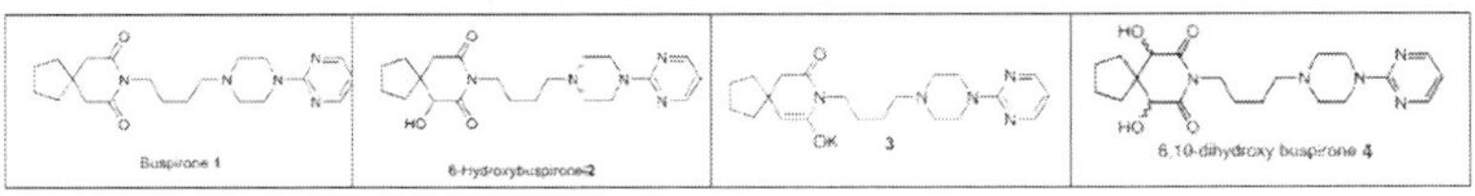

Scheme 11.1 6-Hydroxybuspirone 2 is made from buspirone 1 via the corresponding potassium enolate 3. 6,10-Dihydroxy-buspirone 4 is a by-product resulting from overoxidation of buspirone 1.

The value of PAT was clearly demonstrated when improving the deprotonation step. The main challenge was to prevent:

- the use of excess KHMDS, the consequence of which was the formation of 6,10-dihydroxy-buspirone **4**
- an undercharge of the base KHMDS, resulting in incomplete conversion of buspirone **1** and final contamination of 6-hydroxybuspirone with unconverted starting material

A typical constraint of larger-scale implementation has to do with variability of raw material quality. In this case, potential variation in base concentration, water content, and phosphite quality from batch to batch resulted in a changing impurity profile, an unacceptable proposition for a scalable process.

The use of real-time in situ reaction monitoring with FTIR quickly became the best option to the conundrum of careful control of KHMDS in a situation where its titer may differ from batch to batch.

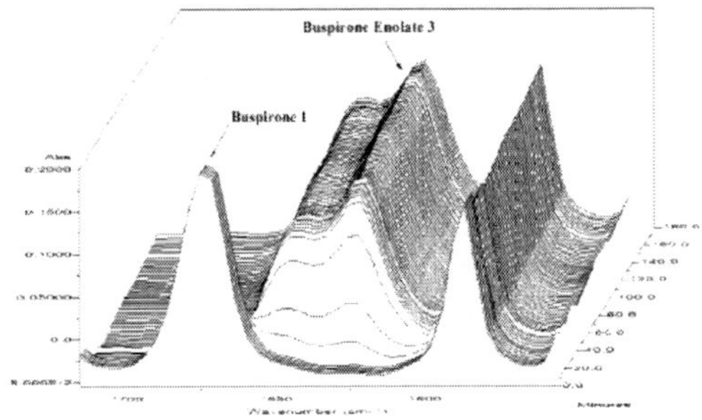

Figure 11.18 A three-dimensional waterfall plot showing the formation of buspirone potassium enolate 3

A promising proof-of-concept study showed that buspirone **1** and buspirone enolate **3** have distinct IR absorbencies (1,677 and 1,627 cm^{-1}, respectively, Fig. 11.18). This three-dimensional waterfall plot shows mid-IR peak intensity (y-axis) against wave number (x-axis) as a function of reaction progression (z-axis, time projecting into the page).

As shown in Fig. 11.19, stepwise addition of the base is visualized by the corresponding stepwise decrease in buspirone concentration and an incremental increase in enolate concentration.

This preliminary investigation also indicated that:

- deprotonation was fast (complete within five minutes)
- it could be conducted at −25°C since the enolate was stable for at least 12 hours at −25°C

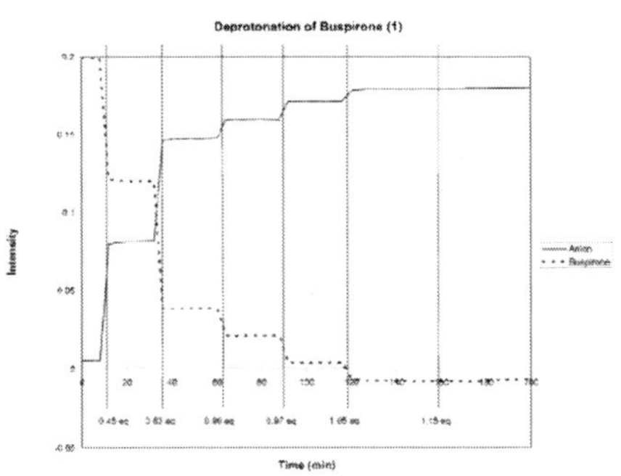

Figure 11.19 Relative concentration trends of buspirone 1 and buspirone potassium enolate 3.

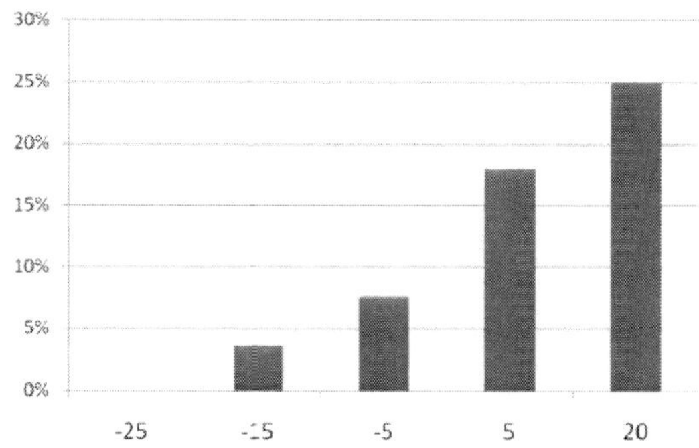

Figure 11.20 Enolate degradation (%) as a function of temperature (°C) over 12 h.

- the addition of triethyl phosphite before addition of the base did not affect FTIR monitoring

A kinetic investigation of enolate degradation versus temperature is summarized in Fig. 11.20. It shows adequate stability of the enolate at −25°C over a 12-hour period. The use of real-time FTIR allowed completion of the study in a time-efficient manner.[18]

To minimize the presence of contaminants **1** and **4** in the final product, that is, finding the correct amount of KHMDS, a strategy was developed where the base was added until the IR band specific to the starting material (ketone function of buspirone **1** at 1,677 cm^{-1}) stopped decreasing. This indicated complete conversion of the starting ketone **1**. At this point, where excess base might have been present, it could be consumed by back-charging a small amount of starting buspirone **1.** The back-charge was continued until 1–3% excess buspirone **1** was observed based on IR absorbance intensity, as illustrated in Fig. 11.21.

This process strategy was first tested at the laboratory scale by running several batches of 6-hydroxybuspirone **2** and then implemented at the pilot plant scale on 13.42 kg of buspirone **1**. Product **2** was obtained in 69% yield and >99 area % purity.

Although FTIR is particularly well suited to the monitoring of unstable intermediates,[19] like enolates, where more traditional methods like high-performance liquid chromatography (HPLC) tend to fail, the FTIR application range is very broad. Figures 11.22 and

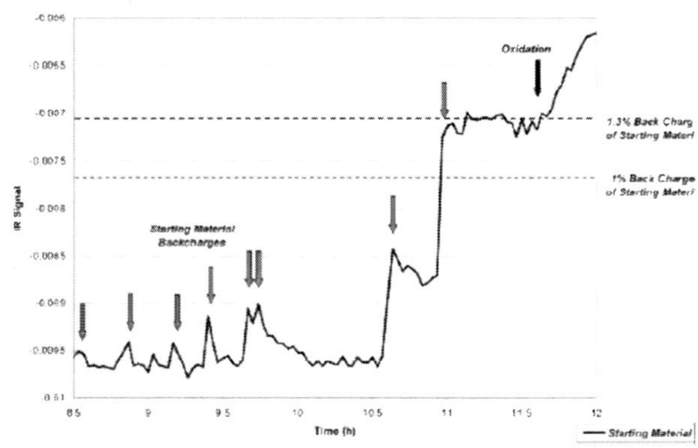

Figure 11.21 Real-time monitoring of the buspirone 1 relative concentration during the back-charge step.

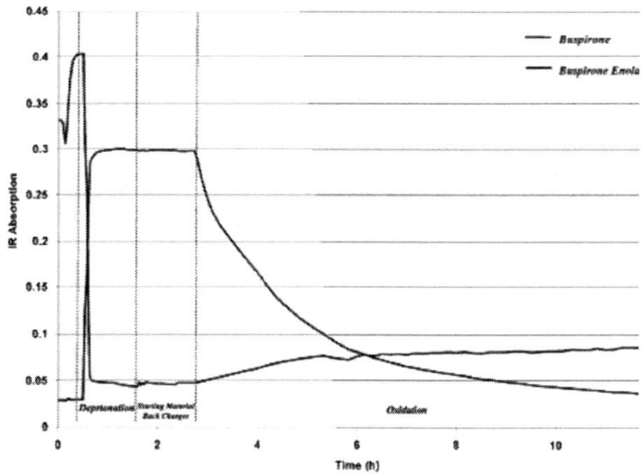

Figure 11.22 6-Hydroxybuspirone whole-process monitoring using FTIR.

11.23 show FTIR monitoring of the process from beginning to end, including the oxidation step.

This case study highlights various aspects of real-time FTIR spectroscopy to design a greener process:

- minimizing contaminants **1** and **4** in the final product, resulting in:

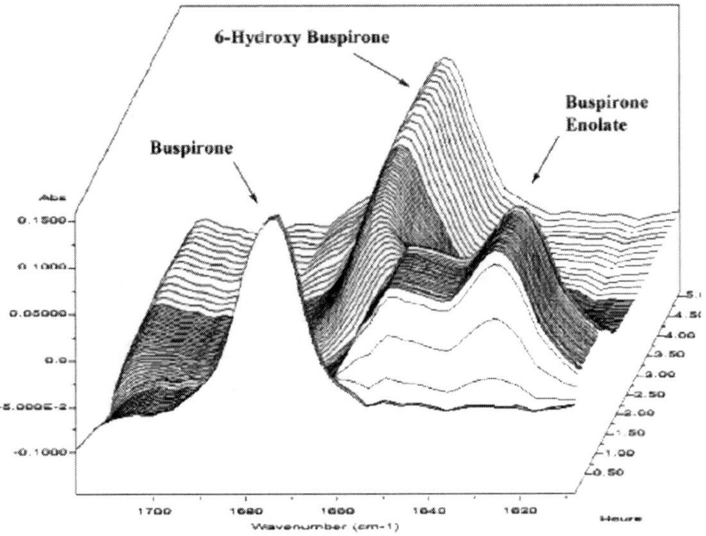

Figure 11.23 6-Hydroxybuspirone whole-process monitoring using FTIR.

- shorter time and fewer resource-consuming final purification stages
- smaller losses to waste streams
- finding the highest possible enolization temperature (−25°C) in a time-efficient manner
- optimizing the amount of KHMDS and accommodating batches of base of ambiguous or different concentration since, more generally, the quality of chemicals in lab quantities often differs from chemicals purchased in bulk for larger-scale manufacturing

The fast response time provided by real-time reaction monitoring (in seconds), applied to the monitoring of a fast enolization process, allows real-time control of process output via constant adjustment of process variables: temperature, excess starting material, and the amount of KHMDS.

A follow-on article[16] described the development of a continuous process to 6-hydroxybuspirone **2** on the basis of the knowledge acquired during the development of the previously described batch process. The strategy discussed before allowed careful control of the

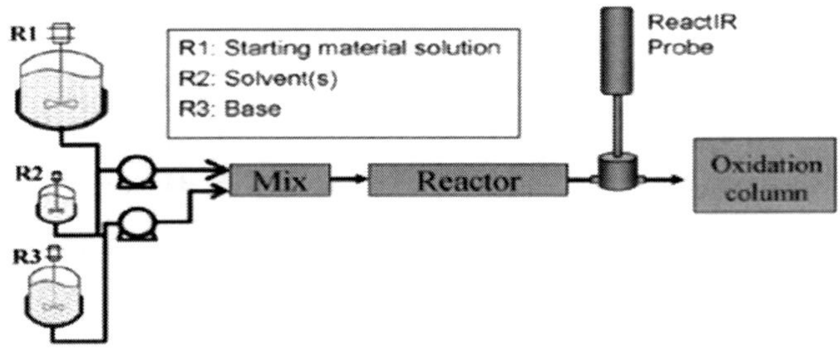

Figure 11.24 Hydroxybuspirone synthesis—a process flow schematic.

KHMDS/buspirone **1** ratio and resulted in enhanced quality of 6-hydroxybuspirone **2**. Even more so than in batch mode, the use of off-line analytical methods to monitor continuous processes is time consuming and wasteful. Indeed, a product of unsuitable quality is often generated, and thus wasted, during the time it takes to obtain analytical results. Conversely, the paper shows that in continuous mode, the use of real-time monitoring allowed shortening of the start-up phase, minimizing the risk of contamination of product streams of different quality and, as a result, optimizing production output.

The previously described strategy to attain the proper base/starting material ratio consisted in a first charge of the base, followed by a careful back-charge of a small amount of buspirone **1** (Fig. 11.24). This strategy was easier to implement in continuous mode as pumping rates of both the base and buspirone **1** were based on real-time feedback from in-line FTIR. A flow cell equipped with a ReactIR™ DiComp probe was used for this purpose.

As displayed in Fig. 11.25, the process could be broken down into the following phases:

1. The solvent and starting material feed were pumped through the column.
2. The solvent was replaced by the KHMDS feed at a flow rate slightly under the calculated stoichiometry.
3. The base flow rate was then incrementally raised until no decrease of buspirone **1** was observed, meaning that all starting material was converted into the corresponding enolate.

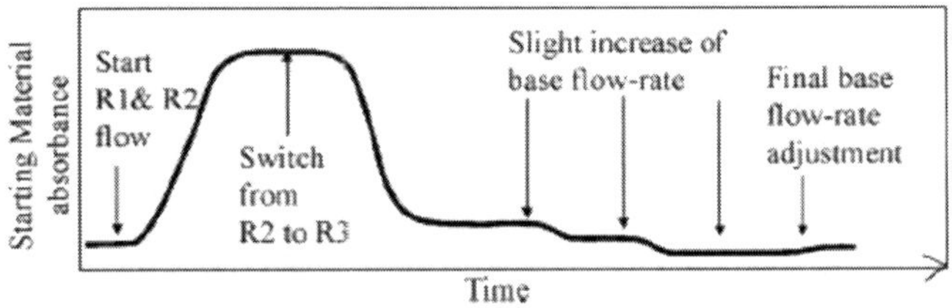

Figure 11.25 Start-up procedure based on the use of ReactIR™.

4. The base feed rate was reduced by 1–3% to keep it slightly undercharged, therefore ensuring the absence of diol impurity **4**.

As can be seen in Fig. 11.25, equilibrium was reached soon after a change in flow rate was implemented, making the strategy easy to implement. The fast enolization rate also facilitated the approach.

After validation at the laboratory scale, the continuous process to 6-hydroxybuspirone was scaled up to the pilot plant and used for the manufacturing of over 100 kg in three different batches.

This publication underscores some of the benefits of real-time process analysis over sampling for off-line analysis when developing and implementing continuous processing. Minimizing waste was shown to be one of the key benefits thanks to:

- shortening of the start-up phase and more rapid attainment of the steady state via real-time detection of phase transitions.
- monitoring of enolization during the steady state in the lab and the pilot plant. As a result, the product stream could be diverted to a waste drum when product quality did not meet specifications, thereby preventing contamination of the active pharmaceutical ingredient (API).

11.2.1.2 Novel innovation systems for a cellular approach to continuous process chemistry from discovery to market

A publication from Cellular Process Chemistry (CPC-Systems) recounts microreaction technology (Fig. 11.26) used for small- and large-scale preparation of APIs and intermediates. It gives an

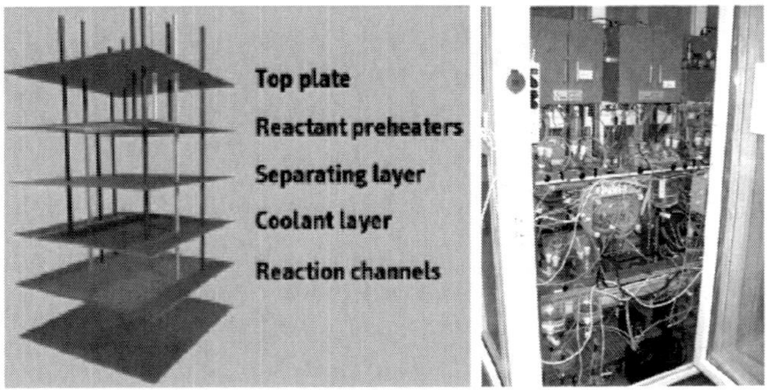

Figure 11.26 Microreactor module (left) and pilot plant scale equipment (right).

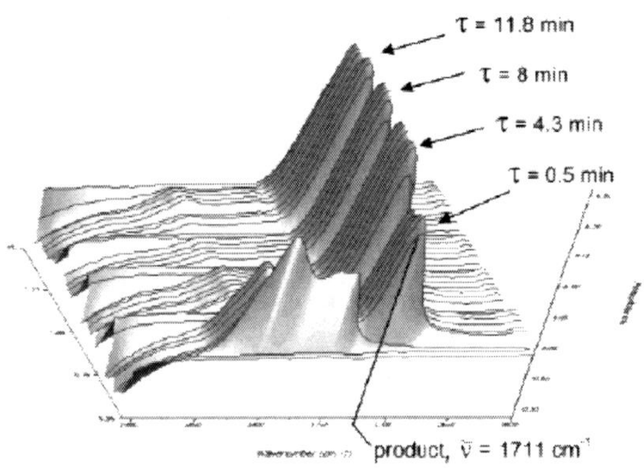

Figure 11.27 In-line IR analysis of the Boc protection (initial experiment). Carbonyl absorbencies: 30 min, 1,711 cm^{-1}; Boc_2O, 1,750, 1,800 cm^{-1}.

overview of the benefits and strategy when using microreactor systems and describes case studies run in collaboration with GlaxoSmithKline (Fig. 11.27) and Clariant where the technology was applied for the preparation of gram- to ton-scale material.

In-line analytics with IR served as a key strategy to minimize compound consumption, shorten process cycles, and maximize throughput. It was used to rapidly screen reaction kinetics and improve reagent concentration and temperature for the preparation of Boc-protected benzylamine (Scheme 11.2).

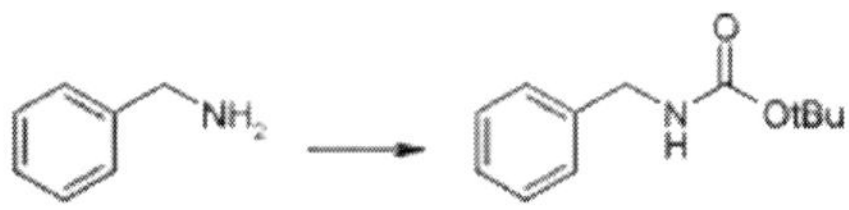

Scheme 11.2 Boc protection of benzylamine.

Figure 11.28 displays the successive IR spectra obtained when screening residence time (τ) by manipulation of flow rates. Change in product concentration at the outlet was monitored using the specific carbonyl absorbance at 1,711 cm^{-1}. The band between 1,750 and 1,800 cm^{-1} showed unconverted Boc_2O reagent. Thanks to the use of real-time FTIR, the whole set of incremental residence times was screened in a much shorter time (less than 30 minutes) than required using alternative analytical methods. Finding the point of maximum conversion or optimum product formation took less time. The methodology can also be applied to the screening of reagent concentration or temperature, resulting in a faster optimization of reaction output.

11.2.1.3 ReactIR™ flow cell: A new analytical tool for continuous-flow chemical processing

A fascinating publication was recently released by Steven Ley *et al.*[20] from the University of Cambridge (U.K.), in collaboration with

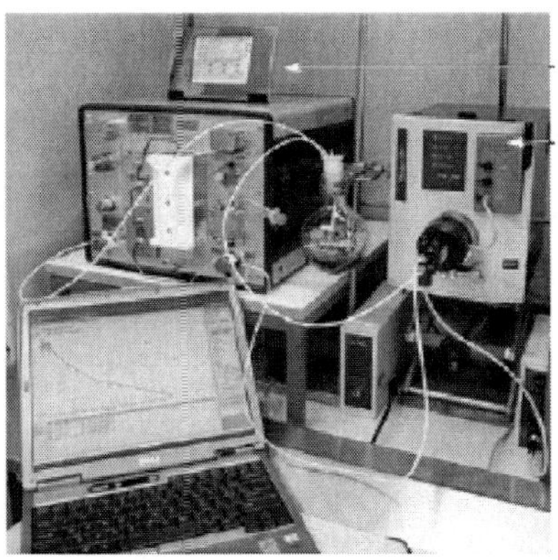

Figure 11.28 Setup for monitoring flow hydrogenation with the IR flow cell and H-Cube Midi™.

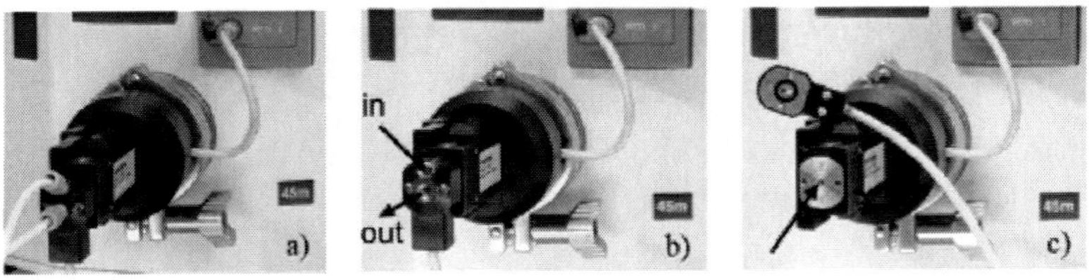

Figure 11.29 ReactIR™ 45 m with prototype IR flow cell. (a) Flow cell in line with OmniFit adapters; (b) flow cell without OmniFit adapters, direction of flow stream through cell indicated; and (c) head unscrewed, free view on diamond window (arrow).

Mettler Toledo Autochem. It shows a wide range of applications of PAT to resolve synthetic issues that concern both discovery and process chemists. Thanks to the use of real-time FTIR, a specially engineered microflow cell head (Figs. 11.29 and 11.30), and a control software designed for reaction monitoring, they were able to monitor reagent consumption and product formation for a variety of chemistries with little need for off-line analytics.

The time to optimize process conditions was shortened, and thanks to the fast response time and structural information contained in each of the many reaction spectra obtained, short-lived intermediates were observed in situ, helping provide a better understanding of the reaction mechanism. Through a number of case studies, various aspects and challenges posed by flow chemistry were investigated by Prof. Ley and his team:

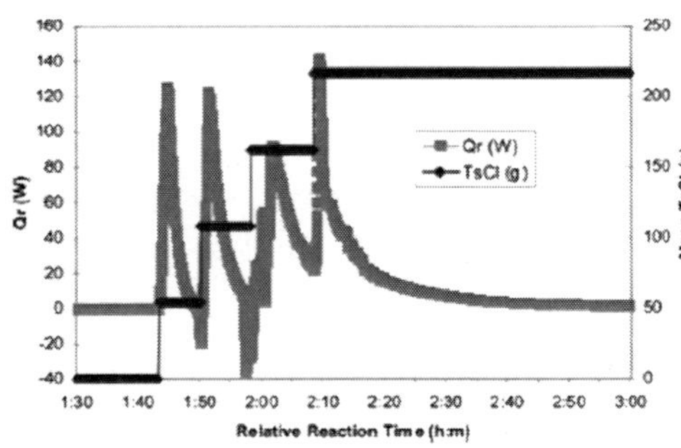

Figure 11.30 Heat flow of 3 ditosylation using solid TsCl.

- Dispersion in the column, impacting time for product to exit. Although ultraviolet (UV) detection had been used for this purpose, two case studies showed that the structural specificity of IR allowed better identification of components.
- Convenience of real-time detection when off-line analysis is challenging. ReactIR™ was coupled to an H-Cube instrument and successfully monitored several hydrogenation reactions.
- Very rapid screening of reaction conditions in one single setup, an alternative to several batch reactions. Case studies involved optimization of an acetal protection reaction and a procedure involving Marshall's chemistry.
- Use of spectroscopic information to identify reactive intermediates. Standard protocols typically involve time and resource-consuming chemical transformations or UV detection, which lacks chemical specificity. A Curtius rearrangement example shows the practicality of in-line IR to track the presence, concentration, and decay of an acyl azide stable enough to be observed in the analytical flow cell but too unstable and hazardous for isolation.
- Monitoring of hazardous azide compounds. In-line monitoring protects the experimenter and can contribute to the development of inherently safer processes. An instance is given where a waste stream was analyzed for the presence of the highly hazardous HN_3.
- Analytical flow cell applied to batch reactions. Although somewhat counterintuitive, the use of ReactIR™ to monitor reactions in a standard vessel using a pump enables simplicity over the use of the fiber probe technology and offers the full mid-IR spectral range (600–4,000 cm^{-1}). This window can be especially valuable early on during the investigational phase of a research project. Prof. Ley and his team applied this idea to a peptide-coupling reaction.

11.2.2 *Process Analytical Technologies for the Greening of Batch Processing*

11.2.2.1 Efficient synthesis of 8-oxa-3-aza-bicyclo[3.2.1]octane hydrochloride

Product and process safety are some of the essential principles of green chemistry and green engineering. A recent publication from Terrence Connolly *et al.*[21] at Wyeth (now Pfizer) developed some aspects of early process safety assessment for the purpose of producing material on a 10 kg scale in the pilot plant.[22] The publication also contained information regarding the use of in situ and real-time mid-IR spectroscopy to enhance process understanding. The sequence of reactions described in the publication is presented in Scheme 11.3.

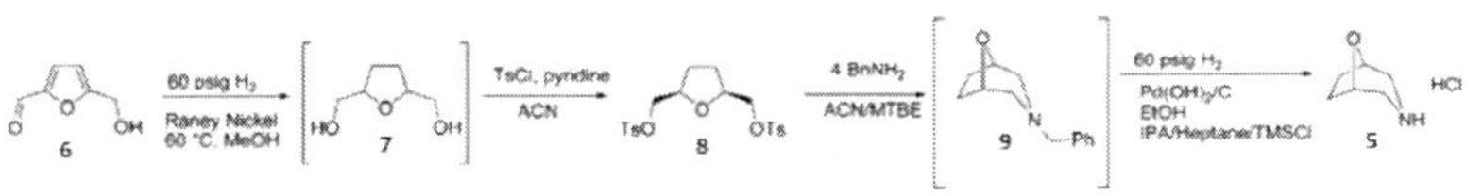

Scheme 11.3 Synthetic pathway to 8-oxa-3-aza-bicyclo[3.2.1]octane hydrochloride.

The conversion of diol **7** into ditosylate **8** was subject to a preliminary heat investigation using an RC1*e*™ calorimeter. The first experiment was run with little or no changes compared to procedures typically used on a small laboratory scale: solid tosyl chloride (TsCl) was added to a mixture of diol **7** in acetonitrile and pyridine at room temperature. As shown in Fig. 11.31, with reaction heat as a function of time, each addition of solid TsCl led to the release of a significant amount of heat. Total heat output for the reaction was 152 J/g of reaction mass, corresponding to a ΔT_{ad} of 151°C. This means that a hypothetical charge of all the TsCl to the reaction vessel in the absence of any cooling (adiabatic conditions) would cause a temperature increase up to 151°C. This theoretical temperature increase does not take into account potential secondary reactions that may be triggered at elevated temperatures. As a result, the actual reaction temperature could rise even higher. This would require further investigation using differential scanning calorimetry (DSC) experiments. The computation of this heat data

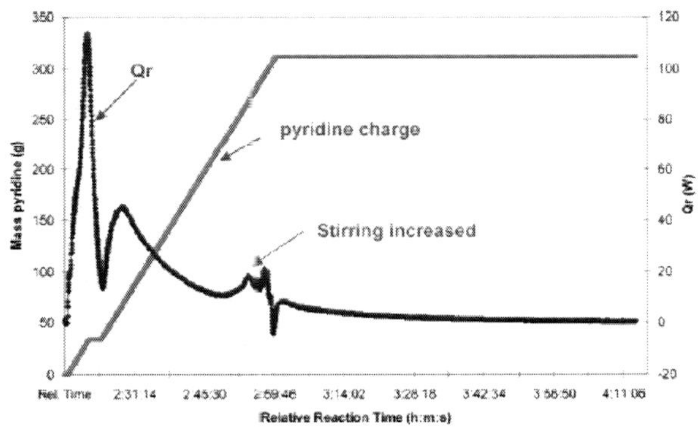

Figure 11.31 Heat flow of **3** ditosylation—addition of pyridine.

into valuable process safety information could be facilitated using modern software technologies (iCSafety™). Despite the significantly high ΔT_{ad}, the heat plot also shows that every portion of TsCl reacts quickly: the heat signal spikes but comes promptly back to the baseline after the addition is completed. One can imagine running such an exothermic process under dose-controlled conditions to make it safer.

More problematic for further scale-up of the procedure were:

- the risk associated with the introduction of a solid reagent to a pilot plant vessel containing a flammable solvent.
- the need for a tin catalyst to achieve reaction completion (mono-diol into diol **7**). In a multipurpose pilot plant, this would require a special cleaning and testing procedure to ensure no tin was carried over to the next batch.

A modified procedure was tested in the same RC1*e*™ system. Pyridine was added to a solution of diol **7** and TsCl. Complete conversion was reached, thanks to an increase in the amount of pyridine from 10 to 13 equiv. and a longer reaction time (16 hours).

Figure 11.32 shows the corresponding heat recording as a function of time. Although pyridine addition was linear, the heat signal showed multiple events:

- A major initial heat spike in excess of 300 W, attributable to activation of diol **7** with TsCl

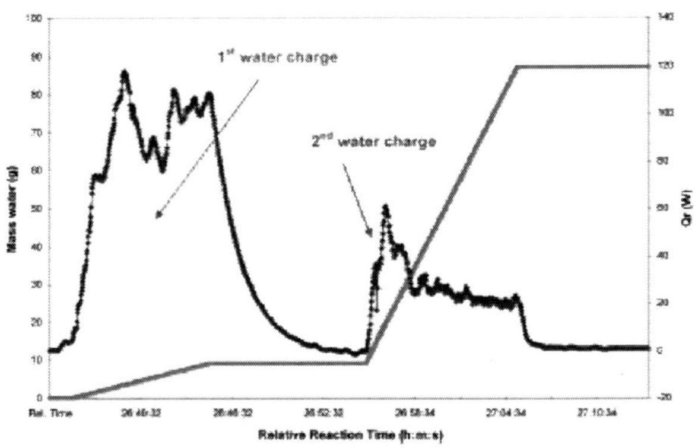

Figure 11.32 Heat flow during aqueous quench of ditosylate **4**.

- A moderate exothermic event from 1:30 to 2:45: simultaneous tosylation and HCl neutralization
- A slight heat increase at 2:50 caused by an increase in stirring speed deemed necessary because of mixture viscosity

Interestingly, the exotherm appeared to be fully dose controlled: when the feed was interrupted (at about 1:00), the heat signal decreased significantly. This means process safety concerns could easily be addressed through feed control. Heat output for the modified procedure was found to be 116 J/g of reaction mass, resulting in a more manageable ΔT_{ad} of 55°C.

As this case study shows, the safety associated with quenching a reactive mixture can be tricky. The high viscosity reached at the end prevented the reaction from being quenched in the safest way—addition of the reaction mixture to water. Instead, water had to be added to the reaction vessel.

Figure 11.33 shows the heat profile and water dosing as a function of time. During the first part of the quench, heat release peaked at 90 W, a significant amount of energy considering the scale (approximately 600 mL). ΔT_{ad} over the entire reaction remained moderate (45°C), but most of the heat was actually released during the first charge, when most of excess TsCl was hydrolyzed. More water was added in a second portion (60 g vs. 8 g), which led to a more moderate exotherm of 20 W.

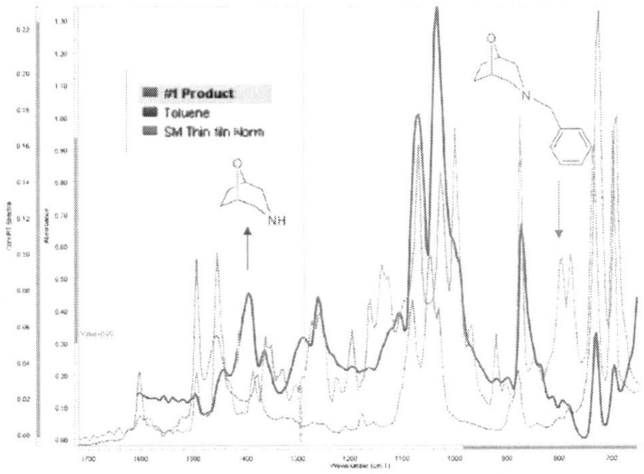

Figure 11.33 Mid-IR peak assignment for benzylamine **9** and product **5**.

The benefit of such an early-on calorimetry investigation using RC1*e*™ technology was to draw attention to operating conditions presenting a risk on a larger scale. As a result, the scale-up procedure could be adjusted accordingly. In this case, pilot plant operation was advised to carefully start adding water at a slow pace (0.2 kg/min rate) and then incrementally increase the rate (up to 6 kg/min) as reaction temperature proved controllable. The process was eventually applied to the preparation of 28.3 kg ditosylate **8**, obtained in 83% yield.

The next step corresponds to the cyclization of ditosylate **8** using double substitution of the tosyl groups with benzylamine, to form compound **9**, and then 8-oxa-3-aza-bicyclo[3.2.1]octane hydrochloride **5** after hydrogenolysis (Scheme 11.3). As is often the case when a chemical reaction needs to be used on a larger scale for batch manufacturing, the following questions arise:

- How long does it take to reach completion?
- Is the product stable if reaction time is extended?

Because of the time it takes to run in-process controls and the potential lack of personnel at night or on weekends, running a process in the plant takes more time than it does in the lab. As a result, product stability in the reaction mixture under extended

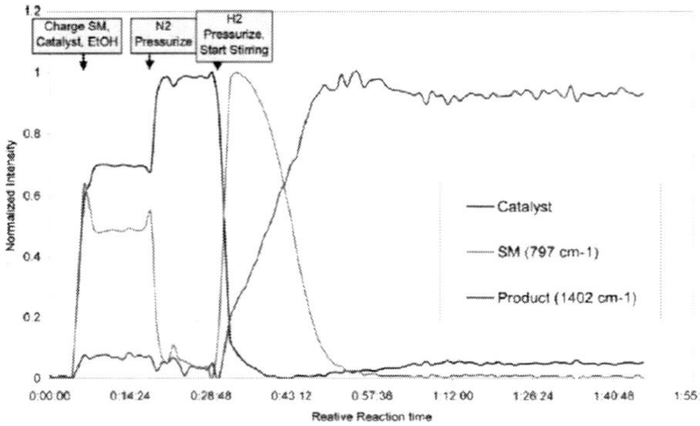

Figure 11.34 Relative concentration trends for starting material, product, and catalyst.

reaction time needs to be examined. In situ FTIR is probably the simplest and the most efficient method to answer these questions. A MonARC™ system with a probe mounted at the bottom of a pressure Parr vessel was used to monitor the hydrogenolysis reaction. Figure 11.34 shows the unique mid-IR absorbance bands for benzylamine **9** (starting material) and the final bicyclic amine **5**:

- 1,455, 1,166, and 797 cm^{-1} for **9**, likely attributable to the aromatic ring
- 1,402 and 1,294 cm^{-1} for amine **5**

Figure 11.34 gives information about component structure and appearance or disappearance from the reaction mixture. Unique bands of interest can then be trended against time to offer kinetic information, as seen in Fig. 11.35. In the absence of sufficient stirring, the catalyst may settle at the bottom of the vessel and make the IR sensor blind. This is why starting material concentration (Fig. 11.35), as measured using the IR peak at 797 cm^{-1}, reached its maximum only when stirring was started (approximately 0:28:00). Then, it decreased as product concentration (band at 1,402 cm^{-1}) grew. Although not demonstrated, it seems that the two trends are complementary, indicating the absence of reaction intermediates. Less than half an hour after stirring was started, starting material and product profiles leveled off and the reaction reached its endpoint. No product degradation was observed. The process was

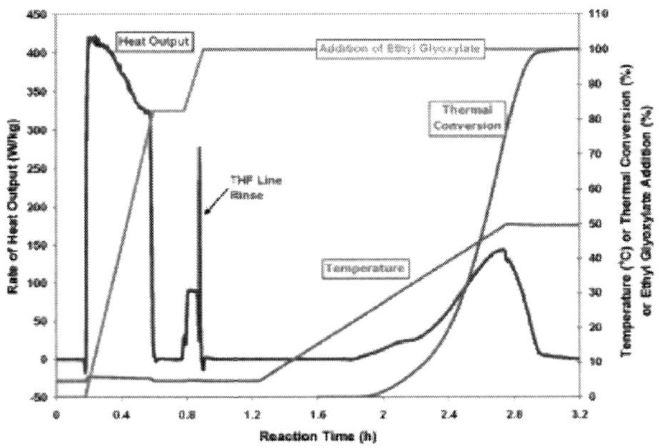

Figure 11.35 Heat output during hydrazone formation using the original procedure.

demonstrated in the pilot plant on 27.5 kg ditosylate **8**. The reaction took twice as much time (48 hours) to reach completion as it did in the lab (24 hours), which is typical of mass transfer-limited reactions under larger-scale conditions.

As often described among professionals, "Process control is environment control." This work is an illustration of the benefits of having a PAT "eye" to monitor the reaction as it occurs in the reaction vessel. That "eye" provided real-time thermodynamic knowledge during the preparation of ditosylate **8**, resulting in modified and safer scale-up conditions in the pilot plant. It also provided spectroscopic information during the debenzylation step, ensuring that:

- the reaction time would not be extended beyond reaction endpoint
- the product was stable
- reaction conditions could successfully be scaled up

Some of the green chemistry principles (real-time analysis for pollution prevention, design of safer chemical syntheses and processes) are properly illustrated in this example.

11.2.2.2 Development of a scalable synthesis of GSK183390A, a PPAR α/γ agonist

Cutting process cycle time and reducing plant personnel's exposure to toxic compounds can clearly be touted as credits to green chemistry principles.[23] This example from GlaxoSmithKline[24] illustrates how to leverage the combined use of calorimetry (RC1e™ and real-time FTIR reaction analysis (ReactIR™) to develop a safer and better-controlled process.[25] The original preparation of GSK183390A, a potent dual agonist of PPAR α/γ and a candidate for treatment of dyslipidemia, was long and convoluted (Scheme 11.4). Their publication reported on the development of a scalable synthetic route that ended up with the preparation of 40 kg of GSK183390A. Although several of the steps were investigated using thermal and spectroscopic techniques, two steps were subjected to a more in-depth investigation and will be the focus of this section.

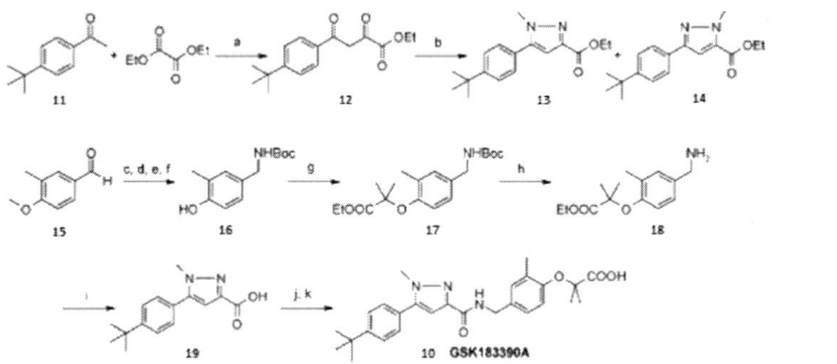

ᵃ (a) NaOEt, 0 to 80 °C, 99%; (b) CH_3NHNH_2, EtOH, 90 °C, **13** 22%, **14** 45%; (c) $NH_2OH \cdot HCl$, NaOAc, EtOH, rt, 93%; (d) ammonium formate, Pd/C, reflux MeOH, 50%; (e) aq HBr, 97%; (f) Boc_2O, CH_2Cl_2, Et_3N, 96%; (g) K_2CO_3, DMF, ethyl bromoisobutyrate, 69%; (h) TFA, CH_2Cl_2, 82%; (i) NaOH, rt, 98%; (j) $SOCl_2$, toluene, 80 °C, then Et_3N, **18**, rt, 96%; (k) NaOH, 80 °C, 73%.

Scheme 11.4 Original synthesis of GSK183390A.

One of the key steps to be modified due to regioselectivity issues was the coupling step leading to the formation of **19**. The scale-up of hydrazonoyl bromide **22**, a coupling partner in the synthesis of **19**, was investigated in detail (Scheme 11.5).

The initial laboratory procedure for the preparation of hydrazone **21** was based on the addition of ethyl glyoxylate in toluene to a tetrahydrofuran (THF) solution of methylhydrazine at 0°C, followed

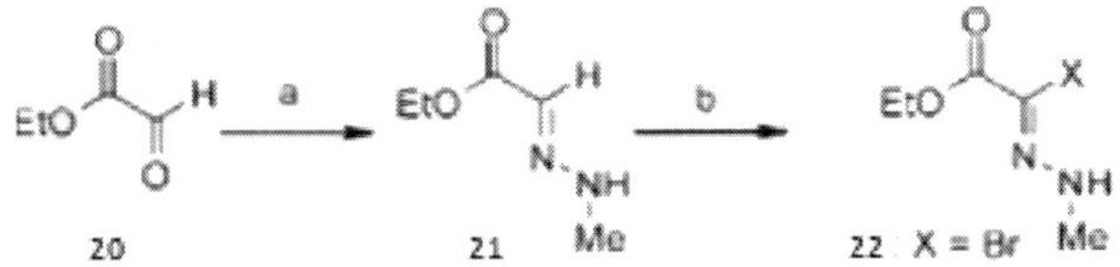

Scheme 11.5 Preparation of hydrazonoyl bromide **22**, the scale-up of which required further investigation.

by heating and holding at 50°C. This procedure presented the following scale-up issues:

- Thermal instability of hydrazone **21** above 75°C.
- Decomposition onset depending on sample purity.
- Reaction heat accumulation (−28 kJ/mol methylhydrazine) visible on the RC1e™ (Fig. 11.36) when the reaction mixture is heated to 50°C (two- to three-hour reaction time). The plot also shows that the heat of reaction during the addition of ethyl glyoxylate is mostly dosing controlled (heat output mostly depending on feed rate) and corresponds to −63 kJ/mol methylhydrazine. This level of heat accumula-

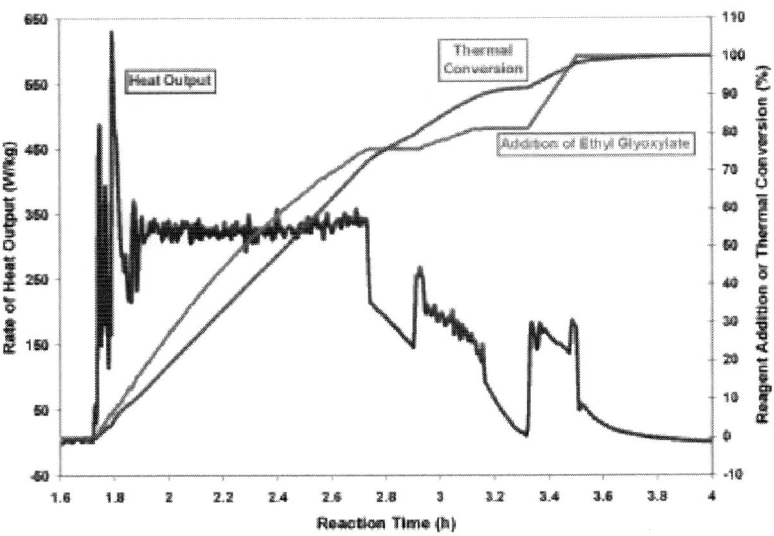

Figure 11.36 Heat output during hydrazone formation using the revised procedure.

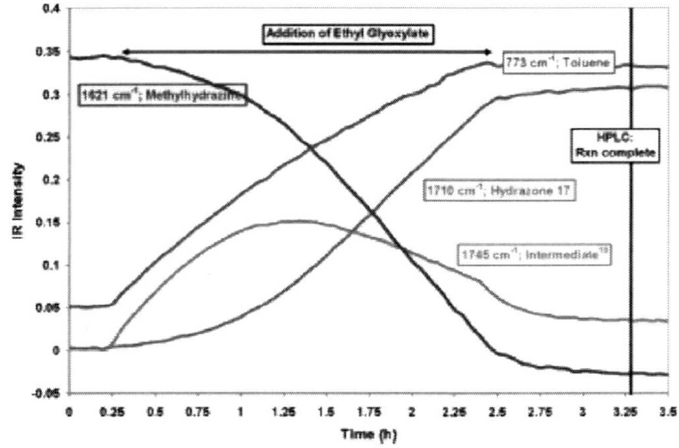

Figure 11.37 ReactIR™ monitoring of hydrazone **21** formation in the pilot plant.

tion together with the known instability of hydrazone **21** was a concern.

A typical strategy is to run the reaction at a higher temperature so it is entirely dosing controlled—in other words completed soon after the addition is finished. Figure 11.37 shows how the process behaves under such conditions: a solution of ethyl glyoxylate in toluene added to a solution of methylhydrazine in methanol at 40°C. Heat release is a higher 75 kJ/mol but entirely addition rate limited. Moreover, the low boiling point of methanol offers an upper boundary to reaction temperature by providing a cooling effect from the condenser in case an uncontrolled exotherm occurs.

As illustrated in previous examples, real-time in situ FTIR spectroscopy, as provided by ReactIR™, provides a deeper insight into reaction evolution at laboratory and plant scale. Figure 11.38 shows the disappearance of methyl hydrazine (band at 1,621 cm^{-1}) over the 2.25-hour addition of ethyl glyoxylate during a pilot plant run monitored by ReactIR™. This disappearance is accompanied by the formation of hydrazone **21** (band at 1,710 cm^{-1}) and an intermediate species at 1,745 cm^{-1}. This intermediate species could be a hemiaminal derivative as the dehydration step is often the rate limiting step. The publication does not further elaborate on

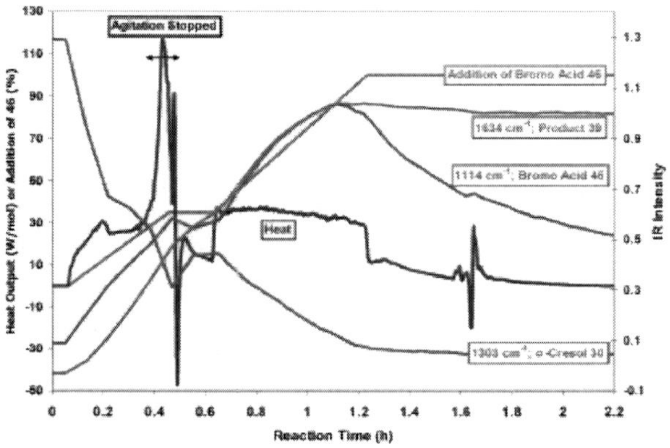

Figure 11.38 Heat output and FTIR trends of the o-cresol alkylation reaction.

the topic. As expected from laboratory runs, the concentration of product **21** levels off soon after the end of ethyl glyoxylate addition, indicative of the reaction endpoint.

This pilot plant run using the enhanced procedure was successful and yielded 39.4 kg hydrazone **21**. The product was deemed of better purity than previously obtained batches because of a higher decomposition onset (124°C instead of 75°C) and lower coloration. Real-time FTIR spectroscopy provided the following benefits:

- The cycle time was reduced, thanks to real-time detection of the reaction endpoint.
- Only one sample was taken from the reaction mixture for final control, which reduced exposure of plant personnel to toxic methylhydrazine.

Alkylation of *o*-cresol to form compound **24**, itself later converted into benzylamine **18** (Schemes 11.4 and 11.6), was also investigated using a combination of real-time FTIR and reaction calorimetry.

Figure 11.39 shows how much information can be obtained in one single run:

- The addition of α-bromoisobutyric acid was accompanied by a large exotherm (−300 kJ/mol), mostly addition controlled. The high viscosity of the reaction mixture caused occasional

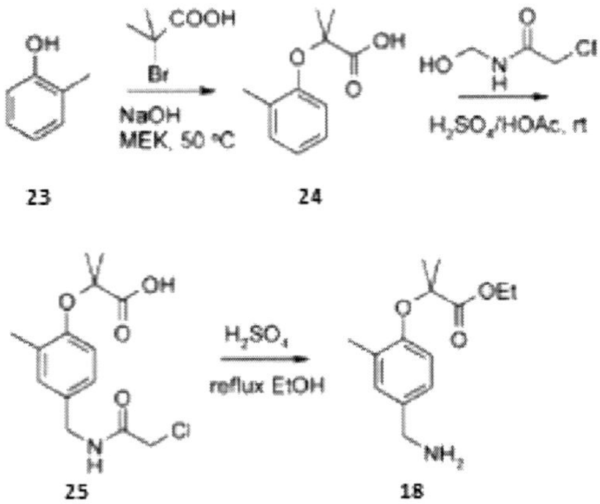

Scheme 11.6 New route to benzylamine **18**.

loss of agitation, which resulted in a large heat spike, clearly a concern for larger-scale implementation.

- FTIR and calorimetry trends are consistent with one another and indicate a reaction endpoint soon after the end of α-bromoisobutyric acid addition.

Thanks to a better understanding of reaction evolution, a modified procedure was devised that involved a slow addition of α-bromoisobutyric acid to the sodium salt of *o*-cresol. The process was finally transferred to the pilot plant and successfully run on 40.3 kg α-bromoisobutyric acid. This example stands out, thanks to the simultaneous use of reaction calorimetry and real-time FTIR that allowed identifying process conditions early in development. As a result, enhanced conditions were developed and successfully applied to the production of 40 kg API. In addition, real-time monitoring at lab and plant scale ensured optimization of cycle time and reduced the number of samples for off-line analysis, thereby minimizing plant operators' exposure to methylhydrazine.

11.2.2.3 Execution of a performic acid oxidation on multikilogram scale

The quest for inherently safer chemical reagents is a worthy goal, fully in accordance with some of the twelve principles of green chemistry. The following example[26] is a good illustration of the choices often available to chemists when deciding for a particular chemical route. It shows how technology can help by providing compelling evidence that the alternative, greener process will be safe to scale up.[27]

En route toward CP-865,569 **27**, a CCR1 antagonist, scientists at Pfizer were faced with the difficult challenge of a synthetic pathway (Scheme 11.7) involving a sulfite displacement generating a large amount of sodium salts.

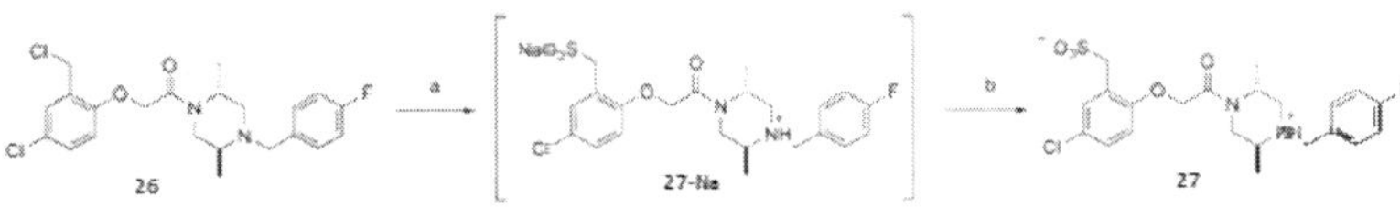

ᵃReagents and conditions: a) Na_2SO_3, MeCN, b) HCl, MeCN, *i*-PrOH

Scheme 11.7 Original synthesis of CP-865,569 using a sulfite displacement.

A cleaner metal ion-free alternative was imagined that would employ an oxidizing step instead. The S-atom could be introduced using a thioacetate group or a bulkier trityl group. The thioacetate option was chosen on the basis of higher atom efficiency and lower cost. Performic acid was selected, thanks to its good environmental profile and its ability to oxidize thioacetate **28** without generating salts (Scheme 11.8).

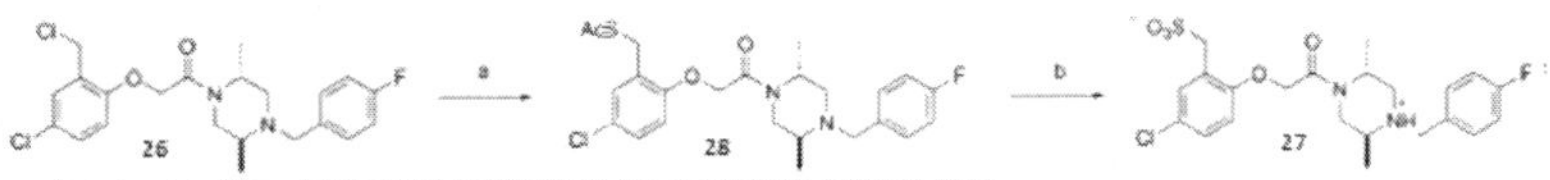

ᵃReagents and conditions: a) KSAc, PhMe, b) HCO_2H, 30% H_2O_2 in H_2O or 30% AcOOH in AcOH

Scheme 11.8 Synthetic route to CP-865,569 involving a thioacetate oxidation step instead.

However, this synthetic route could not be considered "greener" unless the thermal stability of performic acid and the associated heat of reaction could be proven safely controllable. Performic acid's inherent stability was carefully tested using DSC followed by accelerating rate calorimetry (ARC). Reaction calorimetry, using RC1*e*™ technology, was used to investigate performic acid formation and thioacetate oxidation together, leading to the formation of the

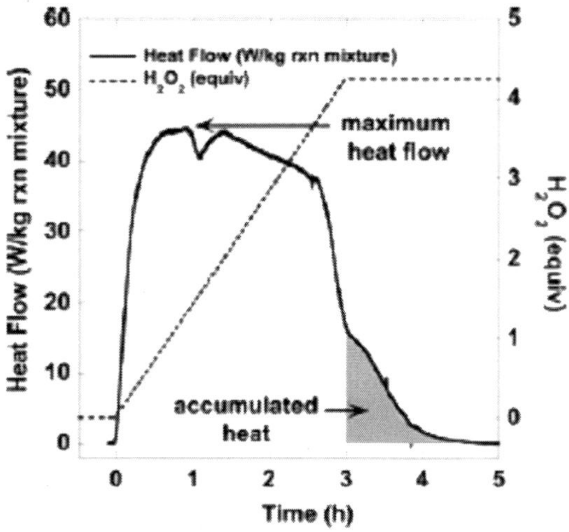

Figure 11.39 RC1e plot for in situ performic acid generation and oxidation of thioacetate **9**.

targeted final product CP-865,569 **27**. The minimum requirements when conducting such a preliminary process safety analysis are the following:

- How much energy does the reaction release?
- What is the instantaneous heat output?
- Is thermal accumulation a challenge, that is, how dosing controlled is the reaction?

The RC1*e*™ experiment, whose plot is shown on Fig. 11.40, gave the answers. The heat generated by adding 4.2 equiv. of 30 wt% H_2O_2 over three hours to a solution of thioacetate **28** in 2.5 volumes of formic acid at 25°C was found to be −975 kJ/mol. The corresponding theoretical adiabatic temperature rise (ATR), the temperature to which the reaction mixture would rise to in the absence of any cooling, is 172°C. More seriously, if the process was to be run under the same conditions on a larger scale, the batch temperature could not be maintained constant at 25°C since the maximum heat output, −44 W/kg, would likely exceed the maximum cooling capacity of the scale-up equipment. However, thermal accumulation, found to be only 9% for a three-hour

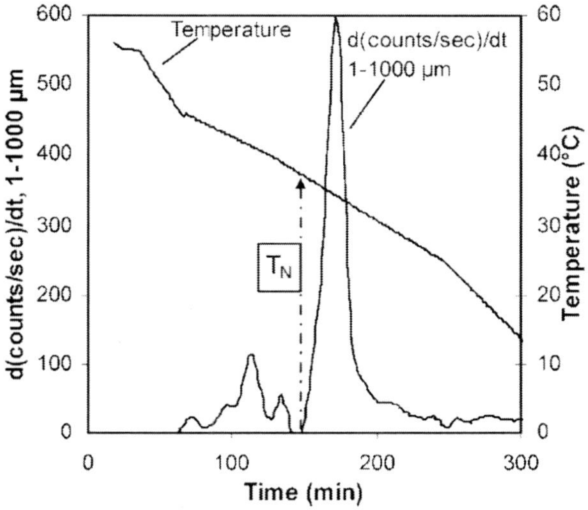

Figure 11.40 First derivative profile of counts/s (1–1,000 μm) by FBRM® and temperature profile. (Seeding occurs when the temperature reaches 46°C—corresponding to the initial detection of particles with FBRM® at a time of approximately 60 min.)

addition time, would easily be minimized using a dosing-controlled procedure. In other words, just slowing down the addition rate would make the process safer.

Reaction quench was also tested using RC1*e*™ technology. The reaction heat and maximum heat output of thioacetate **28** were found to be similarly manageable at −135 kJ/mol and −37 W/kg, respectively. Further calculation and modeling were necessary to address a potential risk of equipment overpressure due to oxygen off-gassing. On the basis of the vent pipe diameter, the equipment available on a larger scale was deemed able to accommodate the anticipated 40 L/h oxygen flow rate that would be generated during the quench phase. The process was eventually transferred to a 300 gal vessel in the pilot plant and used to safely and successfully prepare five batches of 30–35 kg of the final product CP-865,569. The pilot plant runs were monitored using a combination of in situ FTIR using a large-scale ReactIR™ instrument[28] (MonARC™) and sampling for an off-line HPLC assay.

11.2.3 *Applying the Principles of Green Chemistry to Crystallization and Downstream Processing*

11.2.3.1 Introduction

Crystallization is an important process for purification and isolation of intermediates and fine chemical products. From a strict green chemistry approach, it may actually be desirable to avoid crystallization and many associated downstream processes due to the vast quantities of solvents that are added and then separated in energy-intensive crystallization and drying operations. However, from a practical standpoint, crystallization remains a widely used process and one of the least expensive ways in which to purify and isolate an organic compound into the desired solid form. While it is generally easy to get a compound to crystallize, it is acknowledged as an extremely difficult process to optimize. There are often many opportunities for improving the design of crystallization processes to minimize energy consumption and maximize yield of the final product. Over the past decade, process analytical tools have become an essential part of developing vastly improved methodologies for optimization and scale-up of crystallization. Good particle-engineering practices for crystallization can play a significant role in ensuring an overall green process.

Why does this matter from the organic chemist's viewpoint? Within green chemistry it is important to consider the entire process required to manufacture the final product. And when one is going to consider the reduction of waste, the avoidance of exposure to hazardous materials, and the energy efficiency of a process, one should certainly consider the processes that occur downstream from the synthesis steps. It would be a shame to design an elegant synthetic process with exceptional atom efficiency, only to have a significant portion or percentage of the product diverted to waste due to poor design of the separation and product formulation steps. Size reduction of a crystallized product by dry milling, for example, can result in a loss of yield of 10% or more due to holdup in the milling equipment. In addition, the generation of very fine particles during milling may result in a dust that is a potential exposure and explosion hazard. Each of the downstream steps can have a similar impact on the final product yield and overall process efficiency.

With many pharmaceutical and fine chemical compounds, a crystallization step is often the dividing line between the organic synthesis and further physical processing necessary to convert a chemical compound into a final formulated product. For example, a pharmaceutical tablet will proceed through a number of processes that may include filtration and drying of the crystal product; milling and sieving to achieve a targeted crystal size distribution; blending of the API crystals with other excipients according to the final formulation recipe; wet or dry granulation (wet granulation would usually be followed by additional drying); a further sieving to achieve the desired granule size distribution; and finally a compacting of the granulated product into a tablet.

The case studies that follow demonstrate the key role that PAT plays in effective crystal and particle engineering. With the complex nature of crystallization, there is no substitute for real-time monitoring of crystallization for the determination and optimization of critical process parameters in the assurance of downstream and final product performance.[29] But first, here's an exploration of what makes crystallization notoriously difficult to optimize and scale up.

What makes crystallization such a complex process?

The efficiency of a crystallization process is directly tied to the crystal size distribution produced in the crystallizer vessel. Problems with solid bulk density, flowability, and crystal size and shape can often be directly related to the operations of the crystallization step. Missing the crystal size distribution specifications may result in costly rework. The production of excessive fines in the crystallizer vessel can dramatically cut production yields and can reduce throughput due to serious bottlenecks in downstream processing, such as filtration and drying. Excessive milling of the crystal product can result in further yield losses and result in potential dust hazards. Therefore, an engineered crystal size distribution that meets particle size specifications consistently, and avoids excessive downstream modification through milling and sieving, can dramatically improve overall production efficiency and profitability.

Crystallization, however, is extremely difficult to transfer directly from the laboratory to pilot and production scale. Scale-up difficulties are compounded by the importance of both thermodynamic and kinetic properties in determining the final crystal size distribution. Supersaturation, the thermodynamic driving force of crystallization, is a critical parameter in determining the final crystal population. In a laboratory vessel that is relatively well mixed, supersaturation may be effectively constant throughout the vessel. At a larger scale, there are undoubtedly gradients of supersaturation throughout the crystallizer—due to the manner in which the supersaturation is created (most often by cooling or antisolvent addition) and due to the mixing configuration (including parameters such as the vessel dimensions, baffles, impeller type, and agitation speed), which determines how effectively the supersaturation is dispersed throughout the vessel. The introduction of the supersaturation gradient plays a very significant role in the difference between laboratory-scale and full-scale crystallization.

Crystallization is further complicated by the fact that it is a multiphase system. The solid crystal product is very often a different density than the liquid phase (mother liquor). Crystals that are denser than the mother liquor have a desire to settle to the bottom of the crystallizer, and the mixing required to keep a liquid-based system well mixed is no longer sufficient to keep the solids suspended. Although a first instinct response might be to increase the agitation speed or to add baffles, one must also be aware of the possibility of crystal breakage and attrition due to the increased energy input. Attrition can be a significant cause of fine crystals or a source of secondary nucleation, which will dramatically alter the final crystal size distribution and cause a number of scale-up headaches.

The kinetics of crystallization add additional complexity to scale-up. Crystallization kinetics is commonly simplified to two parameters, nucleation (birth of new crystals) and growth (rate of increase of crystal dimensions). (More complex crystallization models may also include rates of dissolution, breakage, attrition, and agglomeration to more fully predict the population balance in the crystallizer, but these are usually factors to avoid in a practical crystallization process.) Nucleation and growth rates are primarily functions of supersaturation. At relatively low levels of supersaturation, growth tends to dominate. Nucleation rates, however, have a higher order relationship with supersaturation, so that if supersaturation reaches a high enough level, nucleation will dominate the crystallization process. In addition, the presence of crystals—and the related crystal size distribution—is another critical factor that interacts with supersaturation in determining how the supersaturation is consumed and therefore also influences the final crystal product. This is because the quantity of crystals present—and specifically the viable crystal surface area available for growth—determines the rate at which supersaturation can be consumed by the existing crystal population.

If avoiding nucleation is desirable, the crystallizer can only generate supersaturation at a rate that the existing crystal surface area can handle at the corresponding growth rate. If the amount of surface area is insufficient to handle the generated supersaturation, then the overall supersaturation level will eventually rise to the point where nucleation becomes a significant factor. However, if the amount of surface area is more than the current growth rate will sustain, the supersaturation will drop (implying that the crystallizer is running at less than full capacity). If that were not complicated enough, having potential gradients of supersaturation and gradients of crystal size distribution throughout the full-scale vessel also means potential gradients of nucleation and growth rates exist, making the final crystal product extremely difficult to predict from simple kinetic models based on laboratory data.

There have been two traditional ways of dealing with this complexity of crystallization—either crashing the solids out of solution and dealing with the headaches of the downstream processing bottlenecks or simply tuning down the crystallizer so that it runs at a low enough level of supersaturation that problems of nucleation are generally avoided. Clearly, neither of these situations is optimized for maximum production yield and throughput, and this in part has fueled the recent push toward real-time monitoring of crystallization using PAT that can directly measure critical parameters such as the crystal size and shape distribution, the crystal form, and even the level of supersaturation.

11.2.3.2 Continuous process improvement: Reduce existing waste and improve process throughput

This case study published by Sepracor (now Sunovion)[30] provides an example of the use of PAT tools to identify, characterize, and resolve serious problems with an existing crystallization process at the manufacturing scale. Making this process greener was achieved through dramatic reduction of wasted energy and resources caused by repeated batch failures and poor initial process design. This study helps illustrate the significant impact of making existing processes greener—where the impact is immediate and large due to an existing level of waste that is often overlooked.

Production campaigns at Sepracor's contract manufacturing organization (CMO) partner facilities highlighted multiple problems during the isolation of a key intermediate by diastereomeric resolution. Most importantly, a significant number of batches were failing optical purity specifications. Many batches also exhibited slow separation performance, with long centrifuge cycle times and repeated manual discharging of the centrifuge. It was reported that the separation problems were greatest within the batches that failed optical purity specifications. The process involved a diastereomeric resolution isolating an enantiomerically pure compound from a chiral mixture. The desired product is referred to as diastereomer A (*S*-isomer), and the undesired compound is diastereomer B (*R*-isomer).

The failed batches and slow separation were identified as significant waste and were systematically addressed through Sepracor's continuous process improvement initiatives. PAT tools played an integral role in identifying the source of the problem and drastically reducing batch-to-batch variability and improving throughput. Scale-down experiments in the laboratory with real-time FBRM® monitoring were used to elucidate the cause of batch-to-batch variability. The crystallization process was designed as a seeded cooling crystallization. During cooling, a small quantity of seed crystals was added as the temperature reached 46°C. Real-time monitoring throughout the crystallization showed that there was very little crystal growth resulting from the addition of the seed crystals, and the system experienced a significant secondary

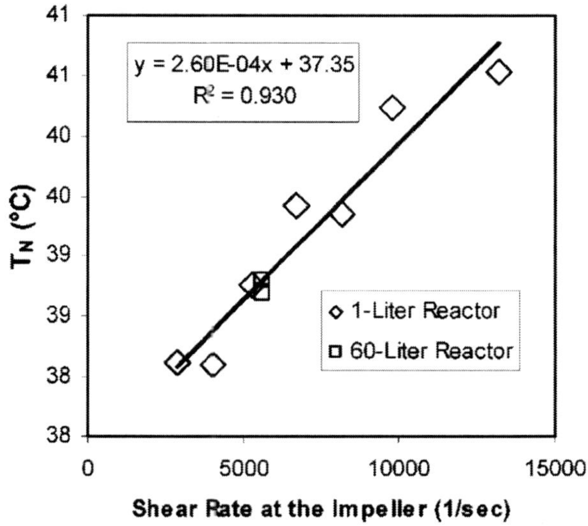

Figure 11.41 Experimental results for T_N at different shear rates in two different-sized vessels.

nucleation event one to two hours after the addition of the seed crystals. Figure 11.41 demonstrates the use of FBRM® for detection of the secondary nucleation event by real-time monitoring of the first derivative of the number of crystals detected. Essentially, it was found that the original seed loading was insufficient to consistently promote crystal growth and the nucleation event was unpredictable and showed a high level of variability.

The temperature at which the nucleation event occurs was identified as T_N, and it was found that the process cycle time in the centrifuge and the product purity were functions of this event. Nucleation occurring at a higher temperature (under a lower level of supersaturation) resulted in higher purity and better separation performance in the centrifuge. With this knowledge it was clear that the nucleation needed to be forced to occur earlier—at a higher temperature—to reduce the batch variability. Experiments were carried out investigating the effect of shear rate at the impeller, seed loading, and seed surface area. Figure 11.42 shows the relationship between shear rate and T_N. Increased seed loading was also found to have a significant impact.

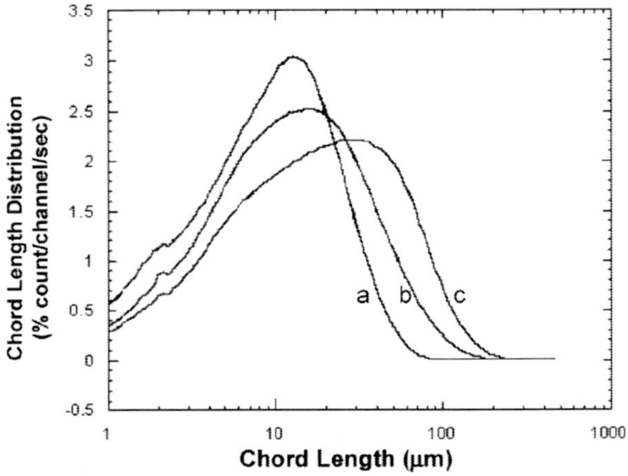

Figure 11.42 Particle size distribution of crystals from (a) the linear addition protocol with 0.5 wt% seeding, (b) the cubic addition protocol with 0.5 wt% seeding, and (c) the cubic addition protocol with 5 wt% seeding.

The crystallization process was scaled up from the laboratory to the pilot plant and finally to the contract manufacturer's site for validation of the redesigned process. The process changes resulting from analysis with the appropriate PAT tools included increasing the speed of the agitator (increasing shear at the impeller tip) and seeding with a much larger seed load. Both of these changes were geared toward improving the predictability of the secondary nucleation event at a higher temperature, that is, earlier in the process. Forcing the secondary nucleation event to occur earlier in the process resulted in a more predictable and robust process with sufficient crystal product providing initial surface area for growth. Results at the CMO, where real-time monitoring was used to confirm that the process was performing as designed, showed a 70% reduction in centrifugation time (from 7.5 hours to 2.2 hours) and dramatically reduced the number of failed batches (from 3 of 18 batches lost before the process changes and 0 of 39 batches lost after the process improvements).

PAT tools played a vital role in identifying the source of variability in Sepracor's process and provided the ability to optimize and scale up the process to remove that variability. The process was greener through significant reduction in waste by eliminating failed batches and also through significant reduction in downstream filtration and drying time to cut energy usage and improve the overall productivity of the existing plant.

11.2.3.3 Particle engineering: Design the crystal product to avoid unnecessary processing

The following case study is based on a publication by Soojin Kim *et al.*[31] (2005), exploring their work in particle engineering at Bristol-Myers Squibb. In the manufacture of any final API, a number of properties related to the crystal product must meet tight specifications to ensure the product has the necessary dissolution and bioavailability in the final formulated drug product. The crystal product must have the necessary crystal purity and correct polymorphic form, as well specified size, shape, and surface properties.[32] At Bristol-Myers Squibb, the application of particle engineering is used to ensure the crystal product meets all of these specifications consistently. In general, particle engineering requires in-depth knowledge of the particulate processes to design the operating conditions that reliably deliver a product within the required specifications. The use of process analytical measurements is standard practice in particle engineering for generating the understanding of the crystallization process that is necessary for optimization.

The drug substance in this investigation is produced by reaction crystallization, with sulfuric acid added to the free base form of the API. The resulting product has low solubility in the solvent system of choice, with the reaction producing the driving force for crystallization. Reactive crystallization can be very tricky to optimize as very high levels of supersaturation can be generated at the point of reactant addition, often resulting in rapid nucleation and very little opportunity for growth. In this Bristol-Myers Squibb example, it was reported that the original process from early process development followed this convention. Rapid generation of supersaturation resulted in excessive fine crystals and a wide crystal size distribution. The poor-quality crystal product filtered very poorly and resulted in caking of the crystal product, which required significant milling before further processing.

Prior to scale-up, it was deemed necessary to improve the crystallization process design to improve process efficiency and to ensure consistent product properties. Particle engineering was applied on the basis of the basic knowledge that the addition rate of sulfuric acid would be a critical process parameter in determining levels of supersaturation, and seeding protocol was a recommended method of providing adequate surface area for growth. Process analytical measurements were invaluable in understanding and documenting the ability to produce crystals of desired size. Figure 11.43 shows the crystal size distribution for operation under different addition and seeding protocols. In situ particle monitoring with FBRM® (Fig. 11.44) provides validation that the crystals are growing over time throughout the sulfuric acid addition so as to reach the desired endpoint. Figure 11.45 provides qualitative verification that the crystals can be controlled by adjustment of the addition rate.

The addition rate has a significant impact on the crystal product, with an optimized crystal population produced with a cubic addition of sulfuric acid. (Cubic addition is characterized by very slow addition during the initial stages and continuing at a gradually increasing rate. The goal is to avoid nucleation and promote crystal growth by only providing as much supersaturation as the existing crystal mass can consume. As the crystal mass in the vessel increases, the rate of addition increases accordingly.)

In the words of the authors, the cubic addition had the following results on the crystal product:

> "The controlled cubic addition crystallization provided a less compressible filter cake, which aids in effective cake deliquoring and washing, as well as a more easily dried product with excellent powder properties than those obtained by uncontrolled or constant addition rate crystallization techniques. API crystals prepared by controlled crystallization also facilitated formulation due to improved bulk flowability, bulk density, and powder properties and handling."

Interestingly enough, the authors then proceed to demonstrate the importance of a holistic viewpoint in particle engineering. The

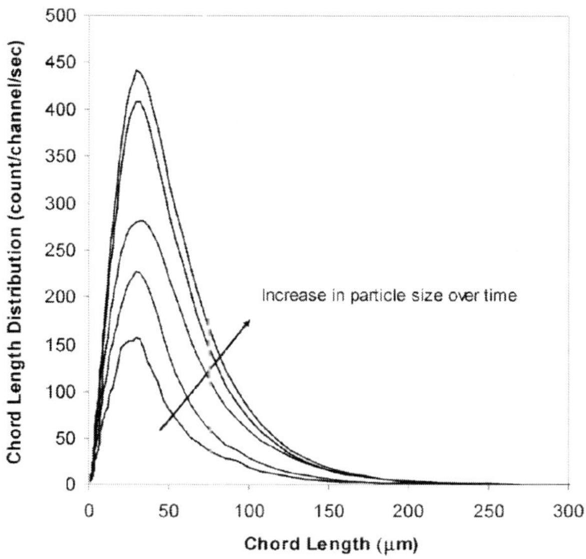

Figure 11.43 Particle size distributions of crystals grown during the cubic addition crystallization of the bisulfate salt, as represented by Lasentec FBRM® CLD.

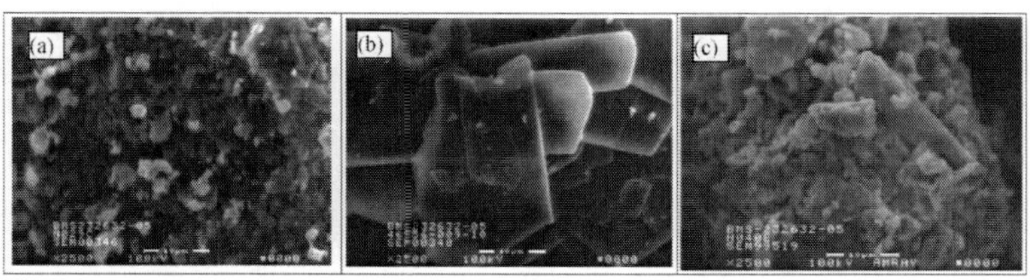

Figure 11.44 Scanning electron micrographs of crystals from (a) uncontrolled crystallization, (b) linear addition crystallization, and (c) cubic addition crystallization.

crystallization process was scaled up as designed to consistently produce crystals with the desired properties for improved filterability and final formulation specifications. However, it was also found that the crystals were sensitive to shear in certain types of agitated dryers that are commonly used at the manufacturing plant. Measurements of the crystal size distribution after drying revealed that significant alteration of the crystal product due to breakage and attrition would result in a final crystal product that did not meet specifications required for the formulated drug product. Crystal breakage and attrition also significantly affected exposed crystal surfaces and total surface area, resulting in significant alteration of dissolution profiles (and therefore bioavailability). To address this problem, a low shear-drying protocol was introduced, and the end result was a crystal product that had "excellent formulation characteristics and drug product dissolution performance."

It is difficult to overstate the role that precise and sensitive real-time measurements of critical process parameters and critical quality attributes play in the effective application of particle engineering. Particle engineering itself is not directly tied to making a process greener. However, delivering the desired crystal form in as efficient manner as possible—and often eliminating unnecessary process steps—shows that particle engineering is an important potential partner to green chemistry and green engineering.

11.2.3.4 Real-time control: Use PAT to ensure optimized process performance

Beyond the basic control of temperature and addition rates, real-time control of critical process parameters and critical quality attributes is almost nonexistent in batch crystallization processes. Most systems are operated according to a batch recipe that is theoretically designed to produce a product within an acceptable range of variability in product specifications. Upon scale-up, many processes routinely have results that fail to meet product specifications, requiring rework and resulting in wasted resources or lost yield.

On the other hand, the implementation of PAT tools from the laboratory through manufacturing scale have the inherent capability of real-time monitoring and control. Despite this, the capability of control is still mostly regarded as outside the scope of standard manufacturing practices. Process analytics, as part of the green chemistry toolbox, also play a significant role in ensuring that crystallization processes deliver product consistently within required specifications.[33]

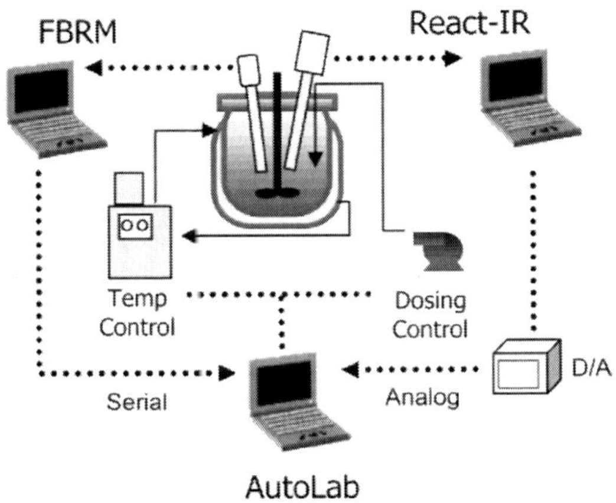

Figure 11.45 Schematic of the equipment and sensors used in this work.

The following case study, published by Liotta and Sabesan[36] (2004), provides a real-world example of designing real-time control of a batch crystallization process in the pharmaceutical industry. As we have seen in many other examples from leading crystallization experts, the researchers from Schering-Plough used probe-based FBRM® technology for monitoring the crystal population in real time. They complemented the crystal size distribution measurement with a real-time measurement of supersaturation using a probe-based mid-IR ATR-FTIR measurement (ReactIR™, as shown in Fig. 11.46).

Supersaturation, as the driving force of the crystallization process, is a critical process parameter that directly impacts the crystallization process and the final crystal product. Determination of the solubility curve and metastable zone width (Fig. 11.47) provide a guide to operation of the crystallization process and should be a first step in any analysis of crystallization. For crystal growth to occur, the crystallizer must operate in a supersaturated state that lies between the solubility curve and the metastable zone width.

The determination of the solubility curve and metastable zone width can be automated using a variety of process analytical measurements to speed up an otherwise monotonous task. It is important to realize that the solubility curve is a thermodynamic

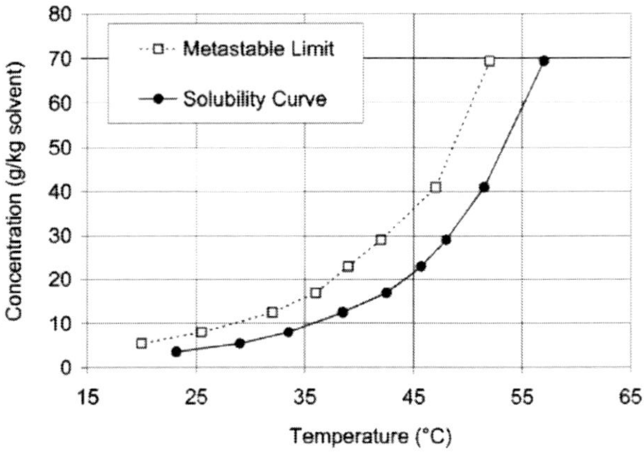

Figure 11.46 Metastable and solubility curves for pharmaceutical N using the AutoMeta technique. *Abbreviation*: AutoMeta, automated metastable zone determination.

function and is dependent on the solute/solvent relationship—that is, it is not a function of scale. The metastable zone width, however, is a kinetic function that will be impacted by factors such as mixing rates and the presence of crystals or contaminant particles. The metastable zone width is also a function of the rate of supersaturation generation. For example, in a cooling crystallization the faster the solution is cooled, the lower a temperature will be reached before spontaneous nucleation occurs. Therefore, the metastable zone width determined on a small-scale screening test may not be

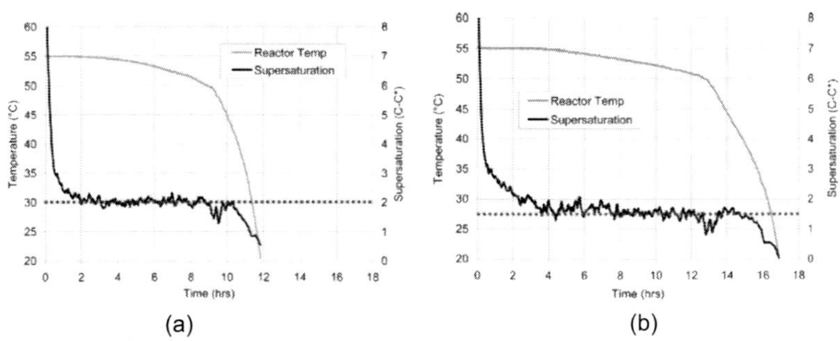

Figure 11.47 Supersaturation and temperature profiles for control experiments with supersaturation set-point values of (a) 2.0 and (b) 1.5.

a reliable indicator when scaling up to a larger volume. Small-scale tests should provide guidance, but it is always advisable to test the metastable zone width at each appropriate scale—which of course requires the application of PAT tools for successful scale-up.

With the mapping of the solubility curve and metastable zone width, how can one be certain they are operating within acceptable limits in order to avoid undesirable spontaneous nucleation? The use of FTIR provides the ability to measure, with high sensitivity, the concentration of most organic solutes—even in the presence of suspended solids. Typically, a calibration must be made in order to convert the FTIR absorbance to a concentration measurement. For the compound being investigated, Liotta and Sabesan selected a wave number region for calibration to the concentration using the method of partial least squares (PLS). Recent work by Barrett *et al.*[35] (2010) demonstrates a calibration-free method for monitoring and control of supersaturation for simpler implementation.

With a model for measuring supersaturation in place, a designed set of experiments was carried out, testing the impact of cooling rate and seeding protocol (seeded or not seeded) on the maximum supersaturation and on the desupersaturation profiles. The use of ReactIR™ to monitor supersaturation in real time allows the researchers to clearly show that rapid cooling (at -0.75°C/min) without seeding results in operation that goes well beyond the earlier determined metastable zone width. The result is the spontaneous nucleation of product, resulting in a large number of fine crystals and potential downstream problems with product filtration and drying. As one might expect, operation at the other extreme—a seeded crystallization with a much slower cooling rate (at -0.075°C/min) results in the maintenance of the supersaturation within the growth-dominated region.

Process analytical measurements have enabled researchers to characterize acceptable limits for operation of this batch crystallization, but true monitoring and feedback control are the only way to be certain the supersaturation remains within an acceptable range. The selected control scheme is based on a cascade controller, with supersaturation providing a master controller with temperature as the slave controller. Effectively, the supersaturation is maintained at a constant level by quickly adjusting the crystallizer cooling rate.

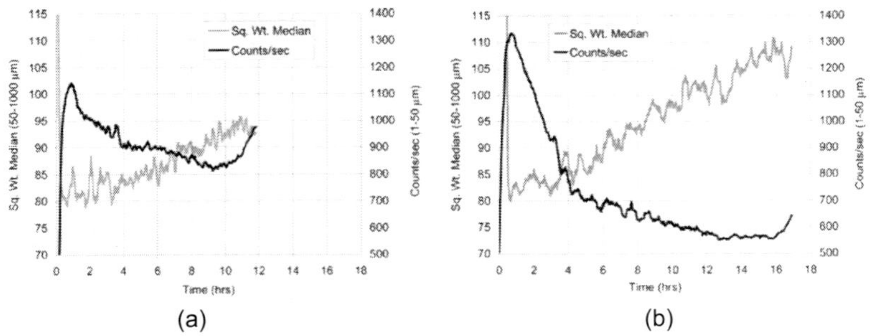

Figure 11.48 Supersaturation large-particle median and small-particle count profiles for control experiments with supersaturation set-point values of (a) 2.0 and (b) 1.5.

The results of this control scheme are shown in Figs. 11.48 and 11.49. Figure 11.48 illustrates the successful real-time feedback control of the supersaturation at set points of 2.0 and 1.5 degrees of supersaturation, respectively. Figure 11.49 shows the corresponding real-time measurements of the crystal population by FBRM® technology. It is the average crystal size (identified as Sq. Wt. Median in the corresponding graphs) that really shows the benefits of this real-time control strategy. Under both supersaturation levels, the average crystal size shows consistent growth throughout the batches. A higher level of supersaturation results in a smaller average crystal size but achieves the batch endpoint sooner. True optimization of the process can be achieved by considering the acceptable crystal size distribution required downstream and setting the supersaturation level to maximize throughput. Real-time control ensures that the end product will meet the necessary specifications even if the process encounters unexpected disturbances—such as the presence of an impurity level that impedes growth.

A recent publication by researchers at Merck[36] (Cote *et al.*, 2009) has shown similar implementation of supersaturation control with an antisolvent crystallization, in which control was necessary to avoid the crystallization of an undesirable polymorph. The system must remain supersaturated with respect to Form II (desired product) and remain undersaturated with respect to Form I (undesirable polymorph), requiring real-time control of supersaturation to reliably achieve these results. Additionally, Raman spectroscopy

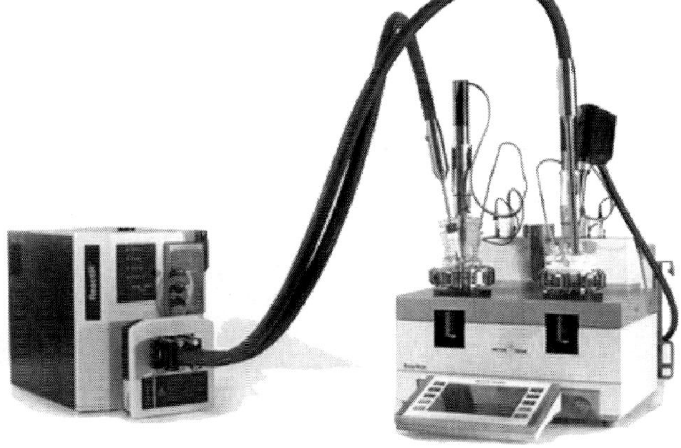

Figure 11.49 Example of setup interfacing PAT (FTIR spectroscopy and FBRM®) and process control instrumentation (EasyMax™).

was also used for real-time verification that the correct crystal form was produced throughout each run.

The widespread implementation of PAT tools for real-time monitoring and control in manufacturing has yet to be realized. However, the data-rich information and detailed process understanding that these tools provide certainly help elucidate the mechanisms of inherently complex crystallization processes. Identifying and measuring the critical process parameters are essential for ensuring maximum efficiency and minimizing batch failures. Developing this process understanding at the laboratory scale is valuable in completing the design of a scalable crystallization process that ensures product delivery within required specifications.

11.2.3.5 Control of crystallization: Which is more important—supersaturation or the crystal size distribution?

In an ideal world, one may want to directly measure the crystal population within the crystallizer (a critical product quality attribute) and measure the supersaturation that is driving the process (a critical process parameter). It is good to know that today's advanced PAT tools allow the measurement of both of these critical parameters in real time. But where should one begin, especially if budget constraints limit implementation to only one advanced measurement?

The crystal population is often the product itself. An online measurement of the crystals gives you the possibility of direct control and the possibility of assurance that the crystals meet final product specifications before actually being discharged from the crystallizer. Supersaturation monitoring and control can provide an optimal path to the final product. However, without actual crystal population information, this only makes sense with tight control of the crystallizer vessel, a precise supersaturation measurement, and a very reliable model of the system, that is, where one can predict nucleation and growth as a function of supersaturation with reasonable accuracy throughout the operating range.

In laboratory-scale research and development this is certainly achievable. However, in a larger-scale crystallizer there are limitations and complications that make control based solely on supersaturation very difficult. As discussed previously, gradients in temperature (and therefore supersaturation) and solid concentrations throughout the vessel can have a dramatic impact on the crystal population. Therefore, to compensate for these potential gradients—without measuring the crystals themselves—the controlled level of supersaturation has to be tuned down to limit the maximum level of supersaturation that might occur. And that is not really control; it is just avoidance.

If one manages to control the crystal product using only supersaturation, it is likely the system is overtuned to stay far away from conditions that might promote nucleation. This likely means the production rate is not optimized. From the viewpoint of process control, one can think of the two measurements as examples of feed-forward and feedback control. Supersaturation provides the ability to predict what is going to happen (feedforward), but one needs a near-perfect model and a high level of measurement precision for this to work. In some batch crystallization applications, this has been shown to be successful using mid-IR (such as ReactIR™ATR-FTIR) as the measurement of supersaturation. Measuring the crystals in process with FBRM® real-time measurement (allowing feedback control) is much more reliable for dealing with disturbances. However, as with any feedback control, one actually depends on a slight disturbance before one can respond and correct the system.

A simple analogy of supersaturation control is trying to get from one location to another using a map (the model of the process) and a compass (the supersaturation measurement). To get to the desired endpoint, the starting point (clear liquor concentration) must be known as well as the effect each step will have along the path. If the map is correct and the measurement has sufficient accuracy, you should reach your destination. But if there are changes to the terrain (such as the presence of impurities), or if the measurement is less than perfect, you can drift off course. Measurement of the crystal population, following the same analogy, is the addition of a global positioning system (GPS). It tells you exactly where you are throughout the course of the batch (within the accuracy of the measurement). The GPS (crystal population measurement), used along with the map (process model), will guide you more effectively than the map (model) and compass (supersaturation). The best results would be achieved with all three components (and one can note that all GPS navigation systems actually do combine a GPS location with a map and a compass).

And in a similar way, many of the best examples of batch crystallizer control actually use both FBRM® crystal population and supersaturation measurements in their models and control algorithms (see academic research from the research groups of Prof. Richard Braatz[37] (UIUC, USA), Prof. Sohrab Rohani[38] (Western Ontario, Canada), Prof. Marco Mazzotti[39] (ETH, Switzerland), and Prof. Brian Glennon[40] (UCD, Ireland). These advanced model-based controllers using neural networks and fuzzy logic rely on measurements of both supersaturation and the crystal population for optimum results. If you have to choose one method of advanced PAT, measuring the crystals themselves is the best option for control and assurance that the product will meet specifications. Supersaturation is very valuable in understanding and modeling the system, but it provides limited monitoring and control capability without real-time confirmation of the crystal size distribution.

11.3 General Conclusions

It is somewhat striking that today, in an era where lofty goals such as productivity gains, the quest for a greener chemistry approach, and better-quality drugs are on the minds of employees at every level in the pharmaceutical industry, too many synthetic chemists still rely on round-bottom flasks and occasional sampling for off-line analysis to conduct everyday research activity. This approach would be the equivalent of relying on outdated maps and the telegraph when embarking on a journey, instead of taking advantage of the GPS and cell phones.

Over the past few years, the miniaturization and simplification of analytical technologies, once considered too bulky and too complicated to use, have allowed PAT and real-time process monitoring to enter most pharmaceutical research laboratories and manufacturing facilities. It is now becoming common for synthetic chemists and process engineers around the world to use calorimetry, mid-IR spectroscopy, and particle size analysis to monitor chemical reactions and crystallizations in real time in order to reduce cycle time, improve process reliability, and more generally pursue quality and efficiency goals often associated with the green chemistry principles. For example, this chapter illustrated that ATR-based FTIR reaction analysis (ReactIR™) and reaction calorimetry (RC1*e*™ are real-time, in-process techniques used to optimize and control chemical reactions—and make chemical processes safer—by directly monitoring reaction progress and preventing the formation of hazardous or contaminating substances.

Real-time particle system characterization (FBRM® and PVM®) can be used to optimize crystallization processes to maximize yield and productivity, while minimizing energy and solvent use—with a complementary goal of an engineered particle distribution that directly meets final particle size specifications and eliminates the risk of dust explosions caused by unnecessary milling. All of these technologies are used to reduce cycle time and increase process throughput, optimizing energy consumption along the way. They are also used to increase the productivity of research and development activities by automating mundane and repetitive tasks, thereby allowing scientists to focus their time and energy toward higher-value activities.

While hardware improvement and simplification have been major drivers for rapid development and implementation of PAT, software for process analytical tools has also seen revolutionary advancement over the past 15 years, the development of user-friendly graphical operating systems has allowed PAT software to evolve from being the restricted domain of a few full-time specialists in safety, analytical, and crystallization laboratories to becoming ubiquitous walk-up tools that do not require any formal training and allow bench chemists to continue to focus on their chemistry.

Going forward, even though further instrument development and miniaturization are expected, a more dramatic evolution is anticipated in the way scientists interact with instrumentation. More specifically, software will provide greatly increased value by shifting from providing basic data to application-specific modules providing meaningful process information for direct use in decision making. This evolution will accelerate the ongoing trend of moving PAT utilization from a small group of specialists to the majority of chemists and engineers. Data processing and visualization technologies that take advantage of increasing computing power already exist and are soon going to become part of the familiar working environment of pharmaceutical scientists.[41]

The trend toward a wider adoption of PAT by pharmaceutical scientists is also fueled by the advancement of networking technologies that provide remote control and an easy interface between various PAT and process control instruments. Today, automated control based on real-time concentration data measured by in situ IR spectroscopy analysis is simple to set up and straightforward to use (Fig. 11.49). Similarly, crystallization processes under the control of an automated laboratory instrument can adjust themselves, without human intervention, on the basis of real-time feedback from a particle size analyzer. Although this may have sounded like science fiction only a few years ago, countless researchers now use these capabilities to automate the design of safe, scalable, and greener processes in much less time and with much less effort than in the recent past.

It will probably not be long before model predictive control becomes a standard laboratory tool. At that point, chemists and engineers will be able to test and simulate kilolab, pilot plant, and full-scale manufacturing operating conditions on a small laboratory scale. Expected process output in the manufacturing plant (yield, purity, impurity profile, particle size, particle distribution, and shape) will be determined at the laboratory scale, and the chemistry will be optimized on a small scale under simulated conditions representative of the specific process vessel scheduled for scale-up. This approach, fully in accordance with green chemistry and green engineering principles, will then enable full implementation of QbD and dramatically improve the efficiency and quality of drug development and manufacturing.

References

1. http://www.fda.gov/AboutFDA/CentersOffices/CDER/ucm128080.htm
2. Rubin, A. E.; Tummala, S.; Both, D. A.; Wang, C.; Delaney, E. J. (2006). *Chem. Rev.*, **106**, 2794–2810; (b) Rubin, A. E.; Tummala, S.; Both, D. A.; Wang, C.; Delaney, E. J. (2005). *Org. Process Res. Dev.*, **9**, 331–364.
3. (a) Barrett, P.; Smith, B.; Worlitschek, J.; Bracken, V.; O'Sullivan, B.; O'Grady, D. (2005). *Org. Process Res. Dev.*, **9**(3), 348–355. DOI: 10.1021/op049783p; (b) Sistare, F.; St. Pierre-Berry, L.; Mojica, C. A. (2005). *Org. Process Res. Dev.*, **9**(3), 332–336. DOI: 10.1021/op0402127
4. Van Loo, M. E.; Lengowski, P. E. (2002). *Org. Process Res. Dev.*, **5**, 833–840.
5. (a) Pollard, M. (2001). *Org. Process Res. Dev.*, **5**, 273–282; (b) Chen, J.; Gervasio, G. J.; Ng, P. N.; Rydzak, J. W.; Kord, A. S. (2010). *Am. Pharm. Rev.*, April-May-June, 32–38.
6. http://us.mt.com/us/en/home/products/L1_AutochemProducts.html
7. (a) http://www.epa.gov/gcc/pubs/principles.html; (b) Tucker, J. L. (2006). *Org. Process Res. Dev.*, **10**, 315–319; (c) Tucker, J. L. (2010). *Org. Process Res. Dev.*, **14**, 328–331.
8. Pharmaceutical Manufacturing. (2010). Greener Processes: PAT & QbD Take Root, **9**(5), 18–24. www.pharmamanufacturing.doc
9. (a) Milosevic, M.; Sting, D.; Rein, A. (1995). *Spectroscopy*, **10**, 44–49; (b) Dozeman, G. J.; Fiore, P. J.; Puls, T. P.; Walker, J. C. (1997). *Org. Process Res. Dev.*, **1**, 137–148.

10. Clegg, I. *Pfizer, PAT: Using Mid-IR Spectroscopy to Monitor a Telescoped Chemical Reaction*. httpv://www.youtube.com/watch?v=EWec0O6WtUU

11. Blackmond, D. G. (2005). *Angew. Chem., Int. Ed.*, **44**, 4302–4320.

12. Landau, R. N. (1996). *Thermochim. Acta*, **289**, 101–126.

13. (a) Barbas, R.; Botija, M.; Camps, H.; Portell, A.; Prohens, R.; Puigjaner, C. (2007). *Org. Process Res. Dev.*, **11**, 1131–1134; (b) Barbas, R.; Botija, M.; Camps, H.; Portell, A.; Prohens, R.; Puigjaner, C. (2009). *Org. Process Res. Dev.*, **13**, 1387–1425; (c) Barbas, R.; Botija, M.; Camps, H.; Portell, A.; Prohens, R.; Puigjaner, C. (2005). *Org. Process Res. Dev.*, **9**, 951–1012.

14. (a) Barbas, R.; Botija, M.; Camps, H.; Portell, A.; Prohens, R.; Puigjaner, C. (2001). *Org. Process Res. Dev.*, 5(6), 612–664; (b) Barbas, R.; Botija, M.; Camps, H.; Portell, A.; Prohens, R.; Puigjaner, C. (2006). *Chem. Eng. News*, **84(**10), 17; (c) Tanaka, K.; Fukase, K. (2009). *Org. Process Res. Dev.*, **13**, 983–990; (d) Valera, F. E.; Quaranta, M.; Moran, A.; Blacker, J.; Armstrong, A.; Cabral, J. T.; Blackmond, D. G. (2010). *Angew. Chem., Int. Ed.*, **49**(14), 2478–2485; (e) McMullen, J. P.; Jensen, K. F. (2010). *Org. Process Res. Dev.*, **14**, 1169–1176.

15. Although calorimetry has been described as well: Roberge, D. M. (2004). *Org. Process Res. Dev.*, **8**, 1049–1053; (b) Roberge, D. M. (2005). *Chem. Eng. Tech.*, **28**(3), 318–323.

16. Watson, D. J.; Dowdy, E. D.; DePue, J. S.; Kotnis, A. S.; Leung, S.; O'Reilly, B. C. (2004). *Org. Process Res. Dev.*, **8**, 616–623.

17. LaPorte, T. L.; Hamedi, M.; DePue, J. S.; Shen, L.; Watson, D.; Hsieh, D. (2008). *Org. Process Res. Dev.*, **12**, 956–966.

18. See, for instance, Connolly T. J., Hansen, E. C., MacEwan, M. F. (2010). *Org. Process Res. Dev.*, **14**, 456–469.

19. Another example where an unstable anion was monitored in real time: Delhaye, L.; Merschaert, A.; Delbeke, P.; Brione, W. (2007). *Org. Process Res. Dev.*, **11**, 689–692.

20. Carter, C. F.; Lange, H.; Ley, S. V.; Baxendale, I. R.; Wittkamp, B.; Goode, J. G.; Gaunt, N. L. (2010). *Org. Process Res. Dev.*, **14**, 393–404.

21. Connolly, T. J.; Considine, J. L.; Ding, Z.; Forsatz, B.; Jennings, M. N.; MacEwan, M. F.; McCoy, K. M.; Place, D. W.; Sharma, A.; Sutherland, K. (2010). *Org. Process Res. Dev.*, **14**, 459–465.

22. Other examples include: (a) Kryk, H.; Hessel, G.; Schmitt, W. (2007). *Org. Process Res. Dev.*, **11**, 1135–1140; (b) Flanagan, R. C.; Xie, S.; Millar, A. (2008). *Org. Process Res. Dev.*, **11**, 1307–1312; (c) Ulysse, L. G.; Yang, Q.; McLaws, M. D.; Keefe, D K.; Guzzo, P. R.; Haney, B. P. (2010). *Org. Process Res. Dev.*, **14**, 225–228; Girgis, M. J.; Liang, J. K.; Du, Z.; Slade,

J.; Prasad, K. (2009). *Org. Process Res. Dev.*, **13**, 1094–1099; (d) Schils, D.; Stappers, F.; Solberghe, G.; van. Heck, R.; Coppens, M.; Van den Heuvel, D.; Van der Donck, P.; Callewaert,. T.; Meeussen, F.; De Bie, E.; Eersels, K.; Schouteden, E. (2008). *Org. Process Res. Dev.*, **12**, 530–536.

23. Example of monitoring of toxic coupounds with reduced exposure: Wiss, J.; Zilian, A. (2003). *Org. Process Res. Dev.*, **7**, 1059–1066.
24. Oh, L. M.; Wang, H.; Shilcrat, S. C.; Herrmann, R. E.; Patience, D. B.; Spoors, P. G. (2007). *Org. Process Res. Dev.*, **11**, 1032–1042.
25. Another example: Akao, A.; Nonoyama, N.; Mase, T.; Yasuda, N. (2006). *Org. Process Res. Dev.*, **10**, 1178–1183.
26. Brown Ripin, D. H.; Weisenburger, G. A.; am Ende, D. J.; Bill, D. R.; Clifford, P. J.; Meltz, C. N.; Phillips, J. E.. (2007). *Org. Process Res. Dev.*, **11**, 762–765.
27. Other examples with scale-up: (a) Ragan, J. A., *et al.* (2003). *Org. Process Res. Dev.*, **7**, 155–160; (b) Merschaert, A. (2007). *Org. Process Res. Dev.*, **11**, 1104–1111.
28. Other examples where in situ FTIR at a large scale with MonARC™ was utilized: (a) Andemichael. Y.; Chen, J.; Clawson, J. S.; Dai, W.; Diederich, A.; Downing, S. V.; Freyer, A. J.; Liu, P.; Oh, L. M.; Patience, D. B.; Sharpe, S.; Sisko, J.; Tsui, J.; Vogt, F. G.; Wang, J.; Wernersbach, L.; Webb, E. C.; Wertman, J.; Zhou, L. (2009). *Org. Process Res. Dev.*, **13**, 729–738; (b) Pesti, J.; Chen, C.-K., Spangler, L.; DelMonte, A. J.; Benoit, S.; Berglund, D.; Bien, J.; Brodfuehrer, P.; Chan, Y.; Corbett, E.; Costello, C.; DeMena, P.; Discordia, R. P.; Doubleday, W.; Gao, Z.; Gingras, S.; Grosso, J.; Haas, O.; Kacsur, D.; Lai, C.; Leung, S.; Miller, M.; Muslehiddinoglu, J.; Nguyen, N.; Qiu, J.; Olzog, M.; Reiff, E.; Thoraval, D.; Totleben, M.; Vanyo, D.; Vemishetti, P.; Wasylak, J.; Wei, C. (2009). *Org. Process Res. Dev.*, **13**, 716–728.
29. Barrett, P.; Smith, B.; Worlitschek, J.; Bracken, V.; O'Sullivan, B.; O'Grady, D. (2005). *Org. Process Res. Dev.*, **9**, 348–355.
30. Mousaw, P.; Saranteas, K.; Prytko, R. (2008). *Org. Process Res. Dev.*, **12**, 243–248.
31. Kim, S.; Lotz, B.; Lindrud, M.; Girard, K.; Moore, T.; Nagarajan, K.; Alvarez, M.; Lee, T.; Nikfar, F.; Davidovich, M.; Srivastava, S.; Kiang, S. (2005). *Org. Process Res. Dev.*, **9**, 894–901.
32. See, for instance, Desikan, S.; Parsons, R. L., Jr.; Davis, W. P.; Ward, J. E.; Marshall, W. J.; Toma, P. H. (2005). *Org. Process Res. Dev.*, **9**, 933–942.
33. Argentine, M. D.; Braden, T. M.; Czarnik, J.; Conder, E. W.; Dunlap, S. E.; Fennell, J. W.; LaPack, M. A.; Rothhaar, R. R.; Scherer, R. B.; Schmid, C. R.;

Vicenzi, J. T.; Wei, J. G.; Werner, J. A.; Roginski, R. T. (2009). *Org. Process Res. Dev.*, **13**, 131–143.

34. Liotta, V.; Sabesan, V. (2004). *Org. Process Res. Dev.*, **8**, 488–494.

35. Barrett, M.; McNamara, M.; Hao, H. X.; Barrett, P.; Glennon, B. (2010). *Chem. Eng. Res. Des.*, **18**, 1108–1119.

36. Cote, A.; Zhou, G.; Stanik, M. (2009). *Org. Process Res. Dev.*, **13**, 1276–1283.

37. (a) Woo, X. Y.; Nagy, Z. K.; Tan, R. B. H.; Braatz, R. D. (2009). *Crystal Growth Des.*, **9**, 182–191; (b) Nagy, Z. K.; Fujiwara, M.; Chew, J. W.; Braatz, R. D. (2008). *J. Process Control*, **18**, 399–407; (c) Togkalidou, T.; Tung, H.-H.; Sun, Y.; Andrews, A.; Braatz, R. D. (2004). *Ind. Eng. Chem. Res.*, **43**, 6168–6181.

38. Trifkovic, M.; Sheikhzadeh, M.; Rohani, S. (2009). *AIChE J.*, **55**, 2591–2602.

39. (a) Lindenberg, C.; Krättli, M.; Cornel, J.; Brozio, J.; Mazzotti, M. (2009). *Crystal Growth Des.*, **9**, 1124–1136; (b) Ruf, A. ; Worlitschek, J.; Mazzoti, M. (2001). *Part Part. Syst. Char.*, **17**, 167–179.

40. O'Grady, D.; Barrett, M.; Casey, E.; Glennon, B. (2007). *Chem. Eng. Res. Des., Trans.*, **85**, 945–952.

41. Tufte, E. R. (2001). *The Visual Display of Quantitative Information* (2nd ed.). Graphics Press.

Chapter 12

Approaches to the Scale-Up of Organic Chemistry Using Microwave Heating

Nicholas E. Leadbeater
Department of Chemistry, University of Connecticut, North Eagleville Road, Storrs, CT 06269-3060, USA
nicholas.leadbeater@uconn.edu

This chapter focuses on the scale-up of synthetic chemistry involving microwave heating. Following an introduction to microwave heating as a tool for synthetic chemistry, there is a discussion of the different approaches to scale-up that can be taken.

12.1 Introduction

"Microwave heating seems useful for the preparation of the first gram or so of our compounds, but is it scalable?" This is a question that is frequently asked, and it certainly has some validity. For microwave heating to be a technology that is useful from an industrial standpoint, the methodologies developed on a small scale need to be scalable to make more significant quantities of material. The objective of this chapter is to address this question by discussion of some case studies from our laboratory. Our work will be placed in

Scalable Green Chemistry: Case Studies from the Pharmaceutical Industry
Edited by Stefan G. Koenig

ISBN 978-981-4316-49-1 (Hardcover), 978-981-4364-81-2 (eBook)
www.panstanford.com

context with other reports in the area and will be preceded by a brief introduction to microwave heating as a tool for synthesis.[1]

12.1.1 *Microwave Heating*

The microwave region of the electromagnetic spectrum, as shown in Fig. 12.1, is defined as that with wavelengths ranging from 1 m down to 1 mm. This corresponds to frequencies of between 0.3 and 300 GHz, a region of the electromagnetic spectrum that is heavily regulated because applications such as wireless devices, satellite radio, and air traffic control operate into this range. Equipment for industrial, scientific, and medical (ISM) use is permitted to operate at only five specific frequencies: 25.125, 5.80, 2.45, 0.915, and 0.4339 GHz. Domestic microwave ovens operate at 2.45 GHz (12.25 cm wavelength), and this same frequency has also been widely adopted by companies manufacturing scientific microwave apparatus for use in preparative chemistry.

Microwave heating is based upon the ability of a particular substance, such as a solvent or a reaction substrate, to absorb microwave energy and convert the electromagnetic energy effectively

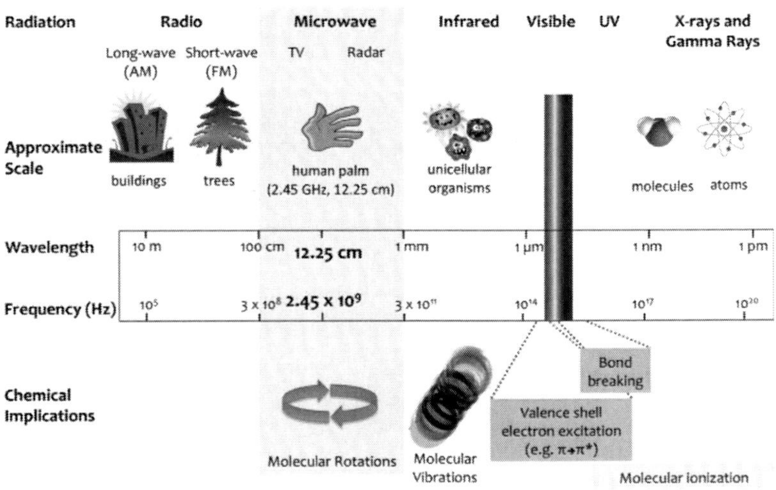

Figure 12.1 Regions of the electromagnetic spectrum with approximate scale as well as chemical implications for selected wavelength regions.

to heat (kinetic energy). Molecules with a dipole moment attempt to align themselves with the oscillating electric field of the microwave irradiation, leading to rotation. In the liquid and solid phases, this rotation is rapidly quenched both by collisions and by translational movement. We have previously described the interaction of microwave energy with a molecule by analogy to baseball or cricket.[2] During the swing, the batter or batsman can be said to be "rotationally excited" and can deliver some amount of rotational force to the incoming pitch (delivery in cricket). At the point of impact, the rotational energy is rapidly converted into translational energy of the ball. Similarly, one water molecule excited rotationally by incident irradiation can strike a second molecule of water, converting rotational energy into translational energy. Under microwave irradiation, a large number of molecules are being rotationally excited and, as they strike other molecules, rotational energy is converted into translational energy and as a consequence heating is observed (Fig. 12.2).

Since microwave heating is dependent upon the dipole moment of a molecule, more polar solvents, such as dimethylsulfoxide, dimethylformamide, ethanol, and water, convert microwave irradiation into heat more effectively than nonpolar ones, such as toluene and hexane. Parameters such as the dielectric constant (δ'), dielectric loss (δ''), or a combination of both, termed as the loss tangent or loss angle ($\tan\delta = \delta''/\delta'$) have been used as measures for the microwave absorbtivity of a material. The dielectric constant describes the polarizability of a molecule in the microwave field, and the dielectric loss expresses the efficiency with which a molecule converts the incident electromagnetic irradiation into molecular rotation and hence heat. The loss angle ($\tan\delta$) is a measure of reactance (resistance in a capacitor) of a molecule.[3] A material that has $\tan\delta = 0$ means that it is completely transparent to microwave irradiation. For a perfectly absorbing material, $\tan\delta = \infty$, and the material shows complete resistance to the incident irradiation. From a preparative chemistry standpoint, materials with a $\tan\delta$ approaching 1 are classed as strong microwave absorbers.

While the dielectric loss or $\tan\delta$ value of a molecule can be used to assess microwave absorbance, the use of these, or any other single parameter, to determine whether a reagent or solvent will heat

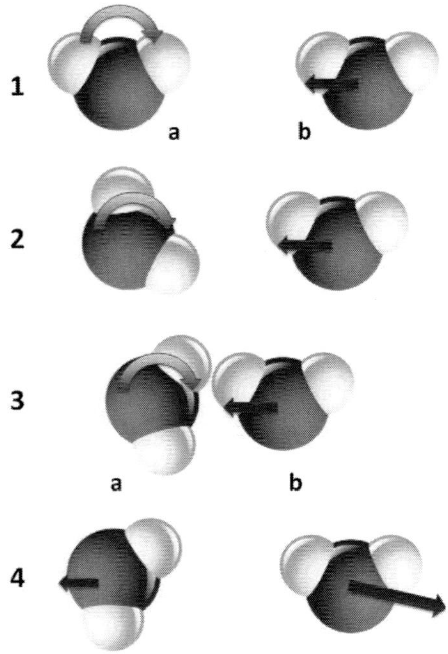

Figure 12.2 Microwave heating. Panels 1–3 show a molecule (**a**) that has been rotationally excited by microwave irradiation being approached by a second molecule (**b**). Upon impact (panel 3), the rotational energy of molecule **a** is converted to translational movement of molecule **b**. In panel 4, notice the increase in translational vector magnitude, the consequence of is an increase in molecular collisions.

efficiently under microwave irradiation is to oversimplify the case. Attributes such as specific heat capacity and heat of vaporization of the substance, as well as the depth to which microwave irradiation can penetrate into the sample, can sometimes have a larger impact upon heating rate, as will be shown later in this chapter.

12.1.2 *Microwave Effects*

There are numerous reports of microwave heating in synthetic chemistry in which comparisons with "conventional" heating are made. As a result, there has been a mystique associated with microwave heating with authors claiming that something special is taking place. However, many of the comparisons are not made under identical, isothermal conditions. Instead, the conventional chemistry is performed at one temperature and the microwave experiment at a higher temperature. Clearly in these cases it is not surprising that a rate acceleration is seen. However, microwave heating can indeed be different than "conventional," solely convection-based heating. For the most part, these differences have been divided into two categories, "specific" microwave effects and "nonthermal" microwave effects.

12.1.2.1 Specific microwave effects

"Specific" microwave effects encompass macroscopic heating events that occur slightly differently under microwave irradiation than when using conventional (convection) heating methods. These effects are often difficult (but not impossible) to reproduce without the use of microwave irradiation. Such examples would include (1) observed heating differences based on microwave absorptivity, (2) macroscopic superheating, and (3) selective heating of substances in heterogeneous (and potentially homogeneous) systems. The first of these has already been addressed—substrates that better convert incident microwave irradiation into heat, heat the bulk faster.

Macroscopic superheating relates to heating solvents above their normal (atmospheric) boiling point without the onset of boiling.[4,5] Solvents only boil when they are in contact with their own vapors.[6] In the absence of nucleation sources such as imperfections in the vessel, solvents are only in contact with their own vapors at the top of the vessel; thus boiling (and hence release of heat) is limited to this relatively small interface. Using microwave irradiation, solvents have been held well above their boiling points for extended periods of time. This phenomenon has been exploited in synthetic chemistry. The acid-catalyzed esterification of 1-hexanol with benzoic acid and the solvent-free cyclization of citronellal (ene reaction) are examples.[7] In the case of the esterification reaction, temperatures of some 38°C above the normal boiling point of 1-hexanol were obtained, and in the case of the ene reaction, it was possible to perform the reaction 35°C above the normal boiling point of citronellal. Accordingly, rate enhancements were observed at these higher temperatures when compared to conventionally heated reactions.

It is sometimes possible to heat very microwave-absorbant substrates and/or catalysts selectively under *heterogeneous* reaction conditions. An example is in the synthesis of CdSe and CdTe nanomaterials using nonpolar hydrocarbon solvents.[8] The precursor substrates are said to absorb the microwave irradiation selectively, leading to more uniform morphology in the resulting nanomaterials as compared to conventional heating methods. It should be noted, however, that the microwave-mediated selective heating at the point of reaction seems to be the exception rather than the rule, existing in only very specific instances or highly manipulated protocols.

12.1.2.2 Nonthermal microwave effects

Numerous attempts have been made over the past 20 years to rationalize perceived enhancements in reaction rates or differences in product distribution when using microwave heating. Proponents of nonthermal microwave effects suggest that when the reaction is performed solvent free or in a nonpolar medium, the presence of an electric field leads to orientation effects of dipolar molecules or intermediates and hence changes the preexponential factor (A) or the activation energy terms in the Arrhenius equation. Additionally, in the case of polar reaction mechanisms where the relative polarity of the reaction is enhanced from the ground state to the transition state, an acceleration due to an increase in microwave absorbance of the intermediates could occur. A range of examples have been cited and suggest that nonthermal effects are most frequently encountered in unimolecular or bimolecular reactions between neutral molecules and anionic reactions of tight ion pairs. However, when nonthermal microwave effects are discussed, they are generally invoked as a result of comparison of the outcome of microwave and conventionally heated experiments. A major flaw in this argument is that the comparisons are often not at isothermal operation. The heating rates of microwave and conventionally heated reactions are very different and thus would be expected to lead to different outcomes. A number of techniques have been used to examine the impact of microwave energy on reaction rates and also to determine where errors may have previously arisen. For instance, multiple fiber-optic probes placed inside a reaction vessel give a clearer picture of temperature gradients and hence inaccuracies in measured and reported data are uncovered.[9,10] Additionally, silicon carbide heating inserts[11] and vessels[12] as well as application of simultaneous cooling of vessel walls[13,14] have been used to probe the impact of microwave power on organic reactions at a constant temperature. Raman spectroscopy has been used to investigate the impact of microwave power input on spectroscopic signatures of molecules, finding no examples of "localized superheating."[15] As we have stated in a previous report,[2] with results continuing to come forward and as previous claims of nonthermal microwave effects are systematically debunked, one thing becomes ever more clear: heating is heating.

12.1.3 *Equipment*

The use of microwave heating in organic synthesis has been widely adopted since seminal publications in 1986.[16,17] The first reports involved the use of domestic microwave ovens. However, as domestic microwave ovens are not designed for performing synthetic chemistry, they are not built to withstand a vessel failure. As a result, there is risk of injury when performing reactions in sealed vessels. Another issue is the lack of direct temperature control or efficient stirring. As a result, reactions often have to be controlled by setting microwave power levels. This, together with the inhomogeneity of the microwave field in a domestic oven and the potential for overheating a reaction mixture, makes both safety and reproducibility major issues. Although costlier, modern scientific microwave apparatus has many advantages. As well as being built to withstand explosions of reaction vessels inside the microwave cavity, temperature and pressure monitoring has been introduced as has the ability to stir reaction mixtures. These features make performing reactions safer, more controllable, more easily monitored, and reproducible.

12.2 Scale-Up of Chemistry Performed Using Microwave Heating

While the use of microwave heating for performing reactions on the mmol scale in sealed vessels is straightforward, there are a number of potential issues associated with scale-up. These issues range from accessing suitable equipment through whether microwave irradiation can effectively penetrate larger reaction volumes. However, as recently addressed by Strauss[18], efficient stirring of reactions should negate any microwave penetration issues. Manufacturers of microwave equipment have for the most part developed equipment to meet scale-up needs using three main approaches: (1) continuous flow, (2) open-vessel batch, and (3) sealed-vessel batch.[19,20] In our laboratory, we have critically assessed all three.

12.2.1 *Continuous-Flow Processing*

There are a number of reasons to adopt a continuous-flow approach for scale-up. Reactions are actually "scaled out" rather than scaled up. Maximum throughput is only a matter of run time and the total number of units operating in parallel. Furthermore, catastrophic loss of a large quantity of valuable substrate can be avoided, as only a small portion of the reaction is subjected to reaction conditions at any given time. Drawbacks to this approach are that reaction mixtures are generally required to be homogeneous before, during, and until they have exited the microwave apparatus. A consequence of this is that extensive reoptimization may need to be undertaken in order to develop appropriate homogeneous reaction conditions and suitable residence times. This in itself may require additional solvent and/or catalyst screening.

For continuous-flow processing where reactants are not heated above their atmospheric boiling points, a flow apparatus that operates at ambient pressure can be used. We took this approach when developing a methodology for the preparation of biodiesel.[21,22] Biodiesel is prepared from vegetable oil by transesterification with methanol using a base catalyst, the by-product being glycerin. We find that the reaction mixture only needs to be heated to around

50°C; the keys to the success being continuous application of significant microwave power to the reaction mixture while it is in the reactor as well as efficient mixing. Using a 4 L reaction vessel it was possible to process feedstock at flow rates of up to 7.2 L/min in flow mode.

The requirement that a reaction be performed under ambient pressure conditions would limit the scope of continuous-flow processing. Dedicated equipment capable of performing chemistry at elevated temperatures and pressures in flow mode has been fabricated and chemistry reported in it.[23,24] Perhaps the most widely used is the Milestone FlowSYNTH (Fig. 12.3). Technical overviews of the unit, together with assessment of its use, have been published recently.[25,26] The reaction chamber of 200 mL working volume is mounted vertically in the microwave cavity. Material is pumped in at the bottom of the reactor and out at the top, and as it comes out of the microwave cavity, it passes through a cooler heat exchanger,

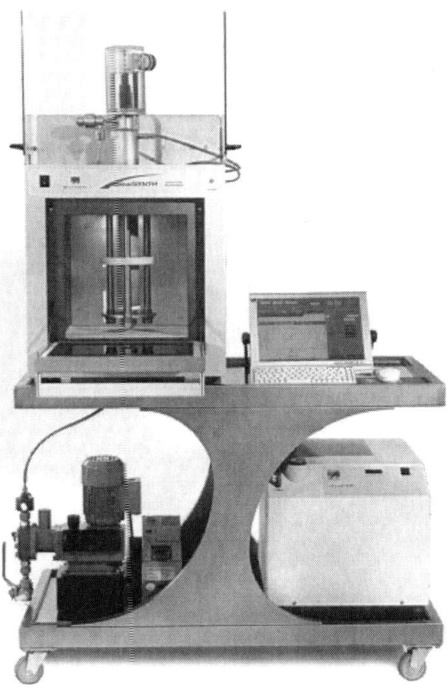

Figure 12.3 The Milestone FlowSYNTH, a continuous-flow microwave unit.

which is attached to a chiller unit. The reaction mixture is stirred mechanically while in the microwave cavity, and the temperature is measured both in the cavity and also after exiting the cooler. The maximum working conditions for the reactor are 200°C and 30 bar. Autogenic pressure upon heating is maintained by means of a variable back-pressure regulator mounted on the exit of the cooler unit. The flow rate is adjustable from 12 to 200 mL/min. We have performed a number of reactions using this unit, starting with a simple esterification.[27] Using a 2:1 stoichiometric ratio of acetic acid to butanol, at a reaction temperature of 150°C and a flow rate of 200 mL/min, we obtained 78% conversion of the desired butyl ester product when employing sulfuric acid as a catalyst and working on the 21 mole scale (Scheme 12.1). We subsequently repeated the reaction but using a 1:3 stoichiometric ratio of butanol to acetic acid, but the conversion improved only slightly (80%).

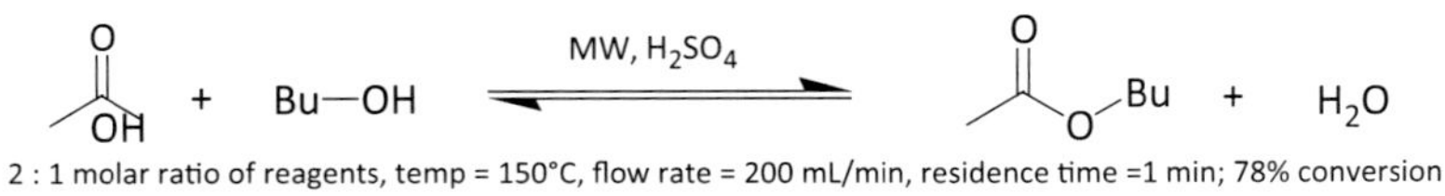

Scheme 12.1 Esterification of acetic acid with butanol using continuous-flow processing.

Although the diameter of the reactor chamber in the microwave cavity is 3.8 cm, there is a narrow bore tube at the inlet of only 2 mm that can easily block with particulate matter. In addition, the exit through the back-pressure regulator is also very narrow and prone to blockage. This became apparent when we attempted to prepare a 1,4-dihydropyridine by means of a Hantzsch methodology (Scheme 12.2). Using a 1:3.2:10 molar ratio of benzaldehyde, ethyl acetoacetate, and ammonium hydroxide, respectively, and a water-ethanol mix as solvent, we passed the homogeneous reaction mixture through the flow unit at a rate of 32 mL/min, corresponding to a residence time of approximately 6 minutes. We found that as we approached the 20-minute mark, the flow rate started to decrease and it was evident that a buildup of the dihydropyridine product in the exit line was responsible for this. The conversion to product of the material we did process was 91%, indicating that the reaction was taking place very efficiently while the unit was running.

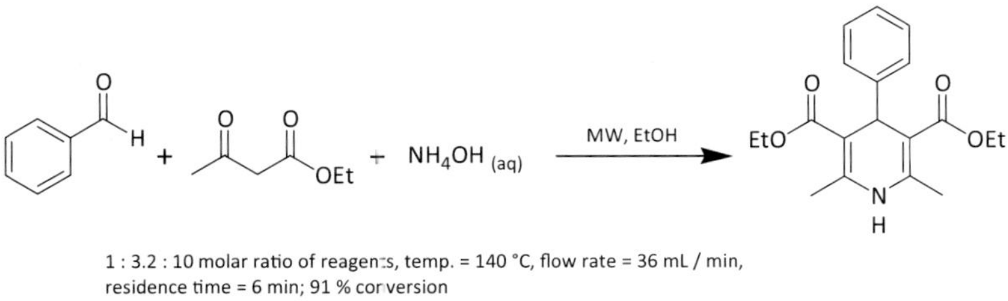

Scheme 12.2 Hantzsch synthesis of a 1,4-dihydropyridine using continuous-flow processing.

We have developed a number of methodologies for palladium-catalyzed coupling reactions using very low catalyst loadings, particular attention being focused on the Suzuki reaction. Indeed, this coupling can be performed using as little as 50 ppb of a simple palladium salt as catalyst. We use either a 1:1 water-ethanol mixture or neat water as solvent, tetrabutylammonium bromide (TBAB) being used as a phase-transfer agent in the case of the latter. Using a water-ethanol mixture is preferable since it offers a cleaner, easier approach and does not require the use of a costly, hard-to-remove additive. However, while the organic substrates readily dissolve in the reaction medium, the sodium carbonate base is only sparingly soluble and, at the end of the reaction, the biaryl product precipitates out of solution even at elevated temperatures. When turning to a flow approach, it was immediately apparent that we would need to find reaction conditions in which the mixtures entering and exiting the unit were totally homogeneous (Scheme 12.3). In small-scale experiments we found that changing

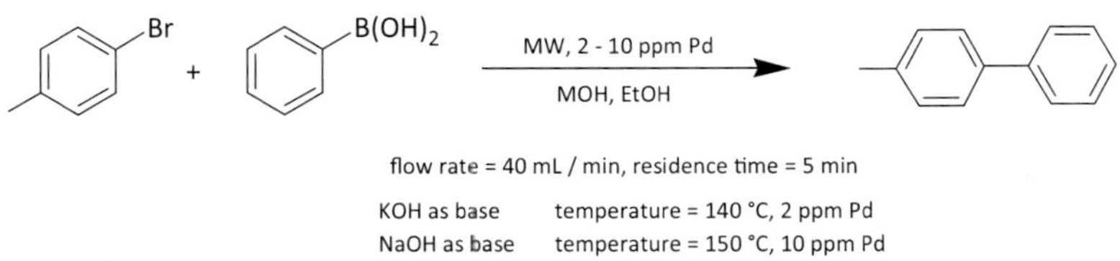

Scheme 12.3 Suzuki reaction using continuous-flow processing.

the base to potassium hydroxide and using ethanol as the only solvent offered a way to process the reaction mixture without loss in product yield. Working at a catalyst loading of 0.008 mol% we attempted to perform the reaction on the 1 mole scale in flow. However, while processing the reaction mixture a pressure rise was observed. Further inspection showed that this was due to a buildup of potassium bromide in the exit line, it being poorly soluble in ethanol. A molar equivalent of potassium bromide is formed during the course of the reaction, originating from the potassium hydroxide base and aryl bromide substrate. Changing the base to sodium hydroxide enabled it and the sodium bromide formed during the reaction to be soluble in ethanol, though the reaction was not as efficient and it was necessary to modify the reaction conditions. This entailed increasing both the reaction temperature and the palladium loading.

Overall, when the reaction and product mixtures are homogeneous, continuous-flow processing is easy and it is possible to scale up with minimal, if any, modification required. However, if any of the starting materials or the product is not completely soluble, then issues can arise while pumping material either in or out. While in some cases this can be overcome by use of additional solvent or a co-solvent, it can lead to a drop in product conversion, depending on the nature of the reaction. The other alternative is to modify the reaction parameters, this taking time and effort.

12.2.2 *Open-Vessel Processing*

When working on a larger scale, performing reactions in standard laboratory glassware at atmospheric pressure can offer operational advantages over a sealed-vessel approach. However, this eliminates one of the greatest attributes of microwave heating, namely, the ability to heat reactions to well above the normal boiling points of solvents in a safe and effective manner. As a result, open-vessel approaches to scale-up have seen a somewhat limited application.[28] That said, when removal of a by-product such as water is key to the success of a synthetic transformation, or if a gas is evolved during the course of the reaction, a large-scale, open-vessel batch microwave reactor may be an effective tool to carry out the procedure.

12.2.2.1 Synthesis using an open-vessel approach

We have scaled up both the Suzuki coupling reaction and the Hantzsch 1,4-dihydropyridine synthesis using an open-vessel approach. In both cases, it was necessary to reoptimize conditions from those previously developed on a small scale in a sealed vessel. In the case of the Suzuki coupling protocol, since the reaction using a 1:1 mix of water and ethanol as solvent was performed at 82°C in an open vessel as opposed to 150°C in a sealed tube, it was necessary to increase both the reaction time and catalyst loading in order to obtain reasonable conversions to the biaryl products.[29] The same was true in the case of the 1,4-dihydropyridine synthesis: time, temperature, and solvent needed modification.[27] The Beckmann rearrangement (acid-catalyzed conversion of ketoximes to *N*-substituted amides) proved much more amenable to translation from small-scale, sealed-vessel processing to a larger, open-vessel approach (Scheme 12.4). In a sealed tube, benzophenone could be converted to *N*-phenylbenzamide in 74% yield on the 5 mmol scale. The reaction mixture was heated to 140°C and held at this temperature for 20 minutes. Moving to an open vessel, the maximum temperature reached was 120°C, corresponding to reflux. Working on the 5 mmol scale, a 95% yield of product was obtained. Scaling further to the 1 mole level using the same conditions was possible without substantial loss in product yield.

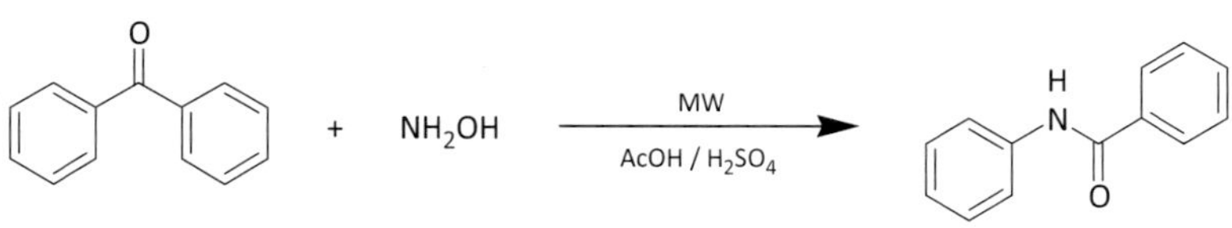

sealed vessel (10 mmol): heat to 140 °C and hold for 20 min - 74 % yield
open vessel (5 mmol): heat to reflux (120 °C) and hold for 20 min - 95 % yield
open vessel (1 mol): heat to reflux (120 °C) and hold for 20 min - 90 % yield

Scheme 12.4 Open-vessel microwave-promoted Beckmann rearrangement.

We scaled up the Heck coupling reaction of iodobenzene and butyl acrylate in open-vessel mode using a solvent-free approach.[30] We chose to use palladium acetate (0.1 mol%) as the catalyst, a combination of sodium carbonate and tributylamine as bases, and TBAB as an additive. These conditions had their root in previous work focused around solvent-free Heck coupling chemistry.[31] Since calculations suggested that the Heck reaction is somewhat exothermic, we wanted to be cautious when heating significant quantities of reagents in the absence of a solvent as a moderator. Working on the 0.5 mole scale, the reaction mixture was heated to 150°C over a three-minute time period, but even after the microwave irradiation was stopped, the contents of the vessel continued to heat to 180°C, demonstrating the exothermic nature of the reaction, once initiated. Repeating the reaction both at the 0.5 and 1 mole scales but heating only to 100°C and holding at this temperature for 10 minutes led to a more controllable approach as well as a quantitative conversion to the desired product. Overall, performing solvent-free reactions on scale using microwave heating in an open-vessel approach is not ideal and some precautions are necessary. A suitably sized reaction vessel is required, and it is advised that the contents should occupy no more than 10% of the vessel volume. In addition, microwave power needs to be very carefully modulated.

12.2.2.2 Reactive distillation

An issue associated with esterification reactions is that in order to drive them to completion, either the ester product or the water generated needs to be removed as the reaction proceeds or else an excess of one of the reagents needs to be used. When using a sealed-vessel approach, only the latter is an option. Using an apparatus that allows for the removal of water generated during the course of the reaction, we scaled up a range of esterification reactions in open-vessel mode.[32] While useful, a simpler alternative is to use a Dean–Stark trap, with which we could run the reaction until such time that the accumulation of water in the trap stopped. The reactions were performed on scales up to 6 moles (Scheme 12.5).[27]

A number of other reactions also benefit from being performed in an open-vessel arrangement with the capacity to collect either a product or a by-product by distillation. Use of microwave heating in conjunction with distillation has found uses in, for example, the purification of ionic liquids,[33] the preparation of pyranoquinolines,[34]

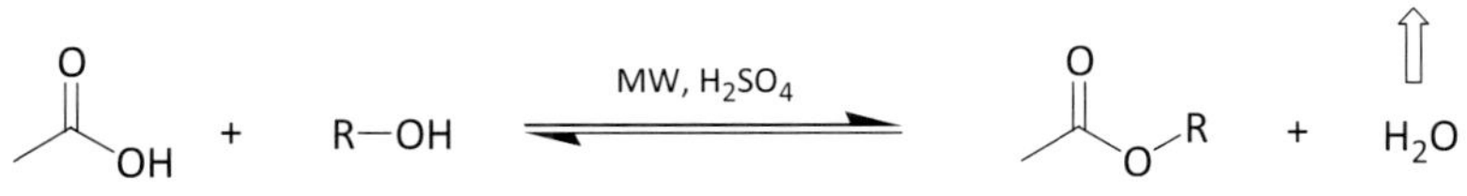

Alcohol = butanol; scale = 2 moles: 82% conversion
Alcohol = butanol; scale = 6 moles: 89 % conversion
Alcohol = cyclohexanol; scale = 2 moles: 80 % conversion
Alcohol = 2-octanol; scale = 2 moles: 86 % conversion

Scheme 12.5 Esterification of acetic acid with primary and secondary alcohols.

and the extraction of components from plants.[35] We have performed transesterification using butanol with diethyl adipate as a substrate, the product being dibutyl adipate (Scheme 12.6). The reaction could be driven toward completion by removing the ethanol by-product in a two-stage process, firstly heating the reaction mixture to 105°C, holding it for 10 minutes, and then raising the temperature to 115°C for another 10 minutes, by which time all the ethanol produced in the reaction had been removed. Raising the temperature to 145°C allowed us to strip off any butanol remaining in the reaction mixture. A 59% yield of dibutyl adipate was obtained.

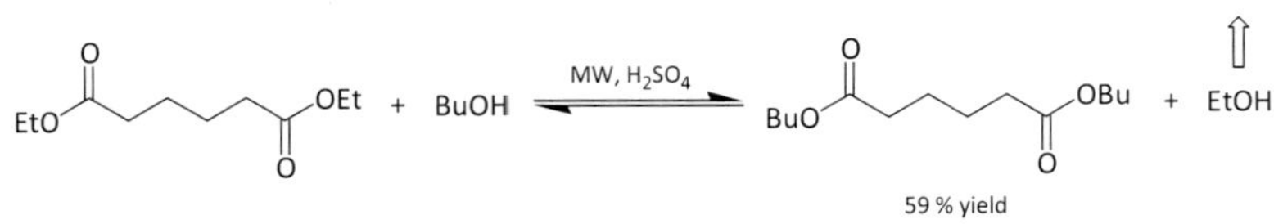

Scheme 12.6 Transesterification reaction of diethyl adipate with butanol.

In the acid-catalyzed dehydration of 2-octanol to yield octenes, the alkene products are considerably more volatile than the alcohol starting material and thus could be collected by direct distillation (Scheme 12.7). Using 1 wt% sulfuric acid as catalyst, alkene formation commenced once the reaction mixture reached a bulk temperature of 145°C. Using a Dean–Stark trap, the octene and water could be collected and separated. A mixture of *cis*- and *trans*-2-octene with a trace of 1-octene was obtained. The batch protocol was converted into a semicontinuous process simply by pumping 2-octanol into the reaction vessel as the octene was produced. The

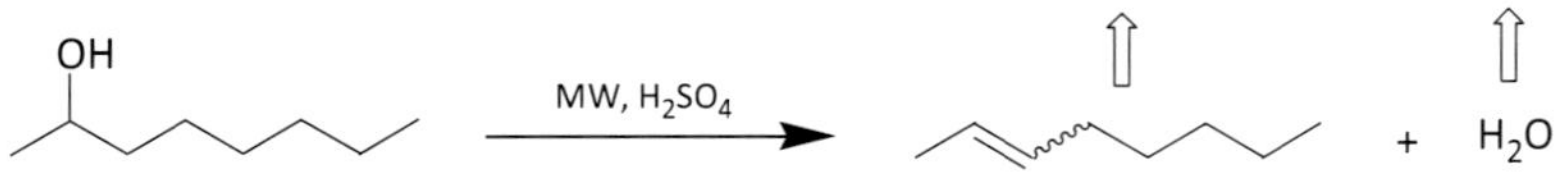

Scheme 12.7 Preparation of octene from octanol.

octanol was introduced at a rate of 15 mL/min, allowing the bulk temperature of 145°C to be maintained throughout.

12.2.3 *Sealed-Vessel Processing*

A sealed-vessel, batch approach represents an attractive choice in the scale-up of microwave-promoted reactions. The primary advantage is that most small-scale reactions are developed under sealed-vessel conditions; thus scale-up is potentially straightforward with little or no reoptimization needed. Disadvantages to this approach are the limits of reaction volume that can be irradiated as well as the safety requirements when working with larger vessels under pressure. There are three options available for sealed-vessel processing (Fig. 12.4), namely, (1) use of multiple smaller reaction vessels, (2) an automated stop-flow approach, and (3) the use of one larger vessel.

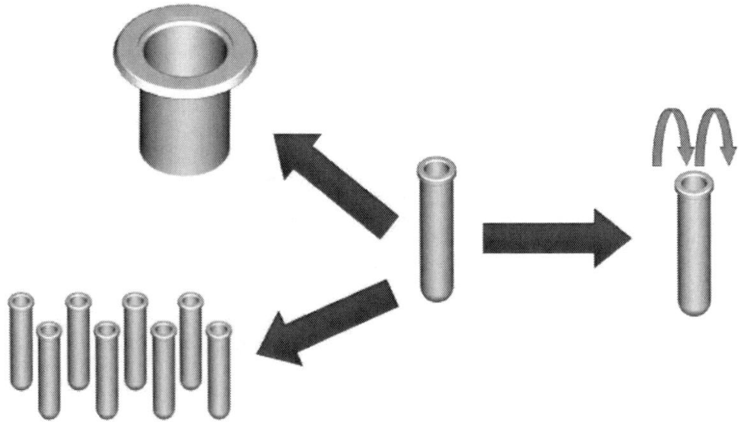

Figure 12.4 Approaches to scale-up using sealed-vessel processing.

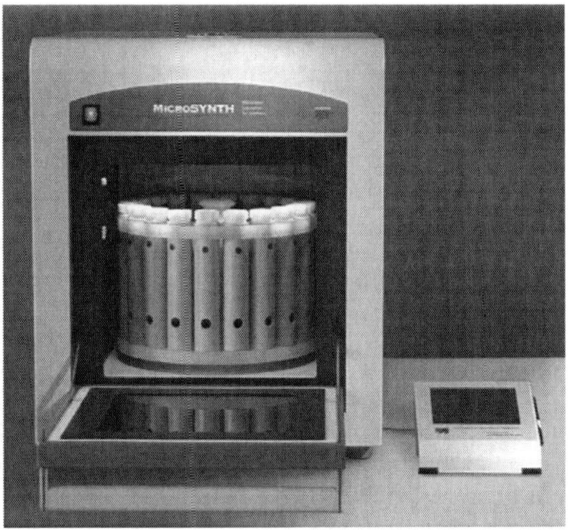

Figure 12.5 The Milestone MultiSYNTH unit, equipped with a 20-vessel carousel.

12.2.3.1 Multiple-vessel processing

When using multiple vessels processed in a carousel, it is generally possible to scale reactions without the need for any reoptimization of conditions developed on small-scale equipment.[36] Most carousels designed for multiple reaction vessels are built such that each individual vessel sits in its own protective cover. Depending on the equipment used, it is possible to monitor the temperature of either one reaction vessel or a number of vessels simultaneously. Pressure monitoring is more complex. Generally the vessels are equipped either with pressure-release valves or with burst discs that will open at a set maximum pressure. Using such equipment, we scaled up the synthesis of N-aryl-functionalized β-amino esters using a Michael addition protocol (Fig. 12.5).[27] Reactions were performed neat at 200°C for 20 minutes, catalyzed by acetic acid.[37] The reaction was highly temperature dependent; insufficient heating led to very poor yields, and too much heating led to side product formation and decomposition. We scaled up the reaction to 2.8 moles by placing 20 mL (0.14 mol) of the reaction mixture into each of the 20 quartz vessels, thus processing a total substrate volume of 0.4 L. Having loaded the reaction vessels into the carousel, we heated the mixtures for 20 minutes using a microwave power of 1,000 W, the final recorded temperature being 198°C. Upon cooling, we analyzed the contents of representative vessels before combining them all. We obtained an overall 78% yield after noting very little variation from vessel to vessel (Scheme 12.8).

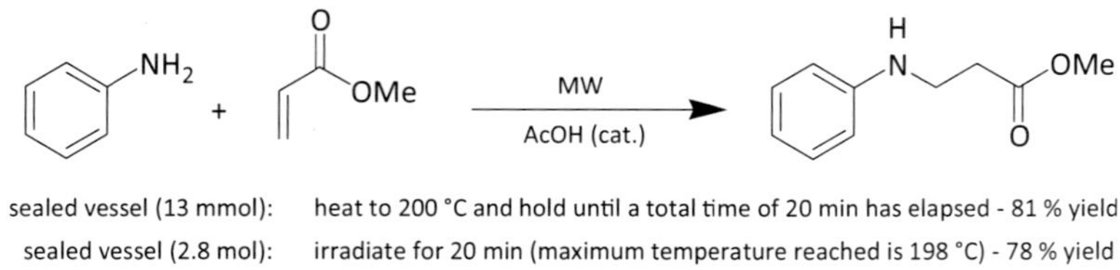

Scheme 12.8 Synthesis of an *N*-aryl-functionalized β-amino ester by an aza-Michael addition reaction.

12.2.3.2 Stop-flow processing

A problem with using a multiple-vessel approach to scale-up is that the loading/sealing of reaction vessels prior to heating and the opening/emptying afterward are all time-consuming processes. In an attempt to overcome this, a stop-flow approach can be used for modest scale-up.[38] Having optimized reaction conditions on a small scale, it is possible in principle to move directly to a stop flow using the same equipment. Reagents are loaded and products pumped out by means of a number of polytetrafluoroethylene (PTFE) lines (Fig. 12.6). While allowing for automation, this can lead to issues if the mixtures going in or coming out of the reaction vessel are not homogeneous, since the PTFE lines are prone to blockage. We have scaled up protocols for the Suzuki and Heck coupling reactions using a stop-flow approach, and in both cases, some reoptimization was required to overcome this issue.[39] Problems were particularly apparent at the end of the reaction, where the products were insoluble in the aqueous solvent mixtures and blocked the exit lines, even at temperatures up to 90°C. To overcome this, it was necessary to add a small charge of organic solvent into the vessel at the end of the reaction in order to solubilize the products and allow them to be pumped out. Ten cycles of each reaction were then run, producing a total of 15–20 g of product. Since the reaction mixtures could be

Figure 12.6 The CEM Voyager stop-flow unit.

heated and cooled fairly rapidly, cycle times were quite short (15 min for Suzuki coupling and 26 min for Heck coupling).

12.2.3.3 Use of a single, larger vessel

A single, larger reaction vessel for scale-up avoids the issues encountered both with using multiple smaller vessels and stop-flow processing. However, a number of new potential problems arise. First, the equipment used must be designed with additional safety measures in mind if larger volumes are going to be heated to elevated temperatures and pressures. Second, significant changes in engineering are required in order to obtain heating profiles that in any way mirror those of smaller microwave units. Third, while it may be possible to heat a reaction mixture effectively using microwave irradiation, it is also important that the contents of the vessel be cooled back to ambient temperature in a timely manner. The increased safety measures needed to withhold a potential vessel failure at elevated temperature and pressure generally mean that thick-walled vessels and microwave cavities are used. As a result, cooling can be very slow. This has been overcome by use of adiabatic

Figure 12.7 The Biotage® Advancer, a single sealed-vessel unit.

cooling, where the contents of the reaction vessel are ejected into a holding tank. Using this approach, the temperature of the mixture can be reduced rapidly to the boiling point of the solvent used. Our initial work in the area focused on using the Biotage® Initiator (Fig. 12.7).[27] With this unit, reactions are performed in a 350 mL capacity Teflon vessel. Temperatures of up to 250°C and pressures of 20 bar can be reached. The contents of the vessel are agitated by means of a mechanical paddle stirrer, and the lid of the reactor also comprises an inlet for gas and three extra available entry/exit ports. With this equipment it was possible to perform the Suzuki coupling reaction at the 50 mmol level using exactly the same conditions as developed in small-scale (1 mmol) optimization trials, with low palladium catalyst loadings of 0.0004–0.01 mol % (Scheme 12.9).

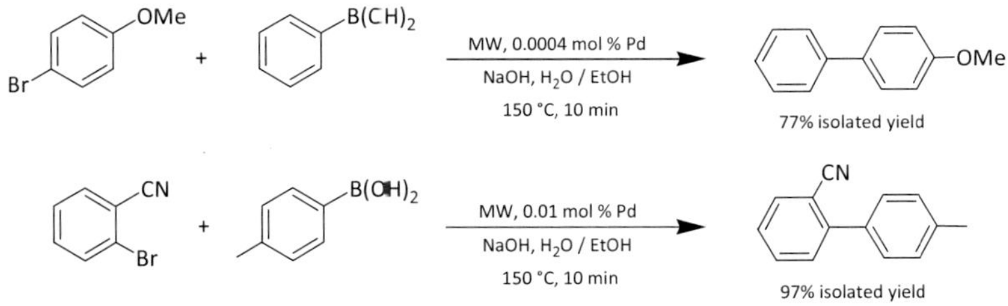

Scheme 12.9 Microwave-promoted Suzuki couplings performed on the 50 mmole scale.

In the case of Heck coupling, we again wanted to use our previously developed conditions for performing the reaction using water as a solvent with low concentrations of the ligandless palladium catalyst (Scheme 12.10). In optimization studies, working

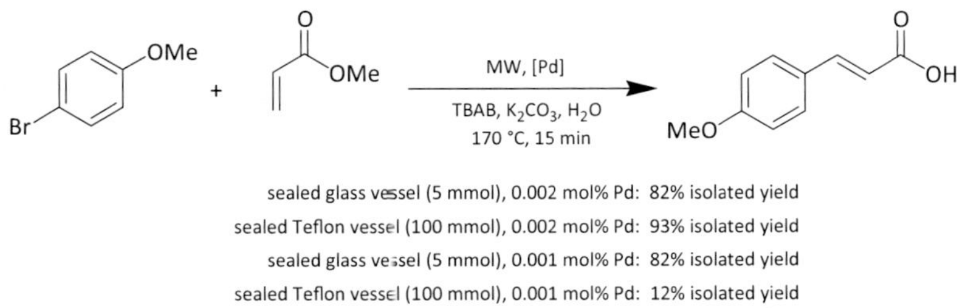

Scheme 12.10 Heck coupling of 4-bromoanisole with methyl acrylate.

on the 5 mmol scale in a sealed glass tube, an 82% isolated yield of 4-methoxycinnamic acid was obtained in the coupling of 4-bromoanisole with methyl acrylate using a catalyst loading of 0.002 mol%. This could be directly scaled to the 100 mmol level in the Advancer, with a slightly improved yield. However, when scaling up a reduced catalyst loading (from 0.002 to 0.001 mol%), a precipitous drop in product yield was observed compared to the small-scale, glass tube trials. Probing this in more detail led us to believe that the Teflon vessel used in the Advancer was slightly permeable to the reaction mixture and adsorbed small quantities of the palladium in solution. Although this was not critical at higher palladium concentrations, when working at the lower limit of catalyst concentrations this adsorption of small quantities of metal resulted in impeding the coupling and poor yields. This highlights the fact that when a reaction is highly dependent on catalyst concentration, particular attention needs to be paid when moving from a glass vessel to one made from Teflon or other such material.

A range of other reactions were scaled up using the Biotage® Advancer (Scheme 12.11). If a reaction mixture is highly nonpolar

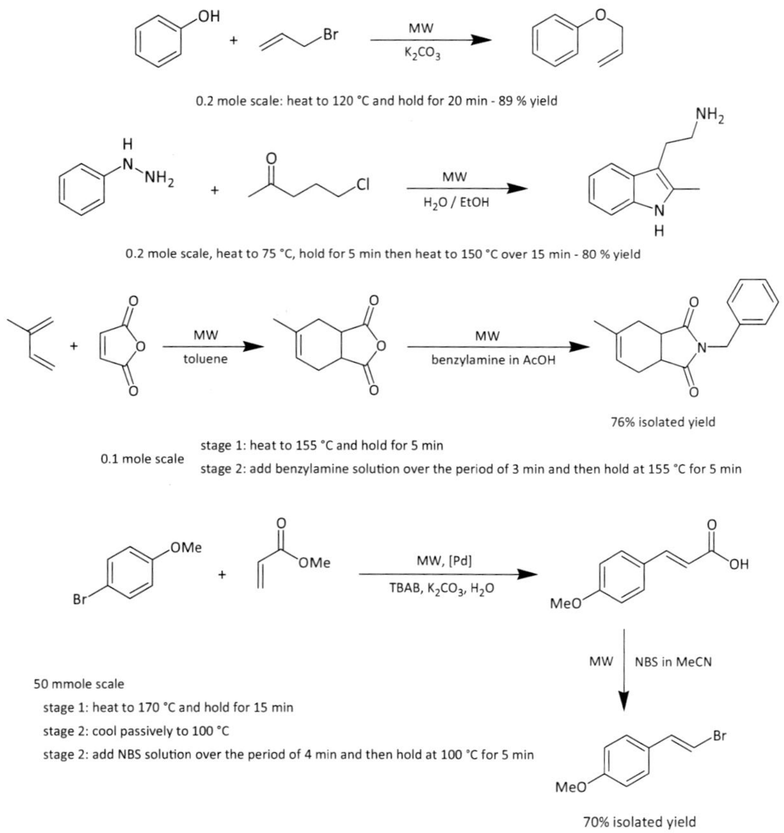

Scheme 12.11 Use of the Biotage® Advancer for scale-up using a single sealed-vessel approach.

then an additive may be required in order to heat it to the desired temperature efficiently. Such effects are magnified when moving from small to larger scales. In addition, borosilicate glass tubes contain small quantities of salt impurities, which can lead to indirect heating of nonpolar reaction mixtures. Teflon on the other hand does not contain such impurities. Thus, the vessel material may also have an effect on the heating profile of nonpolar reaction mixtures. This was the case with the Claisen rearrangement of allylphenyl ether where the addition of a small quantity of TBAB was needed as a heating aid to reach the target temperature of 247°C in the Teflon vessel but not required on the mmol scale in a glass tube. In cases where the reaction was exothermic, we found that judicious use of a multistage heating protocol ensured smooth and controllable heating of the reaction mixture. This was particularly the case in the Grandberg synthesis of 2-methyltryptamine, in essence a modified Fisher indole reaction. When heating the reaction mixture we observed the onset of an exotherm at approximately 120°C. The unit is designed to sense for such occurrences and released the contents of the reaction vessel into the cooling collection pot, thus quenching the reaction. In revising our protocol, we found that by ramping to 75°C and then very gradually increasing the reaction temperature to 150°C over a period of 15 minutes, it was possible to heat the mixture controllably. It was also possible to perform two-step, one-pot reactions using the unit, this being achieved by the addition of the second set of reagents into the reaction vessel while still at elevated temperature and pressure. In one trial, we performed the Diels–Alder reaction of isoprene and maleic anhydride, followed by conversion of the anhydride product to an imide using benzylamine. In another we prepared 4-methoxy-β-bromostyrene via Heck coupling between 4-bromoanisole and methyl acrylate, followed by a decarboxylative bromination.

12.2.3.4 Reactions involving gaseous reagents

A class of reactions that can only be performed in a sealed vessel is those that involve the use of elevated pressure of a reactive gas, such as hydrogenation or carbonylation. We have developed methodologies for the hydroxy- and alkoxycarbonylation of aryl iodides using palladium acetate as a catalyst.[40] In the case of alkoxycarbonylation, we developed a procedure for performing the reaction using a near-stoichiometric loading of carbon monoxide.[41] This has operational advantages when working on larger scales. Not only is the overall inventory of gas smaller, but at the end of the reaction, the majority of the carbon monoxide has been consumed, meaning that there is little left to be vented when depressurizing the reactor to access the vessel. We performed the reaction on scales ranging from 1 mmol to 1 mole (Scheme 12.12). Using a carousel of

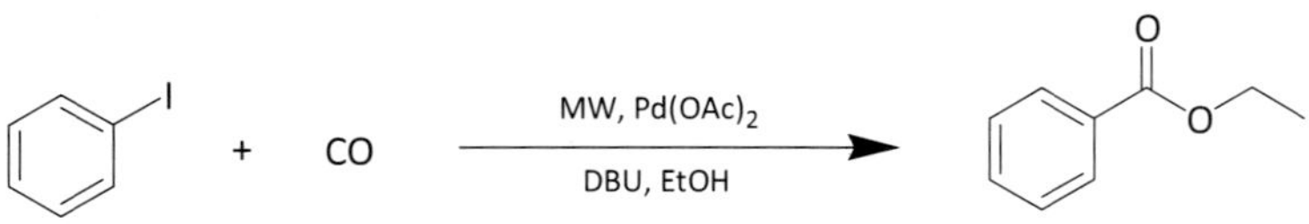

heat to 125 °C and hold for 30 min

1 mmol scale, 0.1 mol% Pd, 10 bar CO (excess CO used):	91%
2 mmol scale, 0.1 mol% Pd, 1 bar CO (1.1 eq):	84%
8 vessels of 32 mmol, total quantity 256 mmol, 0.1 mol% Pd, 17 bar CO (1.4 eq):	81%
100 mmol scale, 0.1 mol% Pd, 17.2 bar CO (1.1 eq):	86 %
1 mol scale, 0.1 mol% Pd, 34 bar CO (2 eq):	79%
1 mol scale, 0.1 mol% Pd, 27 bar CO (1.5 eq):	80%

Scheme 12.12 Alkoxycarbonylation of iodobenzene.

eight reaction vessels we used ethoxycarbonylation of iodobenzene as a test reaction and, starting with 0.1 moles of iodobenzene across the eight reaction vessels, obtained an overall conversion of 91%. Using the Biotage® Advancer, we performed the reaction on scales up to 100 mmol, obtaining >95% conversion to product.

We were not able to scale the reaction further since, at the 0.1 mole level, we were approaching the pressure limits of the apparatus. To transition to larger scales, we moved to another microwave unit. The Milestone UltraCLAVE was originally designed for sample digestion and its design based on a high-pressure autoclave.[42] It has a single, large reaction chamber of 3.5 L volume and is capable of operating at temperatures of up to 200°C and pressures of up to 200 bar. Using this equipment, the alkoxycarbonylation of iodobenzene could be performed on the 1 mole scale.[43] In addition, we could scale up six individual ethoxycarbonylation reactions simultaneously, each at the 50 mmol level.

12.2.3.5 Increasing the scale

A batch microwave reactor capable of running reactions from 2 to 12 L recently became available (Fig. 12.8).[44] Apparatus such as this that is capable of processing material at the kilogram scale could help bridge the gap between a small-scale protocol and larger kilo- or pilot-plant scale. In the development of new chemical entities (NCEs), the transition from the medicinal chemistry route to a scale where enough material can be prepared to carry out initial in vivo toxicology studies is often the most difficult.[45] Medicinal chemists frequently use microwave-mediated transformations that could be difficult to scale to the level necessary to obtain 1–5 kg of an NCE. As a result, process chemists are faced with either running multiple reactions to achieve the desired throughput or having to develop modified conditions to avoid microwave heating. Since the probability of any one of 20–50 promising NCEs becoming the active pharmaceutical ingredient in a new drug is low, companies have a vested interest in pushing off route reoptimization until as late as possible, ideally only after screening and toxicology tests have narrowed the field to a few promising candidates.

Figure 12.8 (A) Engineering renderings of the prototype reactor used and (B) engineering renderings of a potential later-generation unit.

There are three interchangeable glass reaction vessels that can be used with the unit: 5 L, 9 L, and 13 L in capacity with working volumes of 2–4 L, 4–8 L, and 7–12 L, respectively. The desired vessel is placed in a mechanically sealed stainless steel reaction chamber capable of operating at pressures up to 350 psi. The chamber is pre-pressurized to 250–300 psi using nitrogen gas, allowing access to reaction temperatures above the normal boiling points of solvents at atmospheric pressure. The reactor employs three 2.45 GHz magnetrons with an accessible power of 2.5 kW each, for a total maximum allowed output of 7.5 kW. Reaction parameters such as time, temperature, pressure, and magnetron power are constantly monitored, and reaction mixtures are agitated by means of a stirring paddle that operates at speeds from 0 to 125 rpm. At the end of a reaction, the contents can be passed through a water-cooled, counterflow heat exchanger. Alternatively, they can be ejected directly into a receiving chamber at ambient pressure, thus offering more rapid cooling. When assessing the unit, we first probed its operating parameters and then performed a range of pharmaceutically relevant synthetic transformations, scaling them up in a linear fashion to the fullest extent possible.

We chose to first heat various solvents in the unit to give us an indication of its effectiveness. These seven solvents had a range of microwave absorptivities: ethanol, water, acetonitrile, ethyl acetate, tetrahydrofuran, 2-butanone, and dichloromethane. In each case, 4 L of the solvent was heated in the 5 L reaction vessel until it reached 150°C using a constant 7.5 kW power delivery from the magnetrons (2.5 kW $\times$ 3), with the stirring set to $\sim$60 rpm. The solvents that performed the best in this study were ones often regarded as "poor" microwave absorbers. On the other hand, those generally considered "good" solvents for microwave chemistry actually performed relatively poorly. It took between four and five minutes to heat 2-butanone, dichloromethane, acetonitrile, tetrahydrofuran, and ethyl acetate from 30°C to 150°C, but heating 4 L of water to 150°C took over nine minutes. A consideration of heat capacity helps explain these results. The amount of energy required to heat 4 L of water to 120°C is approximately 2.01×10^3 kJ. The same volume of ethanol requires less than half that at 924 kJ, and all other solvents require below 900 kJ (4 L, Δ =

120°C). Also, because the vessel is charged with 280 psi N_2 prior to heating, energy loss due to solvent vaporization is minimized. This said, since the solvents are being heated using microwave irradiation, their relative microwave absorptivities must also have some impact. While the ability of a solvent to be heated using microwave irradiation is generally proportional to its loss tangent and its dielectric loss constant, this is only true if the entire sample cross section can effectively be irradiated. On the small scale this is the case, and solvents that are highly microwave absorbing heat faster than those that have low microwave absorptivities. On this significantly larger scale, the depth to which the microwave energy can penetrate the contents of the vessel varies depending on the solvent. Less absorbent solvents have a larger cross section absorbing the microwave energy as compared to more absorbent solvents. Microwave absorbtivity and the heating cross section are interlinked and inversely proportional, the result being that dichloromethane ($\tan \delta = 0.042$) can be heated to 150°C from 30°C ($\Delta T = 120°C$) in less time than ethanol ($\tan \delta = 0.941$) across the same temperature range using the same applied microwave power. This is shown diagrammatically in Fig. 12.9. The results show that with efficient stirring and properly sized and engineered magnetrons, it should be possible to design batch reactors capable of effectively heating reactions on significantly larger scales than currently used.

In turning to synthetic chemistry, we selected a range of reactions as candidates, including homogeneous condensations, heterogeneous reactions, and phosphine-free palladium-catalyzed transformations carried out in water. Overall, the unit proved an effective tool for the development of kilo-scale microwave chemistry, as shown in Table 12.1. As well as performing individual synthetic transformations, we also developed a four-step sequence with the objective being the simulation of a situation where multiple sequential microwave steps were employed in order to reach a desired target compound (Scheme 12.13). We first optimized conditions on the >10 mmol level, obtaining a 39% overall yield for the four steps. The scale-up sequence employed three microwave steps and afforded 473 g of the final product in 38% overall yield for four steps. In one step (a $POCl_3/Et_3N$ deoxychlorination)

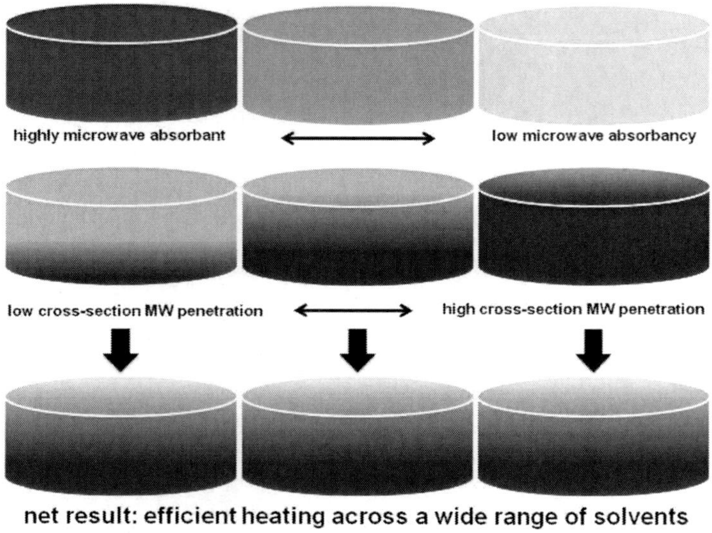

Figure 12.9 Illustration showing that microwave irradiation is effective at heating a wide range of solvents on a large scale, regardless of microwave absorptivity.

conventional heating was used in the scale-up. This was because calorimetry studies (at the 9.0 mmol scale) indicated an exotherm of approximately 120 kJ/mol upon addition of the amine to the reaction mixture and also because the chlorination reaction proceeds smoothly at lower temperatures, albeit with slightly longer reaction times.

12.2.4 *Energy Efficiency*

When using microwave heating for scale-up of synthetic chemistry, the issue of energy efficiency has to be considered.[46] Transitioning from conventional heating approaches to microwave heating often involves moving away from laboratory infrastructure that has been designed over the years to maximize efficiency. For example, many chemical production facilities generally have easy access to cheaply generated steam heating. In their report on comparing the energy efficiency of various heating methods, including microwave heating, Clark *et al.* report a significant reduction in energy demand on

Table 12.1 Reactions performed in a large-scale batch microwave unit. For comparison, small-scale reactions and corresponding yields are included. The scale-up reactions were performed using identical conditions of time, temperature, reaction concentration, and catalyst loadings, differing only in reaction scale and ramp time. All yields represent isolated yields.

Entry	Reaction	Small -scale results	Large -scale results
1	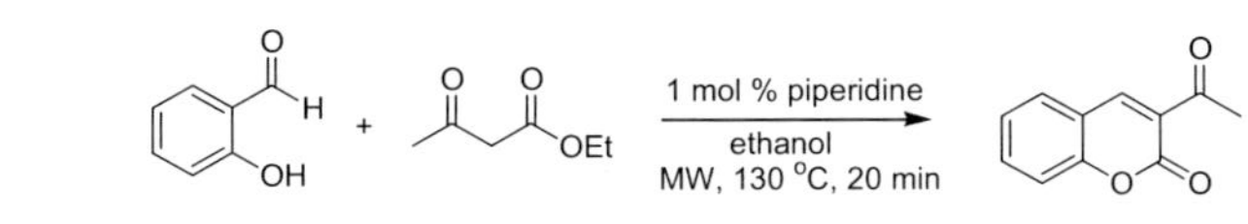 (1)	5 mmol, 72% 300 mmol, 71%	3.0 mol, 66.6% 12.0 mol, 74.0% 21.0 mol, 80.7%
2	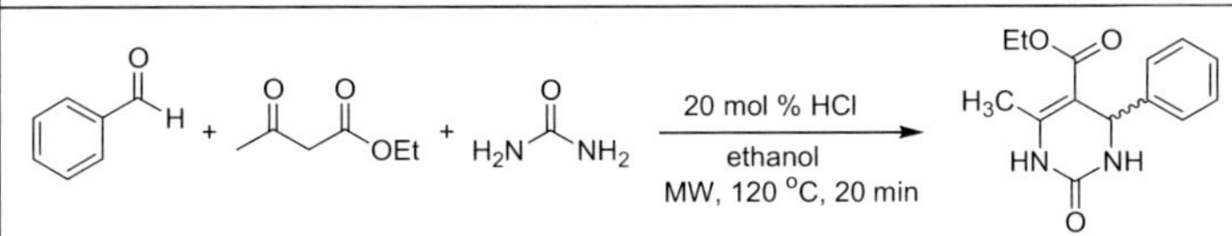 (2)	2 mmol, 54%	4.0 mol, 55.0%
3	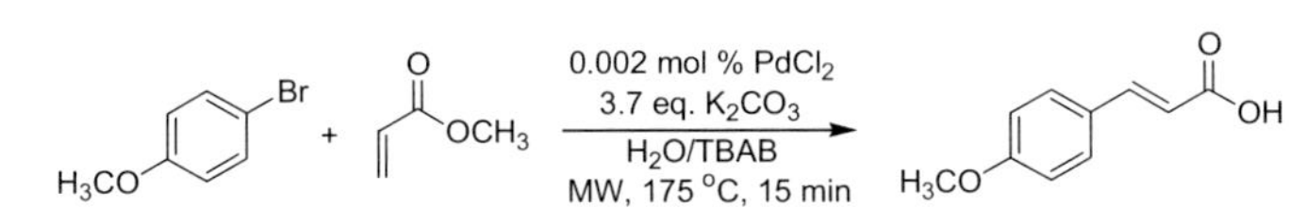 (3)	1 mmol, 73% 100 mmol, 93%	2.0 mol, 95%
4	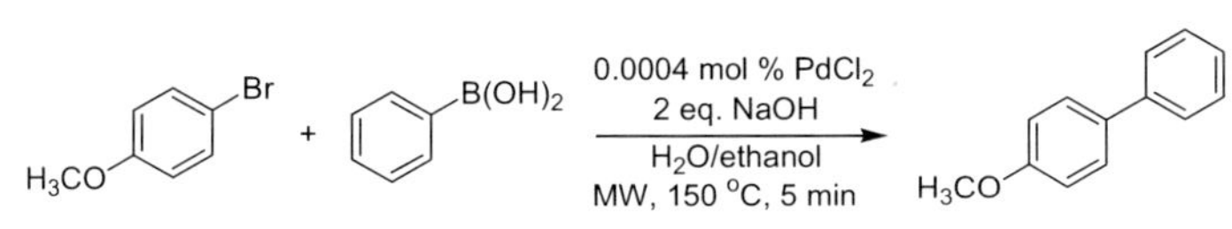 (4)	1 mmol, 93% 50 mmol, 90%	4.0 mol, 91.8%

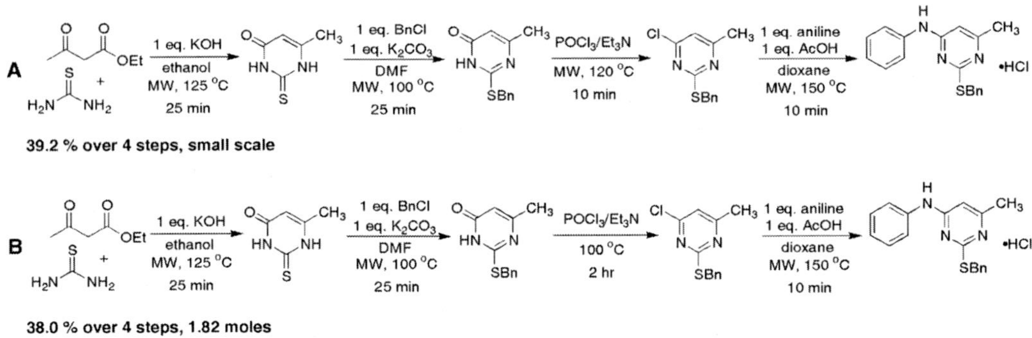

Scheme 12.13 Four-step sequence.

switching from oil bath to microwave for a Suzuki reaction.[47] In 2008, Razzaq and Kappe reported a study of the energy efficiency of four reactions under a number of microwave conditions including the use of sealed vessels and open glassware. They concluded that the efficiency of microwave heating is dependent on the conditions used but is only superior to conventional heating when using a superheated solvent in a sealed vessel.[48] In 2009, there were a number of somewhat contradictory reports. Hoogenboom *et al.* probed the energy efficiency of microwave heating for a number of solvents on a small scale and then deionized water on scales from 5 to 400 mL.[49] They report "maximum average heating efficiencies of 10% for small-scale vessels (5 mL), 20% for medium-scale (50 mL), and 30% for large-scale microwave heating (400 mL)." Moseley and Woodman reported on the energy efficiency of microwave heating at a 1–3 L scale.[50] Four scientific microwave units were compared with each other and against a conventionally heated, jacketed reaction vessel for four organic reactions under identical conditions. The conclusions were that microwave heating can be more energy efficient than conventional heating but that it is highly dependent on the microwave unit used and the scale at which the reaction is performed. Komorowska *et al.* have studied the influence of microwave irradiation on a polyesterification reaction.[51] They suggest that energy loss in the magnetron and the cavity make microwave heating less energy efficient and hence less economical than conventional heating. A similar assertion was made by Dressen *et al.* in a 2010 report.[52] They stated that microwave heating loses at least two times more energy in its fore track and that the energy bill is at least a factor of eight times higher as compared to steam heating for industrial applications. Very recently, Moseley *et al.* screened a number of synthetic transformations and found that on a kilo Joule per mole basis, the energy efficiency of laboratory-scale microwave units was higher than that of pilot-scale reactors under all reaction conditions studied.[53] Also, the most efficient microwave equipment operating under optimal conditions was significantly more energy efficient than pilot-scale reactors even up to 40 L.

12.3 Concluding Remarks

An overview of the approaches that can be taken for the scale-up of reactions using microwave heating has been presented here. Both batch and continuous-flow processing are possible. When using a batch approach there is the option of running reactions in open-vessel mode. While this has some operational advantages, it does preclude reactions occurring well above the normal boiling points of solvents. However, when removal of a product or by-product is key to the success of a synthetic transformation, it can prove very effective. Sealed-vessel approaches can be split into three categories. When using multiple vessels processed in a carousel, reactions can often be scaled without the need for any reoptimization of conditions, but the loading, sealing, opening, and emptying of the vessels can be time consuming. This hurdle can be overcome using a stop-flow approach, though it is replaced by a homogeneity challenge that arises due to the fact that, to take advantage of automated processing, reagents have to be pumped in and products pumped out of the reaction vessel through somewhat narrow tubing. Ease of loading and emptying, together with the ability to run heterogeneous reactions and reactions under an atmosphere of a reactive gas, make performing a reaction in a single, larger, sealed vessel attractive. The increased engineering required to allow larger volumes of material to be heated both effectively and safely could impact parameters such as heating rate and also cooling at the end of the reaction. However, studies show that penetration depth is not an issue when heating larger volumes of solvents.[44] This, together with the ability to scale reactions linearly from small to large without detrimental effects on product yield, and the fact that the contents of a reaction vessel can be ejected directly into a receiving chamber at ambient pressure, thereby affording rapid cooling, make larger, sealed-vessel processing an attractive option.

In using a continuous-flow approach, homogeneity of the reaction mixture is key. When the mixture is homogeneous it is often possible to move from small-scale, sealed-vessel conditions to continuous-flow operation without any modification of reaction conditions or loss in product yield. When either the starting materials or the product mixture contain particulate matter, however, continuous processing can prove a challenge, though reoptimization of reaction conditions as well as reduction of the concentration may allow these difficulties to be overcome.

With energy balance still being a topic of debate, there are areas where larger-scale microwave heating does have clear advantages from a standpoint of greener chemistry. The ease in which reactions can be performed and the ability to use low catalyst loadings and alternative solvents offer avenues for developing cleaner approaches to synthesis. In addition, the application of microwave heating is not limited to organic synthesis. The use of biorenewable feedstocks for microwave-assisted polymerizations has been found to be an excellent route for producing new or improved polymeric materials and biodegradable polymers. In addition, plastic waste can be efficiently recycled using a microwave-promoted chemical depolymerization. Microwave heating is proving an enabling technology in materials chemistry as well as in the biosciences. The reports from our laboratory as well as many others show that microwave heating is not restricted to the small-scale research chemistry laboratory. While not yet perhaps at the stage where tons of a desired compound can be synthesized quickly and easily, kilo-scale synthesis is certainly well within the capability of current commercially available apparatus.

Acknowledgments

Our work would not have been possible without the support of microwave equipment manufacturers. We thank AccelBeam Synthesis, Anton Paar, Biotage, CEM Corporation, and Milestone. The students who worked on the scale-up project all deserve acknowledgment. Three people particularly stand out. Dr Matthew Bowman spearheaded much of our early efforts, and Drs. Chad Kormos and Jason Schmink undertook the majority of the latter studies, and their work forms the basis of this chapter. We thank the National Science Foundation (CAREER award CHE-0847262), the American Chemical Society Petroleum Research Fund (45433-AC1), and the University of Connecticut for funding.

References

1. A number of useful books have been published on the subject of microwave heating in preparative chemistry. See, for example, (a) Leadbeater, N. E. (Ed.). (2010)). *Microwave Heating as a Tool for Sustainable Chemistry*. Boca Raton FL: CRC Press; (b) Loupy, A. (Ed.). (2006). *Microwaves in Organic Synthesis, 2nd Ed.* Weinheim: Wiley-VCH; (c) Kappe, C. O.; Stadler, A. (2005). *Microwaves in Organic and Medicinal Chemistry*. Weinheim: Wiley-VCH.
2. Schmink, J. R.; Leadbeater, N. E. (2010). Chapter 1, in *Microwave Heating as a Tool for Sustainable Chemistry* (N. E. Leadbeater, Ed.). Boca Raton FL: CRC Press.
3. For an excellent discussion of microwave absorptivity and theory from first principles, see Craig, D. Q. M. (1995). *Dielectric Analysis of Pharmaceutical Systems*. Bristol, PA: Taylor and Francis.
4. Saillard, R.; Poux, M.; Berlan, J. (1995). *Tetrahedron*, **51**, 4033.
5. Baghurst, D. R.; Mingos, D. P. M. (1992). *J. Chem. Soc., Chem. Commun.*, 674.
6. Lienhard, J. H., IV; Lienhard, J. H., V. (2008). *A Heat Transfer Textbook, 3rd Ed.* Cambridge MA: Phlogiston Press.
7. Chemat, F.; Esveld, E. (2001). *Chem. Eng. Technol.*, **24**, 735.
8. (a) Washington, A. L.; Strouse, G. F. (2009). *Chem. Mat.*, **21**, 2770. (b) Washington, A. L.; Strouse, G. L. (2008). *J. Am. Chem. Soc.*, **130**, 8916.
9. For a discussion of temperature measurement in microwave chemistry, see Nüchter, M.; Ondruschka, B.; Weiss, D.; Beckert, R.; Bonrath, W.; Gum, A. (2005). *Chem. Eng. Technol.*, **28**, 871.
10. (a) Obermayer, D.; Kappe, C. O. (2010). *Org. Biomol. Chem.*, **8**, 114; (b) Herrero, M. A.; Kremsner, J. M.; Kappe, C. O. (2008). *J. Org. Chem.*, **73**, 36.
11. (a) Kremsner, J. M.; Kappe, C. O. (2006). *J. Org. Chem.*, **71**, 4651; (b) Razzaq, T.; Kremsner, J. M.; Kappe, C. O. (2008). *J. Org. Chem.*, **73**, 6321.
12. Obermayer, D.; Gutmann, B.; Kappe, C. O. (2009). *Angew. Chem., Int. Ed.*, **48**, 8321.
13. Leadbeater, N. E.; Pillsbury, S. J.; Shanahan, E.; Williams, V. A. (2005). *Tetrahedron*, **61**, 3565.
14. Hosseini, M.; Stiasni, N.; Barbieri, V.; Kappe, C. O. (2007). *J. Org. Chem.*, **72**, 1417.
15. Schmink, J. R.; Leadbeater, N. E. (2009). *Org. Biomol. Chem.*, **7**, 3842.
16. Gedye, R.; Smith, K.; Westaway, H. (1986). *Tetrahedron Lett.*, **27**, 279.

17. Giguere, R. J.; Bray, T. L.; Duncan, S. M.; Majetich, G. (1986). *Tetrahedron Lett.*, **27**, 4945.

18. Strauss, C. R. (2009). *Org. Process Res. Dev.*, **13**, 915.

19. For recent reviews, see: (a) Moseley, J. D. (2010). Chapter 5, in *Microwave Heating as a Tool for Sustainable Chemistry* (N. E. Leadbeater, Ed.). Boca Raton FL: CRC Press; (b) Moseley, J. D. (2009). *Chim. Oggi/Chem. Today*, **27**, 6; (c) Wolkenberg, S. E.; Shipe, W. D.; Lindsley, C. W.; Guare, J. P.; Pawluczyk, J. M. (2005). *Curr. Opin. Drug Discov. Devel.*, **8**, 701.

20. For a recent evaluation of a wide range of equipment, see Moseley, J. D.; Lenden, P.; Lockwood, M.; Ruda, K.; Sherlock, J.-P.; Thomson, A. D.; Gilday, J. P. (2008). *Org. Process Res. Dev.*, **12**, 30.

21. Barnard, T. M.; Leadbeater, N. E.; Boucher, M. B.; Stencel, L. M.; Wilhite, B. A. (2007). *Energy Fuels*, **21**, 1777.

22. Leadbeater, N. E.; Barnard, T. M.; Stencel, L. M. (2008). *Energy Fuels*, **22**, 2005.

23. A flow unit has been developed in a partnership between pharmaceutical supplier Cambrex and microwave engineers C-Tech Innovation (www.cambrex.com and www.ctechinnovation.com).

24. For example, see: (a) Bagley, M. C.; Jenkins, R. L.; Lubinu, M. C.; Mason, C.; Wood, R. (2005). *J. Org. Chem.*, **70**, 7003; (b) Wilson, N. S.; Sarko, C. R.; Roth, G. P. (2004). *Org. Process Res. Dev.*, **8**, 535; (c) Shieh, W.-C.; Lozanov, M.; Repic, O. (2003). *Tetrahedron Lett.*, **44**, 6943; (d) Khadilkar, B. M.; Madyar, V. R. (2001). *Org. Process Res. Dev.*, **5**, 452; (e) Marquie, J.; Salmoria, G.; Poux, M.; Laporterie, A.; Dubac, J.; Roques, N. (2001). *Ind. Eng. Chem. Res.*, **40**, 4485; (f) Cablewski, T.; Faux, A. F.; Strauss, C. R. (1994). *J. Org. Chem.*, **59**, 3408.

25. Moseley, J. D.; Lawton, S. J. (2007). *Chem. Oggi/Chem. Today*, **25**, 16.

26. Bergamelli, F.; Iannelli, M.; Marafie, J. A.; Moseley, J. D. (2010). *Org. Process Res. Dev.*, **14**, 926.

27. Bowman, M. D.; Holcomb, J. L.; Kormos, C. M.; Leadbeater, N. E.; Williams, V. A. (2008). *Org. Process Res. Dev.*, **12**, 41.

28. Barnard, T. M.; Vanier, G. S.; Collins, M. J. (2006). *Org. Process Res. Dev.*, **10**, 1223.

29. Leadbeater, N. E.; Williams, V. A.; Barnard, T. M.; Collins, M. J. (2006). *Org. Process Res. Dev.*, **10**, 833.

30. Leadbeater, N. E.; Williams, V. A.; Barnard, T. M.; Collins, M. J. (2006). *Synlett*, 2953.

31. (a) Jeffery, T. (1996). *Tetrahedron*, **52**, 10113; (b) Kaufmann, D. E.; Nouroozian, M.; Henze, H. (1996). *Synlett*, 1091; (c) Perosa, A.; Tundo, P.; Selva, M.; Zinovyev, S.; Testa, A. (2004). *Org. Biomol. Chem.*, **2**, 2249.
32. Amore, K. A.; Leadbeater, N. E. (2007). *Macromol. Rapid Commun.*, **28**, 473.
33. Guerrero-Sanchez, C.; Lobert, M.; Hoogenboom, R.; Schubert, U. S. (2007). *Macromol. Rapid. Commun.*, **28**, 456.
34. Razzaq, T.; Kappe, C. O. (2007). *Tetrahedron Lett.*, **48**, 2513.
35. See, for example, (a) Perino-Issartier, S.; Abert-Vian, M.; Petitcolas, E.; Chemat, F. (2010). *Chromatographia*, **72**, 347; (b) Farhat, A.; Ginies, C.; Romdhane, M.; Chemat, F. (2009). *J. Chromatogr. A*, **1216**, 5077; (c) Veillet, S.; Tomao, V.; Visinoni, F.; Chemat, F. (2009). *Anal. Chim. Acta*, **632**, 203.
36. (a) Lehmann, H.; LaVecchia, L. (2005). *JALA*, **10**, 412; (b) Nüchter, M.; Ondruschka, B. (2003). *Mol. Diversity*, **7**, 253; (c) Stadler, A.; Yousefi, B. H.; Dallinger, D.; Walla, P.; Van der Eycken, E.; Kaval, N.; Kappe, C. O. (2003). *Org. Process Res. Dev.*, **7**, 707; (d) Stadler, A.; Pichler, S.; Horeis, G.; Kappe, C. O. (2002). *Tetrahedron*, **58**, 3177; (e) Alcázar, J.; Diels, G.; Schoentjes, B. (2004). *QSAR Comb. Sci.*, **23**, 906.
37. Amore, K. M.; Leadbeater, N. E.; Miller, T. A.; Schmink, J. R. (2006). *Tetrahedron Lett.*, **47**, 8583.
38. (a) Moseley, J. D.; Woodman, E. K. (2008). *Org. Process Res. Dev.*, **12**, 967; (b) Pitts, M. R.; McCormack, P.; Whittall, J. (2006). *Tetrahedron*, **62**, 4705; (c) Loones, K. T. J.; Maes, B. U. W.; Rombouts, G.; Hostyn, S.; Diels, G. (2005). *Tetrahedron*, **61**, 10338.
39. Arvela, R. K.; Leadbeater, N. E.; Collins, M. J. (2005). *Tetrahedron*, **61**, 9349.
40. (a) Kormos, C. M.; Leadbeater, N. E. (2006). *Synlett*, 1663; (b) Kormos, C. M.; Leadbeater, N. E. (2007). *Org. Biomol. Chem.*, 65.
41. Kormos, C. M.; Leadbeater, N. E. (2007). *Synlett*, 2006.
42. For a description of the apparatus, see Borowski, K. J.; Schoenfeld, C. (2006). *Am. Lab.*, **38**, 26.
43. Iannelli, M.; Bergamelli, F.; Kormos, C. M.; Paravisi, S.; Leadbeater, N. E. (2009). *Org. Process Res. Dev.*, **13**, 634.
44. Schmink, J. R.; Kormos, C. M.; Devine, W. G.; Leadbeater, N. E. (2010). *Org. Process Res. Dev.*, **14**, 205.
45. Federsel, H.-J. (2009). *Acc. Chem. Res.*, **42**, 671.
46. Devine, W. G.; Leadbeater, N. E. (2011). *ARKIVOC*, 127.

47. Gronnow, M. J.; White, R. J.; Clark, J. H.; Macquarrie, D. J. (2005). *Org. Process Res. Dev.*, **9**, 516. [Corrigendum: Gronnow, M. J.; White, R. J.; Clark, J. H.; Macquarrie, D. J. (2007). *Org. Process Res. Dev.*, **11**, 293.]
48. Razzaq, M.; Kappe, C. O. (2008). *ChemSusChem*, **1**, 123.
49. Hoogenboom, R.; Wilms, T. F. A.; Erdmenger, T.; Schubert, U. S. (2009). *Aust. J. Chem.*, **62**, 236.
50. Moseley, J. D.; Woodman, E. K. (2009). *Energy Fuels*, **23**, 5438.
51. Komorowska, M.; Stefanidis, G. D.; Van Gerven, T.; Stankiewicz, A. I. (2009). *Chem. Eng. J.*, **155**, 859.
52. Dressen, M. H. C. L.; van de Kruijs, B. H. P.; Meuldijk, J.; Vekemans, J. A. J. M.; Hulshof, L. A. (2009). *Org. Process Res. Dev.*, **14**, 351.
53. Godwin, D. R.; Lawton, S. J.; Moseley, J. D.; Welham, M. J.; Weston, N. P. (2010). *Energy Fuels*, **24**, 5446.

Chapter 13

Challenges Faced and Future Directions

Stefan G. Koenig
Small Molecule Process Chemistry, Genentech, Inc., a Member of the Roche Group, 1 DNA Way, South San Francisco, CA 94080, USA
koenig.stefan@gene.com

13.1 Introduction

The preceding chapters have provided a series of case studies illustrating to readers the subject of green chemistry—minimized use and generation of hazardous materials—and its current application in the pharmaceutical and associated industries. Although there are numerous other examples in the literature and in practice not covered here, the hope is that the stories gathered in this compendium will inspire readers to seek out more information, become practitioners in their own right, and encourage the next generation of contributions. After 20 years, green chemistry is alive and growing, with more solutions coming about each year.[1] Every branch of chemistry can incorporate the ideas inherent in the 12 principles of the philosophy (see Chapter 1), and every chemist can contribute to making the collective enterprise more sustainable and the planet more stable. Billions of people depend on our efforts.

Scalable Green Chemistry: Case Studies from the Pharmaceutical Industry
Edited by Stefan G. Koenig

ISBN 978-981-4316-49-1 (Hardcover), 978-981-4364-81-2 (eBook)
www.panstanford.com

Green chemistry is built on the accomplishments of generations of chemists that preceded its first articulation. It is a philosophical distinction over traditional modes of thinking rather than a separate branch and follows the path of chemistry as the central science by combining with many other scientific disciplines to advance its goals.[2] The field has also been called sustainable chemistry, to keep the ultimate aim in mind. The focus of this book is on *scalability*, or practical implementation, which is best captured by the terms sustainable development and sustainable process chemistry.[3] Whatever nomenclature is used, the pharmaceutical industry stands to benefit from these applications because it has (a) historically been more wasteful per kilogram of desired product than other chemical businesses due to product complexity, providing greater opportunity for efficiency and cost containment, and (b) the principles align well with the industry's primary objective: the betterment of human health.[4] While it is not always possible to be fully "green" and follow all 12 principles simultaneously, many of the tenets have been harnessed along the way to promote the goal of sustainability.

The pharmaceutical industry has been successful in creating innovative therapies for the treatment of patients because of the committed efforts of individuals involved in this pursuit. This is particularly true of the legion of dedicated chemists that have worked to produce safe and effective medicines in an efficient, robust, and cost-effective manner for decades. Still, new and better methods for discovery and development are becoming available, and the value of a technology is now measured not simply by its performance but also tied to its origin and fate in the form of a life cycle assessment (LCA).[5] The phrases "cradle to grave," meaning resource extraction ("cradle") through usage and ultimate disposal ("grave"), and "cradle to gate," an abbreviated analysis ending at the manufacturer's door, emphasize that a more comprehensive perspective is required in an age where resources are constrained and waste poses myriad problems.[6] Sourcing, using, and disposing of materials in an innocuous fashion are key to limiting undesired side effects, while reducing, reusing, and recycling ("cradle to cradle") are values inherent to a sustainable future.

13.2 Successful Beginnings

This book has focused on real-world, scalable examples of green chemistry in the pharmaceutical industry. The presented applications have been elevated to a point of practicality and thus are of significant utility. The bulk of contributions featured have demonstrated the clever rethinking of conventional chemical synthesis to make the manufacture of drugs and intermediates more efficient and nonhazardous, tools chemists can implement right now (Chapters 2, 5–9). Further green chemistry practices of biocatalysis (Chapter 3), evaluation metrics to enable benchmarking and improvements (Chapter 4), continuous processing (Chapter 10), real-time analytics (Chapter 11), and microwave chemistry (Chapter 12) represent cutting-edge applications that can be adopted to make chemical production more sustainable. All told, these case studies provide a glimpse into the modern methods the pharmaceutical industry and its affiliates have put in place with respect to the manufacture of medicines.

Green chemistry has only existed as a distinct concept since the early 1990s. However, interest in this aspiring area spread quite rapidly to a dedicated core of interested parties. Following the inauguration of the US Environmental Protection Agency's (EPA) Presidential Green Chemistry Awards in 1996, the Green Chemistry Institute® (GCI) was incorporated as a nonprofit organization "devoted to promoting and advancing" the field.[7] Its guiding vision has been that "green chemistry is a fundamental building block in the efforts to create a sustainable economy." In 2001, the GCI was merged into the American Chemical Society (ACS) so the combined effort could better "pursue . . . joint interests in the discovery and design of chemical products and processes that eliminate generation and use of hazardous substances."

Though the ACS GCI is the most visible organization dedicated to growing the green chemistry undertaking, many other groups around the world are pursuing a similar mission.[8] Among the various organizations focused on the subject are the US EPA, both in the Office of Pollution Prevention and Toxics and as the administrator of the Presidential Green Chemistry Awards;[9] the

Green & Sustainable Chemistry Network in Japan;[10] the Green Chemistry Network, launched by the Royal Society of Chemistry in the U.K.;[11] the Canadian Green Chemistry Network;[12] and others included in the remainder of this chapter. The GCI has partnered with several groups to cement ties and share knowledge across the globe, vital steps in a more interconnected scientific enterprise.

In 2005, the GCI partnered with several companies to form the pharmaceutical roundtable (GCIPR). Originally composed of 3 major pharmaceutical producers, the GCIPR has grown to more than 15 members, including generics firms and other affiliated corporations.[13] With their finger on the pulse of pharmaceutical research and development (R&D), this group is in the best position to gauge what green chemistry improvements might be necessary to improve pharmaceutical discovery and development. As mentioned in Chapter 1, the GCIPR published a paper in 2007, outlining the challenges in the field and research areas of interest, that is, where improvements to current methods would be useful.[14] These enhancements included more efficient oxidations, reductions, and coupling reactions.

Beyond the call for improved technologies to standard reaction paradigms, the GCIPR also laid out aspirational transformations as well as improvements in solvent applications. With the growing complexity inherent in modern pharmaceutical molecules, such as the inclusion of multiple chiral centers, fluorine atoms in challenging positions, and novel chemical scaffolds, it became clear that new chemistries would be required. Moreover, after extensive measurement and comparison between companies, it became obvious that solvents were the biggest contributor of waste to any pharmaceutical process. Improved selection, particularly with regard to chlorinated and polar aprotic solvents, and more efficient usage of solvents would greatly impact the environmental profile of a process based simply on volumes. The outlined "grand challenges" addressed the needs of modern, and more complex, pharmaceutical structures and the gap between current abilities and green chemistry aims.[14]

Alongside industry efforts, laws have emerged to encourage or even mandate more environmentally focused research.[15] In 2007, the European Union enacted the *Registration, Evaluation,*

Authorisation, and Restriction of Chemicals (REACH) *Act*, requiring manufacturers to provide data on the safety of their products.[16] The U.S. has similar, though outdated, regulations resulting from the *Toxic Substances Control Act* in 1976, yet updates to this law are expected.[17] Among the vanguard of US states pressing ahead, California approved the *California Green Chemistry Initiative*, set to go into effect in 2013, which requires prioritization of "chemicals of concern" on the basis of the degree of peril posed to public health or environment.[18] In sum, these rulings point to a greater utilization of the 12 principles.

From this overview, it is clear that the philosophy of green chemistry has become accepted and is championed by many participants in the realm of science in government and industry. Green chemistry is the preeminent toolbox for companies seeking to achieve dramatic environmental improvements as set forth in corporate sustainability reports and their stated goals of achieving widely accepted ISO 14000 guidelines established by the International Organization for Standardization,[19] as well as in meeting the triple bottom line, a measure of a firm's tally with respect to profit, people, and the planet.[20] The remaining challenge is for the 12 principles to be put into practice more broadly, from academic instruction and research, through continued industrial application and commercialization, to national guidance and regulation, all with the counsel of advocacy groups.[21] This chapter proposes steps to make green chemistry more impactful and enduring to better tackle the needs of sustainability.

13.3 The Future of Green Chemistry

With a multitude of efforts underway, it is difficult to chart the exact course of this exciting subject. What is paramount is that as many practitioners as possible begin to incorporate the 12 principles into their work and innovate past the expectations of current thinking. Applying groundbreaking concepts and atom-economical strategies,[22] most notably novel modes of catalysis—biotransformations, organocatalysis, and methods based on benign metals[23]—will extend the sustainability initiative. Several pillars should serve to bring about the growth of green chemistry. These foundations of support are (1) undertaking broad-based **education** efforts, (2) fostering **competition** to uncover the best possible solutions, and (3) enabling **collaboration** across boundaries to develop these tools and share the results with the rest of the community. These three elements are reinforcing ideals to a common goal, the expanded utilization of green chemistry philosophy to further the goals of sustainability.

13.3.1 *Education*

First and foremost among the list of factors to grow green chemistry is for chemistry students to have exposure to the 12 principles and their underpinnings, including instruction in toxicology. Only with widespread education will further translation to practice take place on a grand scale. Current practitioners continue to invent new approaches to creating green solutions. However, for the principles to be implemented consistently by more chemists, there simply have to be more green chemists, particularly in the chemical industries.[24] Outreach programs for school children and more rigorous instruction for advanced students and professionals will help spread the field's fundamentals more widely.

Green chemistry education programs have been initiated on many levels and on several continents. At the earliest stages, programs like Beyond Benign, a nonprofit organization that acts to "create tools, opportunities and partnerships to support ... K–12 education resources," will expose students to the benefits at a young age.[25] In addition, several other institutions are dedicated to interacting with local populations to increase awareness, including the Green Chemistry Network Centre at Delhi University.[26] These initiatives facilitate sharing of the 12 principles with a wide number of students. By introducing all students to the subject of sustainability through green chemistry, more will choose chemistry careers and the general public will benefit from improved awareness of contemporary scientific issues.[27]

For those that pursue higher education in science, green chemistry is emerging in academic institutions worldwide. Courses are offered in many settings, with select universities integrating green chemistry across their curricula.[28] With chemistry being an experimental science, full-scale research centers have been established, a prime example of which is the Institute for Green Science at Carnegie Mellon University in Pittsburgh, Pennsylvania.[29] In addition, numerous programs can help educators in their mission of instruction, such as the one offered at the University of Oregon. Within the school's collection of resources, the Green Chemistry Education Network (GCEdNet) works to "support opportunities to research, develop, implement and disseminate green educational materials."[30]

Numerous universities around the world now also offer advanced degrees in this cutting-edge field. At the master's level there are numerous established programs,[31] including one at the University of York in the U.K., as part of its Green Chemistry Centre of Excellence, and emerging programs, including one at Sichuan University in China.[32] Several graduate programs even offer PhD degrees in green chemistry, with the University of Massachusetts, Boston, at the forefront of this level of training.[33] The PhD track was also recently adopted at a sister campus, UMass Lowell, in parallel with its Center for Sustainable Production and Commerce Council (GC3), which provides a forum for business members to "share information and experiences relating to advancing green chemistry and design for environment in commerce and with regard to sustainable supply chain management."[34]

Besides spreading the word to a new generation, engaging experienced professionals can facilitate the transformation even sooner. Many green chemistry conferences have been initiated to enable the presentation and sharing of knowledge.[35] In addition, continuing education programs and short courses can also help the current workforce gain exposure to the 12 principles.[36] Still other resources are available on the Internet for all. For example, the University of Oregon's Green Education Materials (GEMs) for Chemists aims to provide an "interactive collection of chemistry education materials."[37] These providers allow practicing chemists to recognize the benefits of newer technological approaches and adapt before regulations are enacted that mandate green practices. Furthermore, this training enables established chemists to adeptly communicate with a younger generation and better understand the different perspective on the objectives and methods of research.

As with any other discipline, ensuring that knowledge is imparted on a new generation is the best method to fostering growth in its ranks. Those areas that have the greatest promise also have the most to gain by the spread of their ideas and practices. Education is the key to providing a foundation for students on issues such as toxicity; safety, health, and environmental guidelines; energy efficiency; and waste disposal and recycling, among other important notions. The concept of pollution prevention—as opposed to remedying the aftermath—is profound and part of the training employers are beginning to expect from their workforce.[38] By sounding the benefits of green chemistry to students at all levels of science education, both future scientists and the general public will become more knowledgeable about the enormous payoff for humanity.

13.3.2 *Competition*

While the ACS GCIPR enables member pharmaceutical companies to share information in a noncompetitive forum, competition to spur innovation and efficiency is still part of the green chemistry ethos. In fact, the focus on more innocuous products and processes has already delivered some very convincing improvements, as outlined in this book. However, incentives will be required to keep the momentum going and make sure that these green solutions are always improving to better compete in the marketplace. Scientists are naturally predisposed to searching for superior answers to difficult problems and turning insightful questions into the next generation of tools. By challenging chemists around the globe to come up with novel green solutions, a wave of more environmentally sound technologies will emerge through healthy competition.

Often, it is not the energy and drive of chemists that must be ramped up because they are usually self-motivated to work hard and uncover new solutions. Instead, it is the tools available to them that can impose limitations on what can be achieved. Increased awareness of the 12 principles and organizational encouragement to pursue laboratory efforts that go beyond the standard research questions will accelerate the discovery of green chemistry. Many large pharmaceutical companies have established awards to honor their best scientists in this realm and tools to conduct this work.[39] These applications and rewards should be expanded and even emulated by smaller organizations that do not typically think they can act green. Every serious effort toward implementing sustainable practices now will make a difference in the industry's shared long-term viability.

Beyond internal awards that improve one enterprise's environmental footprint, several external organizations have begun rewarding individual company's efforts. By selecting from the best current industry practices, these accolades shine a light on what works and could be replicated at other institutions to elevate the industry's approach and encourage the next great discoveries. Examples of these prizes include the Business Commitment to the Environment (BCE) Leadership Awards, the longest-running award program focused on environmental concern;[40] the Green Chemistry Challenge Awards, presented by the Royal Australian Chemical Institute (RACI),[41] and the Green & Sustainable Chemistry Awards, conferred by the GSC Network of Japan.[10] These awards are provided annually to individuals or groups in business or academic settings for the successful incorporation of green chemistry into their everyday practices.

In the field of chemistry, science fairs and student competitions have become a time-tested platform for encouraging the next generation to excel in the subject matter, a model epitomized by the International Chemistry Olympiad (IChO).[42] A similar methodology has been applied to students in the subdiscipline of green chemistry. The Berkeley Institute for the Environment (University of California, Berkeley) hosts an annual "Big Ideas" prize competition to inspire innovative projects directed at solving the world's most urgent challenges.[43] Students are provided with seed money to foster creative environmental research and development projects. New-thought paradigms are instilled in the participants by use of a competitive but supportive environment, leading to next-generation approaches.

On the regulatory side, environmental law—at the federal, state, and local levels—though considered by some a burden on industry, can actually have the opposite effect and provide fertile areas for growth.[44] By changing the competitive landscape, flexible regulations can spur innovation as companies seek to take advantage of new rules. As applied to the chemical industries, even regulatory schemes such as REACH can benefit those companies that leverage the latest incentives. Pharmaceutical and biotech companies that focus on a targeted product portfolio and manage their suppliers and waste streams carefully can outmaneuver competitors that choose to ignore these new environmental directives. A competitive advantage does not just derive from products with greater sales; implementing efficiency programs throughout the firm also contributes to improved profitability.

Finally, government agencies can take further steps to encourage green behavior. Beyond providing grants to fund research, they can provide credits to parties that implement and document their environmental improvements, thereby emphasizing the critical value these solutions provide. An example of such a program is the Green Technologies Pilot Program of the United States Patent and Trademark Office (USPTO).[45] This undertaking enables a patent aspirant to accelerate review of his or her application, potentially gaining approval of the invention within one year as opposed to the typical three-year cycle. Another option is to create public-private partnerships or preferential tax treatment to foster the growth of new companies dedicated to green chemistry endeavors. While some may fail, the success stories will be disruptive technologies that elevate the industry to the next level. Taken together, these forms of support can help invigorate promising advances and make them more marketable.

Having national agencies, nongovernmental organizations, academic institutions, and the diverse chemical industries recognize and reward significant green chemistry accomplishments will provide these feats broader exposure and enable their adoption on a grander scale. Getting the 12 principles incorporated into the next generation of technologies will bring us closer to the sustainable future the planet requires. Competition, in its various forms, is the best way to enable ultimate success, as it determines what works and what does not. However, even in the realm of a competitive landscape, alliances facilitate rapid progress and greater accomplishment to be shared amongst a larger group. The last pillar of support, collaboration, reminds us of what we are ultimately looking to achieve.

13.3.3 *Collaboration*

With a thriving green chemistry community growing around the globe, collaboration will be key to spreading the achievements. Competition between practitioners keeps driving new discoveries; however, having joint agreements and collaborative organizations will keep ideas flowing with greater overall improvement as a result. A shining example is the accord that pharmaceutical companies have reached in recognizing the value of cooperation on the topic of green chemistry, setting aside their product competition to participate in the pharmaceutical roundtable. As described earlier, the GCIPR encourages the benefits of the discipline to be exchanged by member companies in a noncompetitive format to make each individual organization more environmentally sustainable.

This information-sharing arrangement is only one form of cooperation to expedite the implementation of the 12 principles. Other cross-disciplinary partnerships, corporate joint ventures, and industry-academic alliances could all help cross-fertilize and infuse scientific research on the road to better processes and products. The Warner Babcock Institute, a nonprofit institute "dedicated to the development of non-toxic, environmentally benign, and sustainable technological solutions for society," partners with industry clients in exclusive relationships.[46] They seek to generate unconventional answers that are "functionally superior, cost effective, and environmentally benign" through a synergistic research model.

Green chemistry has opened the door to new business opportunities and provides entrepreneurs with tremendous growth prospects. In many cases, by teaming up with established businesses, start-ups can demonstrate the value of their approach and provide further openings. Examples of companies formed around innovative technologies include Codexis and SiGNa Chemistry. Codexis customizes biocatalysts for processes in the pharmaceutical and energy industries. Chapter 3 details how an initial application can lead to continued success with subsequent partners. Another business that solves challenges in a green manner is SiGNa Chemistry.[47] This company turns reactive alkali metals into stable, free-flowing solids to replace outdated and dangerous transformations with practical and safer alternatives. SiGNa partners with traditional chemical suppliers to enable more downstream customers to practice green chemistry.

Further examples continue to emerge as the marketplace demands a need for better solutions. Collaboration presents a compelling path by which innovations can advance to market. The most revolutionary instances of green chemistry have involved the affiliation of chemical scientists with those in the arena of material-, nano-, and biotechnology and engineering, as well as others across the more unconventional borders of environmentalism, economy, and public policy. For pharmaceuticals, these interdisciplinary interactions are exemplified by the enzyme evolution techniques, making biocatalysis a viable manufacturing option, and life cycle management practices to make optimum decisions from the earliest stages of design through the outsourcing of materials for commercial production. Additional cooperation from the overlap of different fields will lead to more interesting applications.

Areas with significant scientific benefits yet to be realized include those of discovery chemistry and engineering. Chemical development divisions within companies can share the green chemistry lessons they have learned with their discovery colleagues.[48] Medicinal chemistry research operates innumerable reactions on very small scale to discover promising compounds. Cumulatively, this work contributes to a still-significant environmental footprint that could benefit tremendously from the improvements in green chemistry solvents and reagents. Likewise, engineering remains an area where applications could incorporate additional environmental considerations. The Center for Green Chemistry and Green Engineering at Yale has been established "to advance sustainability by catalyzing the effectiveness of . . . Green Engineering."[49] The need for such efforts was also addressed in a paper published by the GCIPR with specific tenets outlined for green engineering,[50] including a focus on chemical processing, solvents, and energy issues.

One critical area where green engineering will have a large impact is in the elevation of chemical production to commercial manufacturing scale. The Novartis-Massachusetts Institute of Technology Center for Continuous Manufacturing strives to transform large-scale pharmaceutical production.[51] As described in Chapter 10 by coworkers at DSM with expertise in the field, continuous processing offers the potential of radically improving manufacturing from the more conventional batch-based practices. Combining Novartis's industrial expertise with the Massachusetts Institute of Technology's (MIT) scientific and technical capabilities could enable more efficient chemical processes in smaller facilities with minimized waste and raw material and energy use.

Similarly, at the University of California, Berkeley, the Consortium on Green Design and Manufacturing, a multidisciplinary partnership between academia, industry, and government, has been created to explore similar questions regarding the future of chemical production.[52] At the same institution, the Berkeley Center for Green Chemistry (BCGC) brings together several disciplines to forge an integrated approach to answer society's looming environmental challenges.[53] The center seeks to build a broad-based "novel academic program . . . through interdisciplinary scholarship" with contributions from "chemistry, public health, engineering, natural resources, law, and business," serving as a model of leadership in green chemistry for others to emulate.

Cooperation can also occur across other boundaries, for example, within national industry groups, as between Japanese chemical organizations in the Green & Sustainable Chemistry Network,[10] or within educational systems, such as with the Interuniversity National Consortium "Chemistry for the Environment" (INCA), an Italian university consortium that facilitates knowledge transfer.[54] Support for sustainability initiatives can come in the form of grants to fund basic research or commercialization efforts for early-stage discoveries. Several of these institutions have taken root over the past several years, including GreenCentre Canada, which fosters academic and industry partnership,[55] and the Centre for Environment and Sustainability, GMV, in Göteborg, Sweden, a partnership between Chalmers University of Technology and the University of Gothenburg, which promotes multidisciplinary programs and alliances with the business community.[56]

What could be a more unifying force than the proposition of a sustainable future? Attaining the equilibrium of an enduring planet, capable of providing a lasting home for our own species and the many others on earth, is the ultimate goal of this sustainability endeavor. Green chemistry is a significant part of this mission.

And it is only fitting that the path to this envisioned stable, long-term existence involves intense collaboration between chemists and numerous other disciplines. Only together will we be able to succeed. With this overarching theme in mind, it makes sense that cooperation should be a necessary element for us to achieve success in our own research projects as well as in the ultimate aspiration for a sustainable future.

13.4 Conclusions

Whereas case studies present past accomplishments, the future of any field is always difficult to navigate and impossible to predict. Having built a solid foundation from which to enable further opportunities, green chemistry is now on the precipice of delivering grand prescriptions to many of society's challenges. With sustainability an acknowledged goal in most every circle, uncovering solutions and putting them into practice is still a challenge. Green chemistry—often referred to as "sustainability on the molecular level"—will play a major role in facilitating new scientific discoveries for the express purpose of creating a more enduring world. At the same time, it is important not to let the hype overwhelm the ultimate goal, to craft new products and processes with a comprehensive environmental perspective.

In a few short years, the philosophy of green chemistry has made a significant impact on science and the pharmaceutical industry, in particular. For this discipline to grow further in influence and make our industries more efficient and benign, several key targets should be addressed. With enhanced (1) education, (2) competition, and (3) collaboration, a virtuous circle of progress could be established from which further rounds of improvement should emerge. Education infused with the 12 principles will enable chemists to better integrate green chemistry into their work. Incentives—part stimulatory and part market driven—will help maintain the momentum by creating a level of competition to evolve the best green solutions. Lastly, collaboration—through agreements and dedicated organizations—will keep ideas flowing back to practitioners to encourage successive waves of improvement.

The pharmaceutical (and biotechnology) industry, from the large multinational companies to smaller niche enterprises, is dedicated to helping people live longer, healthier lives. Their main aim, to treat disease and infection, will always be a worthwhile endeavor. However, by contributing to a healthier planet in the form of green chemistry, the industry will also help prevent many maladies from becoming epidemic later. Green chemistry empowers scientists to solve problems in a new way and make the planet more sustainable. Providing chemists with the right tools will help define our generation on the issue of sustainability. Participating in this quest will establish the sciences as the vocation of leadership for mankind.

The chapters of this book have provided case studies to explain green chemistry and its impact in the pharmaceutical industry. With additional discoveries and solutions being uncovered every year, this specialty continues to grow, but there remains more to be done. Every branch of chemistry can benefit from the 12 principles, and all chemists can contribute to this progress to make the enterprise and planet more sustainable. The onus is on us to bring green chemistry into the standard practice of chemical research and development at every level. The future is certainly bright for green chemistry innovation as the world begins to face its challenges. By incorporating the 12 principles into standard scientific training and having chemists routinely practice them, we will be well on our way to a stable future.

References

1. Anastas, P. (2011). *Chem. Eng. News*, **89**(26), 62–65.
2. Ghosh, A.; Gupta, S. S.; Bartos, M. J.; Hangun, Y.; Vuocolo, L. D.; Steinhoff, B. A.; Noser, C. A.; Horner, D.; Mayer, S.; Inderhees, K.; Horwitz, C. P.; Spatz, J.; Ryabov, A. D.; Mondal, S.; Collins, T. J. (2001). *Pure Appl. Chem.*, **73**, 113–118.
3. Appell, R.; Gala, D.; Sanghvi, Y. S. (2011). *Org. Process Res. Dev.*, **15**, 898–899.
4. Tucker, J. L. (2006). *Org. Process Res. Dev.*, **10**, 315–319.
5. http://www.epa.gov/nrmrl/lcaccess/ (accessed September 4, 2011).

6. Cue, B. W.; Zhang, J. (2009). *Green Chem. Lett. Rev.*, **2**, 193–211.
7. http://www.acs.org/greenchemistry (accessed September 4, 2011).
8. http://www.incaweb.org/transit/iupacgcdir/links.htm (accessed September 4, 2011).
9. http://www.epa.gov/greenchemistry/ (accessed September 4, 2011).
10. http://www.gscn.net/indexE.html (accessed September 4, 2011).
11. http://www.greenchemistrynetwork.org/ (accessed September 4, 2011).
12. http://www.greenchemistry.ca/ (accessed September 4, 2011).
13. http://portal.acs.org/portal/PublicWebSite/greenchemistry/industriainnovation/roundtable/index.htm (accessed September 4, 2011).
14. Constable, D. J. C.; Dunn, P. J.; Hayler, J. D.; Humphrey, G. R.; Leazer, J. L., Jr.; Linderman, R. J.; Lorenz, K.; Manley, J.; Pearlman, B. A.; Wells, A.; Zaks, A.; Zhang, T. Y. (2007). *Green Chem.*, **9**, 411–420.
15. Walker, D. (2008). The environment, in *The Management of Chemical Process Development in the Pharmaceutical Industry* (Vol. 1). USA: Wiley-AIChE, 87–108.
16. http://ec.europa.eu/environment/chemicals/reach/reach_intro.htm (accessed September 11, 2011).
17. Hogue, C. (2011). *Chem. Eng. News*, **89**(7), 30–31.
18. http://www.dtsc.ca.gov/PollutionPrevention/GreenChemistry Initiative/index.cfm (accessed September 11, 2011).
19. http://www.iisd.org/greenstand/default.htm (accessed September 11, 2011).
20. (Nov. 17, 2009). Idea: triple bottom line—it consists of three Ps: profit, people and planet. *The Economist*. http://www.economist.com/node/14301663 (accessed September 11, 2011).
21. Tucker, J. L. (2010). *Org. Process Res. Dev.*, **14**, 328–331.
22. Li, C.-J.; Trost, B. M. (2008). *Proc. Natl. Acad. Sci. USA*, **105**, 13197–13202.
23. Crabtree, R. H.; Anastas, P. T. (Eds.). (2009). *Handbook of Green Chemistry, Green Catalysis* (vols. 3). Weinheim, Germany: Wiley-VCH.
24. Kang, E. (2010). *Tetrahedron*, **66**, 1029–1030.
25. http://www.beyondbenign.org/ (accessed September 4, 2011).
26. http://greenchem.du.ac.in/ (accessed September 11, 2011).
27. Burmeister, M.; Jokmin, S.; Eilks, I. (2011). *CHEMKON*, **18**(3), 123–128.
28. Pratt, M. (Oct. 9, 2007). Green chemistry joins college curriculum. *The Washington Post*. http://www.washingtonpost.com/wp-dyn/

content/article/2007/10/09/AR2007100900247.html (accessed September 11, 2011).

29. http://www.chem.cmu.edu/groups/Collins/about/ (accessed September 4, 2011).
30. http://cmetim.ning.com/ (accessed September 4, 2011).
31. http://portal.acs.org/portal/PublicWebSite/greenchemistry/education/academicprograms/index.htm (accessed September 4, 2011).
32. http://school.cucas.edu.cn/HomePage/76/2009-12-23/Program_8532.shtml (accessed September 11, 2011).
33. http://www.greenchemistry.umb.edu/ (accessed September 4, 2011)
34. http://www.greenchemistryandcommerce.org/about.whatisGC3.php (accessed September 11, 2011).
35. http://portal.acs.org/portal/PublicWebSite/meetings/greenchemistryconferences/index.htm (accessed September 4, 2011).
36. Wilson, E. K. (2011). *Chem. Eng. News*, **89**(4), 34–35.
37. http://greenchem.uoregon.edu/gems.html (accessed September 11, 2011)
38. van Arnum, P. (2009). *Pharm. Technol. Sourc. Manage.*, **5**(9). http://pharmtech.findpharma.com/pharmtech/Green-Chemistry-in-Pharmaceutical-Applications/ArticleStandard/Article/detail/622983 (accessed September 11, 2011).
39. Goodman, S. (Mar. 25, 2009). "Green chemistry movement sprouts in colleges, companies. *The New York Times/Greenwire*. http://www.nytimes.com/gwire/2009/03/25/25greenwire-green-chemistry-movement-sprouts-in-colleges-c-10287.html (accessed September 4, 2011).
40. http://www.bceawards.org/ (accessed September 25, 2011).
41. http://www.raci.org.au/events-awards/national-awards-2011/green-chemistry-challenge-awards (accessed September 25, 2011).
42. http://www.icho.sk/ (accessed September 25, 2011).
43. http://bie.berkeley.edu/studentcompetitions (accessed September 25, 2011).
44. Porter, M. E.; Van Der Linde, C. (1995). *Harvard Bus. Rev.*, **73**(5), 120–134.
45. http://www.uspto.gov/patents/init_events/green_tech.jsp (accessed September 4, 2011).
46. www.warnerbabcock.com/ (accessed September 11, 2011).

47. http://signachem.com/ (accessed September 25, 2011).
48. Alfonsi, K.; Colberg, J.; Dunn, P. J.; Fevig, T.; Jennings, S.; Johnson, T. A.; Kleine, H. P.; Knight, C.; Nagy, M. A.; Perry, D. A., Stefaniak, M. (2008). *Green Chem.*, **10**, 31–36.
49. http://greenchemistry.yale.edu/ (accessed September 11, 2011).
50. Concepción Jiménez-González, C.; Peter Poechlauer, P.; Broxterman. Q. B.; Yang, B.-S.; am Ende, D.; Baird, J.; Bertsch, C.; Hannah, R. E.; Dell'Orco, P.; Noorman, H.; Yee, S.; Reintjens, R.; Wells, A.; Massonneau, V.; Manley, J. (2011). *Org. Process Res. Dev.*, **15**, 900–911.
51. http://engineering.mit.edu/research/labs_centers_programs/novartis.php (accessed September 25, 2011).
52. http://cgdm.berkeley.edu/ (accessed September 25, 2011).
53. http://bcgc.berkeley.edu/ (accessed September 11, 2011).
54. http://www.incaweb.org/ (accessed September 11, 2011).
55. http://www.greencentrecanada.com/ (accessed September 25, 2011).
56. http://www.chalmers.se/gmv/EN/ (accessed September 25, 2011).

Index